L. G. PHILLIPS

MATHEMATICS
FOR HIGHER NATIONAL
CERTIFICATE

L. G. PHILLIPS.

MATHEMATICS
FOR HIGHER NATIONAL CERTIFICATE

BY

S. W. BELL, B.Sc., F.I.M.A.

Head of the Department of Mathematics and Computing

AND

H. MATLEY, B.Sc., A.F.I.M.A.

Principal Lecturer in Mathematics Norwich City College

VOLUME II

(ELECTRICAL)

SECOND EDITION [METRIC]

CAMBRIDGE
AT THE UNIVERSITY PRESS
1971

Published by the Syndics of the Cambridge University Press
Bentley House, 200 Euston Road, London, N.W. 1
American Branch: 32 East 57th Street, New York, N.Y. 10022

ISBN: 0 521 08081 9

First edition 1958
Second edition 1963
Reprinted 1965 1967
Third edition [metric] 1971

Printed in Great Britain
at the University Printing House, Cambridge
(Brooke Crutchley, University Printer)

CONTENTS

Chapter 5

Further complex variables

Chapter 6

Further first-order differential equations

Chapter 7

Further second-order differential equations

Chapter 8

Some electrical applications of linear differential equations

CHAPTER 9

Simultaneous linear differential equations and some applications

CHAPTER 10

Solution in series. The Laplace transformation

CHAPTER 11

Introduction to partial differential equations

CHAPTER 12

Further Fourier series

CHAPTER 13

Determinants. Solution of simultaneous linear equations

CHAPTER 14
Double integrals and some applications

CHAPTER 15
Elements of matrix algebra

CHAPTER 16
An introduction to the application of matrices to network analysis

APPENDIX
Inversion and current loci

PREFACE

This volume is primarily intended to cover the work done in the final year of the Higher National Certificate in Electrical Engineering, and to give sufficient topics to form the basis of an A.3 Mathematics course for electrical engineers.

Since the publication of Volume I, there have been important changes in technical education with the establishment of Colleges of Advanced Technology and Regional Colleges. This has meant a marked increase in full-time and sandwich courses. We hope that this book, with Volume I, can be used to advantage by students following such courses in electrical engineering. Students taking mathematics in Part II for the examination of the Institution of Electrical Engineers may also find the book useful. Although some of the text is not of the rigour demanded by University courses, engineering degree students would probably find helpful the practical applications we have included.

Enquiries at a number of technical colleges have indicated that at this level the main need is for a mathematics course for electrical rather than mechanical engineers. This, and because we wished to cover as wide a variety of applications as possible, led us to restrict Volume II to electrical work. A number of topics normally found in books on electrical technology have been used freely, as we firmly believe that the student should not consider mathematics as isolated from his other subjects.

In selecting the topics to form a basis for this book, we have been guided by the report of the Mathematical Association on 'The Teaching of Mathematics in Technical Colleges'; and we hope we have interpreted its suggestions correctly.

Our thanks are due to the Heads of the Mathematics Departments of the Technical Colleges at Cardiff, Derby, Edinburgh, Kingston, Gloucester and Salford, who kindly supplied syllabuses and helpful suggestions. Special thanks are due to Mr W. G. L. Sutton, Vice-Principal of Leeds

College of Technology, for his helpful comments and advice; and to Mr H. Burnip, B.A., for his comments, help in proof-reading, and checking many of the answers to the exercises.

We are indebted to the following authorities for permission to use questions from their examinations: the Senate of the University of London; the Syndics of the Cambridge University Press; and the Institution of Electrical Engineers. In this connexion we wish to point out that any solutions of questions from these examination papers appearing in the text are our own.

In conclusion we desire to thank the printer and his readers for their courtesy and care whilst seeing this book through the press.

S. W. BELL
H. MATLEY

NORWICH
January 1958

PREFACE TO THE SECOND EDITION

We have corrected some errors and made a few minor alterations in the text. The main change has been the addition of more questions, both in the text and in the miscellaneous exercises at the end of the book. Many of these are from recent mathematics papers for Part II of the examination of the Institution of Electrical Engineers.

S. W. BELL
H. MATLEY

NORWICH
1962

PREFACE TO THE THIRD EDITION

We have amended the text to incorporate S.I. units and altered the units given in examination questions where necessary.

S. W. BELL
H. MATLEY

NORWICH
1970

SOME ABBREVIATIONS USED IN THE TEXT

$\neq$	not equal to	C.F.	complementary function
$\bumpeq$	approximately equal to	e.m.f.	electromotive force
$\equiv$	identically equal to	e.s.u.	electrostatic units
$\sum$	the summation of	G.S.	general solution
$\rightarrow$	approaches	i.p.	imaginary part
$<$	less than	lim	limit of
$>$	greater than	M.V.	mean value
$\leqslant$	less than or equal to	P.I.	particular integral
$\geqslant$	greater than or equal to	p.d.	potential difference
a.c.	alternating current	r.p.	real part
A.E.	auxiliary equation	s.h.m.	simple harmonic motion

SOME LETTERS OF THE GREEK ALPHABET IN COMMON USE IN MATHEMATICS

α	alpha	ν	nu
β	beta	π	pi
γ	gamma	Π	capital pi
δ	delta	ρ	rho
Δ	capital delta	σ	sigma
ϵ	epsilon	Σ	capital sigma
η	eta	ϕ	phi
θ	theta	ψ	psi
λ	lambda	ω	omega
μ	mu	Ω	capital omega

CHAPTER 1

FURTHER DIFFERENTIATION AND EXPANSION IN SERIES

1·1. Taylor's theorem

This is a more general theorem than Maclaurin's theorem; in fact, the form of Maclaurin's expansion can be obtained as a special case of Taylor's theorem.

One form of the theorem gives the expansion of $f(a+h)$ in powers of h. This is the form which will now be derived, using integration by parts.

1·11. Derivation of the theorem

Let $f(x)$ have a continuous nth derivative in the interval

$$x = a \text{ to } x = a + h.$$

$$\int_a^{a+h} f'(x)\,dx = [f(x)]_a^{a+h} = \underline{f(a+h) - f(a)}\,. \tag{1}$$

Now $$dx = -\,d(a+h-x). \tag{2}$$

Using integration by parts, treating $f'(x)$ as 'u' and

$$dx = -\,d(a+h-x)$$

as 'dv',

$$\int_a^{a+h} f'(x)\,dx = \int_a^{a+h} -f'(x)\,d(a+h-x)$$

$$= [-(a+h-x)\,f'(x)]_a^{a+h} + \int_a^{a+h} (a+h-x)f''(x)\,dx$$

$$= hf'(a) - \int_a^{a+h} (a+h-x)\,f''(x)\,d(a+h-x),$$

(using (2) again).

Using integration by parts on the right-hand integral, taking $f''(x)$ as 'u' and $(a+h-x)\,d(a+h-x)$ as 'dv',

$$\int_a^{a+h} f'(x)\,dx = hf'(a) - \left[\frac{(a+h-x)^2}{2} f''(x)\right]_a^{a+h}$$

$$+ \int_a^{a+h} \frac{(a+h-x)^2}{2} f'''(x)\,dx$$

$$= hf'(a) + \frac{h^2}{2!} f''(a) - \int_a^{a+h} \frac{(a+h-x)^2}{2!} f'''(x)\,d(a+h-x),$$

using (2) again and also $2 = 2!$.

Again integrating the right-hand integral by parts,

$$\int_a^{a+h} f'(x)\,dx = hf'(a) + \frac{h^2}{2!}f''(a) - \left[\frac{(a+h-x)^3}{3!}f'''(x)\right]_a^{a+h}$$

$$+ \int_a^{a+h} \frac{(a+h-x)^3}{3!} f^{\text{iv}}(x)\,dx$$

$$= hf'(a) + \frac{h^2}{2!}f''(a) + \frac{h^3}{3!}f'''(a)$$

$$+ \int_a^{a+h} \frac{(a+h-x)^3}{3!} f^{\text{iv}}(x)\,dx. \qquad (3)$$

Integrating by parts $(n-1)$ times will give

$$\underline{\int_a^{a+h} f'(x)\,dx = hf'(a) + \frac{h^2}{2!}f''(a) + \dots}$$

$$\underline{+ \frac{h^{n-1}}{(n-1)!}f^{n-1}(a) + R_n,} \qquad (4)$$

where $$\underline{R_n = \int_a^{a+h} \frac{(a+h-x)^{n-1}}{(n-1)!} f^n(x)\,dx.} \qquad (5)$$

From (1) and (4), equating the values for $\int_a^{a+h} f'(x)\,dx$,

$$\underline{f(a+h) = f(a) + hf'(a) + \frac{h^2}{2!}f''(a) + \dots}$$

$$\underline{+ \frac{h^{n-1}}{(n-1)!}f^{n-1}(a) + R_n.} \qquad (6)$$

R_n is called the remainder after n terms.

If n is increased indefinitely, the series in (6) becomes an infinite series. This series truly represents the function $f(a+h)$ if the series converges *and* if $\lim_{n\to\infty} R_n = 0$.

Cases in which these conditions are not satisfied are beyond the general scope of this book. Unless otherwise stated the student may assume that any function met with here is capable of expansion by Taylor's series. That is

$$\underline{f(a+h) = f(a) + hf'(a) + \frac{h^2}{2!}f''(a) + \dots + \frac{h^n}{n!}f^n(a) + \dots.} \qquad (7)$$

Note that if a is put equal to zero and h equal to x Maclaurin's series is obtained:

$$f(x) = f(0) + xf'(0) + \frac{x^2}{2!}f''(0) + \dots + \frac{x^n}{n!}f^n(0) + \dots .$$

Putting $a = x$, another form of the series is

$$f(x+h) = f(x) + hf'(x) + \frac{h^2}{2!}f''(x) + \dots + \frac{h^n}{n!}f^n(x) + \dots . \quad (8)$$

1·2. Small changes in a function

In volume I the mean-value theorem was used as a means of finding an approximation to the new value of a function due to a small change in the variable. Equation (7) may be similarly used to any required degree of accuracy. In general, an idea of the accuracy of the approximation can be found by considering the term following that at which the calculations were stopped.

Example 1

Find the value of sin 31° correct to six places of decimals.

$$\sin 31° = \sin(30° + 1°) = \sin\left(\frac{\pi}{6} + \frac{\pi}{180}\right).$$

Thus $f(a+h) = \sin\left(\frac{\pi}{6} + \frac{\pi}{180}\right)$, where $a = \frac{\pi}{6}$ and $h = \frac{\pi}{180}$.

$$f(a) = \sin\frac{\pi}{6} = 0{\cdot}5,$$

$$f'(a) = \cos\frac{\pi}{6} = \frac{\sqrt{3}}{2},$$

$$f''(a) = -\sin\frac{\pi}{6} = -0{\cdot}5,$$

$$f'''(a) = -\cos\frac{\pi}{6} = -\frac{\sqrt{3}}{2},$$

$$f^{\text{iv}}(a) = \sin\frac{\pi}{6} = 0{\cdot}5, \text{ etc.}$$

Using (7) of § 1·11,

$$\sin\left(\frac{\pi}{6}+\frac{\pi}{180}\right) = 0{\cdot}5 + \frac{\pi}{180}\frac{\sqrt{3}}{2} + \frac{1}{2!}\left(\frac{\pi}{180}\right)^2(-0{\cdot}5)$$

$$+\frac{1}{3!}\left(\frac{\pi}{180}\right)^3\left(-\frac{\sqrt{3}}{2}\right) + \frac{1}{4!}\left(\frac{\pi}{180}\right)^4(0{\cdot}5) + \ldots$$

$$\simeq 0{\cdot}5 + 0{\cdot}0151150 - 0{\cdot}0000761 - 0{\cdot}0000008$$

$$+ 0{\cdot}000000002 + \ldots$$

$$\simeq 0{\cdot}5150381.$$

As the last term calculated is 0·000000002, the figure given above is certainly correct to six decimal places.

Correct to 6 decimal places $\sin 31^\circ = 0{\cdot}515038.$

EXAMPLE 2

Find an expansion for $\cosh(a + x)$ in ascending powers of x.

Putting $h = x$ in equation (7) of § 1·11

$$f(a+x) = f(a) + xf'(a) + \frac{x^2}{2!}f''(a) + \ldots.$$

Now if $f(a + x) = \cosh(a + x)$,

$$f(a) = \cosh a,$$

$$f'(a) = \sinh a,$$

$$f''(a) = \cosh a, \text{ etc.}$$

$$\therefore \cosh(a+x) = \cosh a + x\sinh a$$

$$+\frac{x^2}{2!}\cosh a + \frac{x^3}{3!}\sinh a + \ldots.$$

EXERCISE 1

1. Given that $\cos 60^\circ = 0{\cdot}5$, $\sin 60^\circ = 0{\cdot}866025$, find the values of $\cos 61^\circ$ and $\sin 61^\circ$ correct to 5 decimal places.

2. Find an expansion for $\log_e(a + x)$ in powers of x, $a > 0$ and $|x| < a$.

3. Given that sin 34° = 0·559193 and cos 34° = 0·829038, find sin 35° and cos 35° correct to 5 significant figures.

4. Expand $\cos(x + h)$ as a power series in h.

5. Show that

$$\tan(\tfrac{1}{4}\pi + x) = 1 + 2x + 2x^2 + \tfrac{8}{3}x^3 + \tfrac{10}{3}x^4 + \dots.$$

Hence evaluate tan 46°30′ correct to 4 decimal places.

[L.U.]

6. By writing $(x - a)$ for h in Taylor's series, verify that

$$f(x) = f(a) + (x - a)f'(a) + \frac{(x-a)^2}{2!}f''(a) + \dots.$$

Deduce an expansion for $\sin x$ in powers of $(x - \frac{1}{2}\pi)$ and evaluate sin 92° correct to 4 decimal places.

Answers

1. 0·48481; 0·87462.

2. $\log_e a + \dfrac{x}{a} - \dfrac{x^2}{2a^2} + \dfrac{x^3}{3a^3} - \dots.$

3. 0·57358; 0·81915.

4. $\cos x - h\sin x - \dfrac{h^2}{2!}\cos x + \dfrac{h^3}{3!}\sin x + \dots.$

5. 1·0538.

6. $1 - \dfrac{(x - \frac{1}{2}\pi)^2}{2!} + \dfrac{(x - \frac{1}{2}\pi)^4}{4!} - \dots$; 0·9994.

1·3. Indeterminate forms

Suppose that it is required to find $\lim\limits_{x\to a} f(x)/F(x)$, where $f(x)$ and $F(x)$ are two continuous functions and $f(a) = 0 = F(a)$. $f(a)/F(a)$ would merely give 0/0 which is indeterminate.

There are two usual methods of finding the limit.

The first consists of writing $\lim\limits_{x\to a}\dfrac{f(x)}{F(x)} = \lim\limits_{h\to 0}\dfrac{f(a+h)}{F(a+h)}$ and then expanding numerator and denominator in powers of h, using Taylor's theorem. If $x \to 0$, Maclaurin's expansion is used as in vol. I, § 9·8.

EXAMPLE

$$\lim_{x\to\alpha} \frac{x(\cos x - \cos \alpha)}{(\alpha^2 - x^2)}$$

$$= \lim_{h\to 0} \frac{(\alpha + h)\{\cos(\alpha + h) - \cos\alpha\}}{\{\alpha^2 - (\alpha + h)^2\}}$$

$$= \lim_{h\to 0} \frac{(\alpha + h)\left\{\left(\cos\alpha - h\sin\alpha - \frac{h^2}{2!}\cos\alpha + \dots\right) - \cos\alpha\right\}}{-h(2\alpha + h)}$$

$$= \lim_{h\to 0} \frac{-(\alpha + h)h\left(\sin\alpha + \frac{h}{2!}\cos\alpha - \dots\right)}{-h(2\alpha + h)}$$

$$= \lim_{h\to 0} \frac{(\alpha + h)\left(\sin\alpha + \frac{h}{2!}\cos\alpha - \dots\right)}{(2\alpha + h)}$$

$$= \frac{\alpha\sin\alpha}{2\alpha} = \underline{\frac{\sin\alpha}{2}}.$$

The second method is to use L'Hôpital's rule in the form

$$\lim_{x\to a} \frac{f(x)}{F(x)} = \lim_{x\to a} \frac{f'(x)}{F'(x)}. \tag{1}$$

This rule can be derived as follows:

Assuming that $f(x)$ and $F(x)$ satisfy the conditions necessary for applying the mean-value theorem, then

$$\frac{f(a+h) - f(a)}{F(a+h) - F(a)} = \frac{hf'(a+\theta_1 h)}{hF'(a+\theta_2 h)} \quad (0 < \theta_1 < 1,\ 0 < \theta_2 < 1)$$

(see vol. I, § 1·9)

$$= \frac{f'(a+\theta_1 h)}{F'(a+\theta_2 h)} \quad (h \neq 0).$$

If $f'(x)$ and $F'(x)$ are continuous, then

$$\lim_{h\to 0} \frac{f'(a+\theta_1 h)}{F'(a+\theta_2 h)} = \lim_{x\to a} \frac{f'(x)}{F'(x)}.$$

Also, as $f(a) = 0 = F(a)$,

$$\lim_{h\to 0} \frac{f(a+h) - f(a)}{F(a+h) - F(a)} = \lim_{x\to a} \frac{f(x)}{F(x)}.$$

(1) then follows.

If $\lim\limits_{x\to a} f'(x)/F'(x)$ is also indeterminate it can be shown in a similar manner that $\lim\limits_{x\to a} \dfrac{f'(x)}{F'(x)} = \lim\limits_{x\to a} \dfrac{f''(x)}{F''(x)}$, and so on.

Example

$$\lim_{x\to \alpha} \frac{x(\cos x - \cos \alpha)}{\alpha^2 - x^2}$$

$$= \lim_{x\to \alpha} \frac{(\cos x - \cos \alpha) - x \sin x}{-2x}$$

$$= \frac{-\alpha \sin \alpha}{-2\alpha} = \underline{\frac{\sin \alpha}{2}}.$$

In general, if successive derivatives of $f(x)$ and $F(x)$ are easily found and lead quickly to a determinate answer, then L'Hôpital's rule is probably the quicker method. When the derivatives are complicated, the expansion method should be used.

Further Examples

(1)
$$\lim_{x\to 0} \frac{\tan x - x}{\sin x - x} = \lim_{x\to 0} \frac{\sec^2 x - 1}{\cos x - 1}$$

$$= \lim_{x\to 0} \frac{-\sec^2 x(1 - \cos^2 x)}{(1 - \cos x)}$$

$$= \lim_{x\to 0} - \sec^2 x(1 + \cos x)$$

$$= \underline{-2}.$$

(2)
$$\lim_{\theta\to 0} \left(\frac{(\cos \pi\theta - 1)}{\tan^2 \pi\theta}\right) = \lim_{\theta\to 0} \left(\frac{-\pi \sin \pi\theta}{2\pi \tan \pi\theta \sec^2 \pi\theta}\right)$$

$$= \lim_{\theta\to 0} \left(\frac{-\sin \pi\theta \cos^3 \pi\theta}{2 \sin \pi\theta}\right)$$

$$= \lim_{\theta\to 0} (-\tfrac{1}{2} \cos^3 \pi\theta)$$

$$= \underline{-\tfrac{1}{2}}.$$

EXERCISE 2

Find the following limits:

1. $\lim_{x\to 0}\left(\dfrac{e^x - 1}{x}\right)$.

2. $\lim_{x\to 0}\left(\dfrac{x - \sin x}{x^3}\right)$.

3. $\lim_{x\to 0}\left[\dfrac{\log_e(1+x) - x}{\sin^2 x}\right]$.

4. $\lim_{x\to 0}\left[\dfrac{e^x - (1+2x)^{\frac{1}{2}}}{\log_e(1+x^2)}\right]$.

5. $\lim_{x\to 1}\left[\dfrac{\log_e x}{(x-1)}\right]$.

6. $\lim_{m\to 0}\left[\dfrac{A}{m^2}\left(\dfrac{\sin mx}{\sin ml} - \dfrac{x}{l}\right)\right]$.

7. $\lim_{m\to 0}\left[\dfrac{1}{m^4}\left(\sec\dfrac{ml}{2} - 1\right) - \dfrac{l^2}{8m^2}\right]$.

8. $\lim_{\omega\to n}\dfrac{\omega E(\cos\omega t - \cos nt)}{L(n^2 - \omega^2)}$.

9. Assuming that L'Hôpital's rule holds for $\lim_{x\to\infty} f(x)/F(x)$, show that $\lim_{x\to\infty} x^n e^{-kx} = 0$, where n is a positive integer and $k > 0$.

ANSWERS

1. 1. 2. $\frac{1}{6}$. 3. $-\frac{1}{2}$.

4. 1. 5. 1. 6. $\dfrac{Ax(l^2 - x^2)}{6l}$.

7. $5l^4/384$.

8. $Et \sin nt/2L$. (This type of limit often occurs in electrical resonance.)

1·4. Derivation of tests for maxima and minima

Taylor's theorem will be used in the form

$$f(a+h) - f(a) = hf'(a) + \frac{h^2}{2!}f''(a) + \frac{h^3}{3!}f'''(a) + \frac{h^4}{4!}f^{\text{iv}}(a) + \dots, \qquad (1)$$

where h is a small quantity, positive or negative.

If h is taken sufficiently small the sign (+ or −) of the right-hand side can be taken as the sign of the first non-zero term.

(a) *Minimum values.* The condition that $f(a)$ is a minimum value of $f(x)$ is that $f(a+h)-f(a)>0$, whether h is positive or negative and however small h may be (see vol. I, § 1·81). Equation (1) then shows that $f'(a)$ must be zero; otherwise $f(a+h)-f(a)$ would change sign with change of sign in h. Thus

$$f(a+h)-f(a)=\frac{h^2}{2!}f''(a)+\frac{h^3}{3!}f'''(a)+\frac{h^4}{4!}f^{\text{iv}}(a)+\dots. \qquad (2)$$

As h^2 cannot be negative, then if $f''(a)>0$,

$$f(a+h)-f(a)>0$$

for h positive or negative.

This gives the usual test for a minimum value, viz.

$$f'(a)=0 \quad \text{and} \quad f''(a)>0. \qquad (3)$$

However, if $f''(a)=0$ equation (2) becomes

$$f(a+h)-f(a)=\frac{h^3}{3!}f'''(a)+\frac{h^4}{4!}f^{\text{iv}}(a)+\dots. \qquad (4)$$

In this case, if $f(a+h)-f(a)$ is not to change sign with h, then $f'''(a)=0$. Also as h^4 is always positive, $f^{\text{iv}}(a)$ must be positive for $f(a+h)-f(a)$ to be positive.

Continuing in this way it is easily seen that a complete test for a minimum value at $x=a$ is: *the first non-vanishing derivative at $x=a$ must be of even order and positive in sign.*

EXAMPLE

If $y=x^4$

$$f'(x)=4x^3, \quad f''(x)=12x^2, \quad f'''(x)=24x, \quad f^{\text{iv}}(x)=+24,$$

$$f'(0)=0, \quad f''(0)=0, \quad f'''(0)=0, \quad f^{\text{iv}}(0)=+24.$$

Thus $x=0$ gives a minimum value of y.

(b) *Maximum values.* For a maximum value of $f(x)$ at $x=a$, $f(a+h)-f(a)<0$ whatever the sign of h and however small h may be.

Proceeding as in (a) above, it is shown that a complete test for a maximum value is that *the first non-vanishing derivative at $x=a$ must be of even order and negative in sign.*

EXAMPLE

Find the turning points of the curve $y = 4\cos x - \cos 2x$.

$$f(x) = 4\cos x - \cos 2x,$$
$$f'(x) = -4\sin x + 2\sin 2x,$$
$$f''(x) = -4\cos x + 4\cos 2x,$$
$$f'''(x) = 4\sin x - 8\sin 2x,$$
$$f^{iv}(x) = 4\cos x - 16\cos 2x.$$

When $f'(x) = 0$,

$$2\sin 2x - 4\sin x = 0.$$
$$4\sin x\cos x - 4\sin x = 0,$$
$$4\sin x(\cos x - 1) = 0,$$

giving $\sin x = 0$ or $\cos x = 1$.

When $\sin x = 0$,

$$x = 0, \pi, 2\pi, 3\pi, \ldots .$$

When $\cos x = 1$,

$$x = 0, 2\pi, 4\pi, \ldots .$$

Thus the values $x = k\pi$ (k any integer) cover all the turning points. (1)

(*a*) *Taking k odd.* Let $k = (2n + 1)$, n any integer.

Then $f''(x) = -4\cos(2n + 1)\pi + 4\cos 2(2n + 1)\pi$

$$= -4\cos\pi + 4\cos 2\pi,$$

as multiples of 2π do not alter the values of a trigonometric ratio.

Thus $f''(x) = -4(-1) + 4(1) = +8$.

Therefore when $x = (2n+1)\pi$, $f(x)$ has a minimum value.

The minimum value is

$$4\cos(2n + 1)\pi - \cos 2(2n + 1)\pi$$
$$= 4\cos\pi - \cos 2\pi$$
$$= -4 - 1$$
$$= -5.$$

The minimum value of y is -5 and occurs when $x = (2n + 1)\pi$. (2)

(*b*) *Taking k even.* Let $k = 2n$, n any integer.

Then
$$f''(x) = -4\cos 2n\pi + 4\cos 4n\pi$$
$$= -4 + 4 = 0,$$
$$f'''(x) = 4\sin 2n\pi - 8\sin 4n\pi$$
$$= 0 - 0 = 0,$$
$$f^{\text{iv}}(x) = 4\cos 2n\pi - 16\cos 4n\pi$$
$$= 4 - 16$$
$$= -12.$$

Therefore when $x = 2n\pi$, $f(x)$ has a maximum value.

The maximum value is
$$4\cos 2n\pi - \cos 4n\pi = 4 - 1$$
$$= 3.$$

The maximum value of y is $+3$ and occurs when $x = 2n\pi$. (3)

EXERCISE 3

These questions include some standard revision examples of maxima and minima.

1. If the first non-vanishing derivative of $f(x)$, when $x = a$, is of odd order higher than the first, what kind of a point is $x = a$?

2. If $f(x) = \cosh x + \cos x$, show that $f(0)$ is a minimum value of $f(x)$.

3. Show that $f(x) = \sinh x - \sin x$ has a point of inflexion at $x = 0$.

4. If in testing for a maximum or a minimum value of $y = f(x)$ it is found that $d^2y/dx^2 = 0$, what further tests must be tried?

A wire 3 m long has to be bent into the form of a rectangle with an external circular loop at one corner, and the rectangle is to have one side double the other. Find the dimensions of the circle and rectangle so that the total area enclosed is a minimum. [L.U.]

5. Find the values of a and b in order that $x^6+ax^5+bx^4$ should have a maximum value when $x=1$ and a minimum value when $x = 3$. What is the nature of the value when $x = 0$?

6. A canning firm makes tins of two sizes in equal quantities. Both sizes have the same diameter but different heights, and the tins may be taken as closed cylinders. The smaller tin is to contain 200 ml and the larger 400 ml. Find the dimensions of each size in order that the total area of sheet-metal used is a minimum.

7. A firm decides to make tins of two varieties, one type a closed cylinder and the other box-shaped with a square cross-section. Each type is to have the same volume. Which type is the more economical from the point of view of tin-plate used, assuming that the dimensions in each case are those for minimum surface area?

8. Find the maximum and minimum values of

$$12\log_e x + 8 + x^2 - 10x.$$

9. Find the points of inflexion of the curve $y = e^{-x^2}$. Also find the maximum value of y, and make a sketch of the curve.

10. A current is given by $i = 6\cos\omega t - \cos 3\omega t$, where ω is a constant. Find all the maximum and minimum values of i within the range $t = 0$ to $2\pi/\omega$. Sketch the graph of the current in this range.

Answers

1. Point of inflexion.

4. Radius 0·196 m; rectangle 0·294 m by 0·589 m.

5. $a = -\frac{24}{5}$, $b = \frac{9}{2}$; at $x = 0, f(x)$ is a minimum.

6. Diameter 72·6 mm; heights 48·4 mm and 96·8 mm.

7. The cylindrical type.

8. Maximum value 0·317, minimum value 0·183 at $x = 2$ and $x = 3$ respectively.

9. Points of inflexion are $(\pm 1/\sqrt{2}, 1/\sqrt{e})$; (0, 1) is the maximum point.

10. $\omega t = 0$, i is a minimum value of $+5$.
$\omega t = \frac{1}{6}\pi$, i is a maximum value of $3\sqrt{3}$.
$\omega t = \frac{5}{6}\pi$, i is a minimum value of $-3\sqrt{3}$.
$\omega t = \pi$, i is a maximum value of -5.
$\omega t = \frac{7}{6}\pi$, i is a minimum value of $-3\sqrt{3}$.
$\omega t = \frac{11}{6}\pi$, i is a maximum value of $3\sqrt{3}$.
$\omega t = 2\pi$, i is a minimum value of $+5$.

1·5. *n*th derivatives

1·51. Simple cases

To find the general nth term in a power series for a function of x, the general nth derivative of the function must be found. In only a few cases is this a simple matter. Some of these are now given.

(*a*) $f(x) = x^m$ (m a positive integer).

$$f'(x) = mx^{m-1},$$

$$f''(x) = m(m-1)x^{m-2},$$

$$f'''(x) = m(m-1)(m-2)x^{m-3},$$

and so on.

Thus, if $m > n$,

$$f^n(x) = m(m-1)(m-2)\dots(m-n+1)x^{m-n}$$

$$= \underline{\frac{m!}{(m-n)!}x^{m-n}}.$$

(*b*) $f(x) = \sin x$.

$$f'(x) = \cos x = \sin(x + \tfrac{1}{2}\pi),$$

$$f''(x) = \cos(x + \tfrac{1}{2}\pi) = \sin\{(x + \tfrac{1}{2}\pi) + \tfrac{1}{2}\pi\} = \sin(x + 2.\tfrac{1}{2}\pi),$$

and so on.

Thus $\underline{f^n(x) = \sin(x + \tfrac{1}{2}n\pi)}.$

Similarly, if $f(x) = \cos x$,

$$f^n(x) = \cos(x + \tfrac{1}{2}n\pi).$$

(*c*) $f(x) = e^{ax}$.

$$f'(x) = ae^{ax},$$

$$f''(x) = a^2e^{ax},$$

and so on.

Thus $$f^n(x) = a^n . e^{ax}.$$

(d) $f(x) = e^{ax} \sin bx.$

$$f'(x) = e^{ax}(a \sin bx + b \cos bx)$$

$$= e^{ax}\sqrt{(a^2+b^2)}\left\{\frac{a}{\sqrt{(a^2+b^2)}} \sin bx + \frac{b}{\sqrt{(a^2+b^2)}} \cos bx\right\}$$

$$= re^{ax} \sin (bx + \alpha),$$

where $\underline{r = \sqrt{(a^2+b^2)}}$ and $\underline{\alpha = \tan^{-1} b/a}$.

Repeating this process it is easily seen that

$$f''(x) = r^2 e^{ax} \sin (bx + 2\alpha),$$

and, in general,

$$f^n(x) = r^n e^{ax} \sin (bx + n\alpha).$$

1·52. Leibniz's theorem

The nth derivative of the product of two functions, say $f(x) = uv$, may be found from a theorem due to Leibniz. The symbols $u_r \equiv d^r u/dx^r$, $v_r \equiv d^r v/dx^r$ will be used.

The theorem then states that

$$\frac{d^n(uv)}{dx^n} = u_n v + n u_{n-1} v_1 + \frac{n(n-1)}{2!} u_{n-2} v_2 + \ldots + u v_n. \quad (1)$$

The coefficients are the same as those in the general binomial expansion.

Thus

$$\frac{d^4(uv)}{dx^4} = u_4 v + 4u_3 v_1 + \frac{4.3}{2!} u_2 v_2 + \frac{4.3.2}{3!} u_1 v_3 + u v_4$$

$$= u_4 v + 4u_3 v_1 + 6u_2 v_2 + 4u_1 v_3 + u v_4.$$

A general proof of the theorem will not be given, but it is easy to satisfy oneself that it is a general extension of the rule for differentiating a product.

Thus

$$\frac{d(uv)}{dx} = v\frac{du}{dx} + u\frac{dv}{dx} = u_1 v + u v_1,$$

$$\frac{d^2(uv)}{dx^2} = \frac{d}{dx}\left(v\frac{du}{dx}\right) + \frac{d}{dx}\left(u\frac{dv}{dx}\right)$$

$$= \left(v\frac{d^2u}{dx^2} + \frac{dv}{dx}\frac{du}{dx}\right) + \left(\frac{du}{dx}\frac{dv}{dx} + u\frac{d^2v}{dx^2}\right)$$

$$= v\frac{d^2u}{dx^2} + 2\frac{dv}{dx}\frac{du}{dx} + u\frac{d^2v}{dx^2}$$

$$= u_2v + 2u_1v_1 + uv_2,$$

$$\frac{d^3(uv)}{dx^3} = \frac{d}{dx}(u_2v) + 2\frac{d}{dx}(u_1v_1) + \frac{d}{dx}(uv_2)$$

$$= (u_3v + u_2v_1) + 2(u_2v_1 + u_1v_2) + (u_1v_2 + uv_3)$$

$$= u_3v + 3u_2v_1 + 3u_1v_2 + uv_3,$$

and so on.

EXAMPLE

Find $$\frac{d^5}{dx^5}(x^4e^x).$$

Take $u = e^x$ and $v = x^4$. Then

$$u = u_1 = u_2 = \ldots = u_5 = e^x$$

and $v_1 = 4x^3, \quad v_2 = 12x^2, \quad v_3 = 24x, \quad v_4 = 24, \quad v_5 = 0.$

Leibniz's theorem then gives

$$\frac{d^5}{dx^5}(x^4e^x) = e^xx^4 + 5e^x4x^3 + 10e^x12x^2 + 10e^x24x + 5e^x24 + e^x0$$

$$= e^x(x^4 + 20x^3 + 120x^2 + 240x + 120).$$

1·53. Use in power series

In a power series expansion of $f(x)$ it is $f^n(0)$ which is required, not $f^n(x)$. If an algebraical relationship can be found between the function and its first, or first two, derivatives, then Leibniz's theorem can be used to find a formula which will give any coefficient required in the expansion. This method is used when it is awkward to find the nth derivative of the function by normal methods.

EXAMPLE

Expand $\dfrac{\sin^{-1} x}{\sqrt{(1-x^2)}}$ as a power series in x, giving the general term.

Let
$$y = f(x) = \frac{\sin^{-1} x}{\sqrt{(1-x^2)}},$$

then
$$\frac{dy}{dx} = f'(x) = \frac{1}{(1-x^2)} + \frac{x \sin^{-1} x}{(1-x^2)^{\frac{3}{2}}}.$$

Thus
$$(1-x^2)\frac{dy}{dx} = 1 + \frac{x \sin^{-1} x}{\sqrt{(1-x^2)}} = 1 + xy,$$

i.e.
$$\underline{(1-x^2)\frac{dy}{dx} = 1 + xy.} \qquad (1)$$

Differentiate both sides of equation (1) n times, using Leibniz's theorem and writing $y_r \equiv d^r y/dx^r$.

$$(1-x^2)y_{n+1} - n.2xy_n - \frac{n(n-1)}{2!}2y_{n-1} = 0 + xy_n + n.1.y_{n-1}.$$

This gives
$$(1-x^2)y_{n+1} = x(2n+1)y_n + n^2 y_{n-1}.$$

Putting $x = 0$, and using functional notation:

$$\underline{f^{n+1}(0) = n^2 f^{n-1}(0).} \qquad (2)$$

Now $f(0) = \sin^{-1} 0 = 0$.
Therefore, from (2),

$$f(0) = f^2(0) = f^4(0) = f^6(0) = \ldots = f^{2r}(0) = 0,$$

i.e. $\qquad$ <u>all even derivatives are zero when $x = 0$.</u> $\qquad$ (3)

From (1), putting $x = 0$,

$$f'(0) = 1.$$

Therefore, from (2),

$$f^3(0) = 2^2.f'(0) = 2^2.1 \qquad \text{(putting } n = 2 \text{ in (2))},$$

and
$$f^5(0) = 4^2.f^3(0) = 4^2.2^2.1 \quad \text{(putting } n = 4 \text{ in (2))}.$$

In general

$$f^{2n+1}(0) = (2n)^2(2n-2)^2 \ldots 4^2 . 2^2 . 1,$$

i.e. $$f^{2n+1}(0) = 4^n(n!)^2. \qquad (4)$$

Using (3) and (4) in Maclaurin's expansion for the function:

$$f(x) = x + \frac{4}{3!}x^3 + \frac{4^2 . (2!)^2}{5!}x^5 + \ldots + \frac{4^n(n!)^2}{(2n+1)!}x^{2n+1} + \ldots .$$

Exercise 4

Find the nth derivatives of:

1. $1/x$. **2.** $\log_e (3x - 4)$. **3.** $\cos 3x$.

4. $e^x \sin x$. **5.** $\log_e \left(\frac{1-x}{1+x}\right)$.

6. If $y = \sin (\log_e x)$, prove that $x^2 \frac{d^2y}{dx^2} + x\frac{dy}{dx} + y = 0$.

Using Leibniz's theorem, show that

$$x^2 y_{n+2} + (2n+1)xy_{n+1} + (n^2+1)y_n = 0.$$

7. If $y = f(x) = e^{x+3x^2}$, prove that $dy/dx = (1 + 6x)y$. Also prove that $f^{n+1}(0) - f^n(0) - 6nf^{n-1}(0) = 0$. Hence expand y as a power series in x as far as x^5. [L.U.]

8. If $y = e^{ax} \cos bx$, prove that $y_n = r^n e^{ax} \cos (bx + n\alpha)$, where $r^2 = a^2 + b^2$ and $\tan \alpha = b/a$. By finding y_n also by Leibniz's theorem prove that

$$r^n \cos (bx + n\alpha) = a^n \cos bx + na^{n-1}\, b \cos (bx + \tfrac{1}{2}\pi)$$
$$+ \frac{n(n-1)}{2!} a^{n-2}\, b^2 \cos (bx + 2 . \tfrac{1}{2}\pi) + \ldots$$
$$+ b^n \cos (bx + \tfrac{1}{2}n\pi).$$ [L.U.]

9. If $y = (1 - x^2)^{\frac{1}{2}} \sin^{-1} x$, show

(i) $(1 - x^2)y_1 + xy = 1 - x^2$;

(ii) $(1 - x^2)y_{n+1} - (2n-1)xy_n - n(n-2)y_{n-1} = 0$ $(n > 2)$.

Hence, or otherwise, expand y in a series of ascending powers of x as far as the term in x^7.

10. Bessel's function of order zero, $J_0(x)$, satisfies the equation $x\dfrac{d^2y}{dx^2}+\dfrac{dy}{dx}+xy=0$, and also the conditions that when $x=0$, $y=1$ and $dy/dx=0$. Find an infinite series in powers of x for $J_0(x)$

11. The function $x/(e^x-1)$ may be expanded in the form

$$\frac{x}{(e^x-1)}=\sum_{n=0}^{\infty}\frac{B_n x^n}{n!}.$$

Determine the coefficients B_n as far as B_4. By showing that $f(x)=\frac{1}{2}x+[x/(e^x-1)]$ is an even function, i.e. that $f(-x)=f(x)$, deduce that B_3, B_5, B_7 ..., are all zero, and derive the expansion

$$\coth\frac{x}{2}=\frac{2}{x}+\frac{2B_2x}{2!}+\frac{2B_4x}{4!}+\dots . \qquad \text{[I.E.E.]}$$

12. Prove that $D^n\cos(\pi x)=\pi^n\cos(\pi x+\frac{1}{2}n\pi)$, and show that the $2m$th derivative of $x^2\cos\pi x$ when $x=1$ has the value $(-1)^{m+1}\pi^{2m-2}(\pi^2+2m-4m^2)$. [L.U.]

Answers

1. $\dfrac{(-1)^n n!}{x^{n+1}}$.

2. $\dfrac{(-1)^{n+1}3^n(n-1)!}{(3x-4)^n}$.

3. $3^n\cos(3x+\frac{1}{2}n\pi)$.

4. $2^{\frac{1}{2}n}e^x\sin(x+\frac{1}{4}n\pi)$.

5. $(n-1)!\left\{\dfrac{(-1)^n}{(1+x)^n}-\dfrac{1}{(1-x)^n}\right\}$. [Use difference of two logs.]

7. $1+x+\dfrac{7x^2}{2!}+\dfrac{19x^3}{3!}+\dfrac{145x^4}{4!}+\dfrac{601x^5}{5!}\dots$.

9. $x-\dfrac{2}{3!}x^3-\dfrac{4\,.\,2^2}{5!}x^5-\dfrac{6\,.\,4^2\,.\,2^2}{7!}x^7\dots$.

10. $1-\dfrac{x^2}{2^2}+\dfrac{x^4}{2^2\,.\,4^2}-\dfrac{x^6}{2^2\,.\,4^2\,.\,6^2}+\dots$.

11. $B_0=1$, $B_1=-\frac{1}{2}$, $B_2=\frac{1}{6}$, $B_3=0$, $B_4=-\frac{1}{30}$.

CHAPTER 2

FURTHER INTEGRATION

It will be assumed that the student is already familiar with all the methods of integration and standard integrals given in volume I. If any were omitted in the A 1 Course, they should be dealt with before proceeding further.

2·1. $\int \frac{1}{x\sqrt{(px^2+qx+r)}}\,dx$

This integral is reduced to the form $\int \frac{1}{\sqrt{(Px^2+Qx+R)}}\,dx$ by the substitution $u = 1/x$; more generally, if the integrand is of the form $\frac{1}{(ax+b)\sqrt{(px^2+qx+r)}}$, the substitution $u = \frac{1}{ax+b}$ will reduce it to a standard form.

EXAMPLE 1

Find $$I = \int \frac{1}{x\sqrt{(5x^2-6x+1)}}\,dx.$$

Let $u = 1/x$, i.e. $xu = 1$.

Then $u\,dx + x\,du = 0$, giving $\frac{dx}{x} = \frac{-du}{u}$.

Also $5x^2 - 6x + 1 = \frac{5}{u^2} - \frac{6}{u} + 1 = \frac{u^2-6u+5}{u^2}$.

$$\therefore \quad I = \int -\frac{u}{\sqrt{(u^2-6u+5)}}\frac{du}{u}$$

$$= \int \frac{-du}{\sqrt{\{(u-3)^2-4\}}}$$

$$= -\cosh^{-1}\left(\frac{u-3}{2}\right) + C$$

$$= -\cosh^{-1}\left(\frac{1-3x}{2x}\right) + C.$$

In the case of a definite integral the limits can be changed to u values, thus avoiding the return to the variable x.

EXAMPLE 2

Evaluate $I = \int_3^4 \frac{1}{(x-2)\sqrt{(x^2-1)}}\,dx.$

Let $$u = \frac{1}{(x-2)},$$

then $(x-2)u = 1$ and $u\,dx + (x-2)\,du = 0$,

giving $$\frac{dx}{(x-2)} = \frac{-du}{u}.$$

Also $$x - 2 = \frac{1}{u},$$

therefore $$x = \frac{1}{u} + 2 = \frac{1+2u}{u}.$$

Thus $$x^2 - 1 = \frac{(1+2u)^2}{u^2} - 1 = \frac{3u^2+4u+1}{u^2}.$$

When $x = 3$, $$u = \frac{1}{3-2} = 1,$$

and when $x = 4$, $$u = \frac{1}{4-2} = \frac{1}{2}.$$

Thus $$I = \int_1^{\frac{1}{2}} \frac{-u}{\sqrt{(3u^2+4u+1)}}\frac{du}{u} = \int_{\frac{1}{2}}^1 \frac{du}{\sqrt{(3u^2+4u+1)}}.$$

$$\therefore \quad I = \frac{1}{\sqrt{3}}\int_{\frac{1}{2}}^1 \frac{du}{\sqrt{(u^2+\frac{4}{3}u+\frac{1}{3})}}$$

$$= \frac{1}{\sqrt{3}}\int_{\frac{1}{2}}^1 \frac{du}{\sqrt{\{(u+\frac{2}{3})^2-\frac{1}{9}\}}}$$

$$= \frac{1}{\sqrt{3}}\left[\cosh^{-1}(3u+2)\right]_{\frac{1}{2}}^1$$

$$= \frac{1}{\sqrt{3}}\left(\cosh^{-1}5 - \cosh^{-1}\tfrac{7}{2}\right)$$

$$= \frac{1}{\sqrt{3}} [\log_e (5 + \sqrt{(25 - 1)}) - \log_e (\tfrac{7}{2} + \sqrt{(\tfrac{49}{4} - 1)})]$$

$$= \frac{1}{\sqrt{3}} \left[\log_e (5 + 2\sqrt{6}) - \log_e \left(\frac{7 + 3\sqrt{5}}{2}\right)\right]$$

$$= \frac{1}{\sqrt{3}} \log_e \frac{2(5 + 2\sqrt{6})}{(7 + 3\sqrt{5})} \simeq \underline{0{\cdot}2124.}$$

2·2. Use of the substitution $t = \tan \frac{1}{2}x$

If $t = \tan \frac{1}{2}x$, differentiating:

$$dt = \tfrac{1}{2} \sec^2 \tfrac{1}{2}x \, dx = \tfrac{1}{2}(1 + t^2)\, dx.$$

Thus
$$dx = \frac{2\,dt}{(1 + t^2)}. \tag{1}$$

From standard trigonometric formulae

$$\sin x = \frac{2t}{1 + t^2}, \quad \cos x = \frac{1 - t^2}{1 + t^2}, \quad \tan x = \frac{2t}{1 - t^2}. \tag{2}$$

Substitutions from (1) and (2) above can be used to advantage when dealing with integrals of the type

$$\int \frac{1}{a + b\cos x}\, dx, \quad \int \frac{1}{a + b \sin x}\, dx.$$

In general, they will reduce to an algebraic form any fraction in which the numerator and denominator contain only first powers of trigonometric ratios of the same angle.

EXAMPLES

(1) $\int \operatorname{cosec} x\, dx.$

Put $t = \tan \frac{1}{2}x$, then

$$dx = \frac{2\,dt}{1 + t^2}, \quad \operatorname{cosec} x = \frac{1}{\sin x} = \frac{1 + t^2}{2t}.$$

$$\therefore \int \operatorname{cosec} x\, dx = \int \frac{(1 + t^2)}{2t} \frac{2\,dt}{(1 + t^2)} = \int \frac{1}{t}\, dt = \log_e t + C$$

$$= \underline{\log_e \tan \tfrac{1}{2}x + C.}$$

Similarly

$$\int \sec x \, dx = \log \left(\frac{1 + \tan \frac{1}{2}x}{1 - \tan \frac{1}{2}x}\right) + C = \underline{\log_e \tan (\tfrac{1}{4}\pi + \tfrac{1}{2}x) + C.}$$

(2) $$\int_{\frac{1}{2}\pi}^{\pi} \frac{dx}{1 - \cos x + \sin x}.$$

Put $t = \tan \frac{1}{2}x$, then

$$dx = \frac{2\,dt}{1 + t^2},$$

and

$$1 - \cos x + \sin x = 1 - \left(\frac{1 - t^2}{1 + t^2}\right) + \frac{2t}{1 + t^2}$$

$$= \frac{1 + t^2 - 1 + t^2 + 2t}{1 + t^2}$$

$$= \frac{2t(1 + t)}{(1 + t^2)}.$$

Also when $x = \frac{1}{2}\pi$, $t = \tan \frac{1}{4}\pi = 1$, and when $x = \pi$, $t = \tan \frac{1}{2}\pi = \infty$.

$$\therefore \quad I = \int_1^{\infty} \frac{(1 + t^2)}{2t(1 + t)} \frac{2}{(1 + t^2)}\, dt = \int_1^{\infty} \frac{1}{t(1 + t)}\, dt$$

$$= \int_1^{\infty} \left(\frac{1}{t} - \frac{1}{1 + t}\right) dt$$

$$= \left[\log_e \left(\frac{t}{t + 1}\right)\right]_1^{\infty}$$

$$= \left[\log_e \left(\frac{1}{1 + 1/t}\right)\right]_1^{\infty}$$

$$= \log_e 1 - \log_e \tfrac{1}{2}$$

$$= \underline{\log_e 2.}$$

Integrals involving $\sinh x$ and $\cosh x$ can be dealt with in a similar manner by putting $t = \tanh \frac{1}{2}x$.

Then $$dt = \tfrac{1}{2}\operatorname{sech}^2 \tfrac{1}{2}x\, dx = \frac{(1-t^2)}{2}\, dx,$$

giving $$dx = \frac{2\, dt}{(1-t^2)}.$$

Also $$\sinh x = \frac{2t}{1-t^2}, \quad \cosh x = \frac{1+t^2}{1-t^2}, \quad \tanh x = \frac{2t}{1+t^2}.$$

EXAMPLE 3

$$I = \int \operatorname{cosech} x\, dx.$$

Let $t = \tanh \frac{1}{2}x$.

Then $dx = \dfrac{2\, dt}{(1-t^2)}$ and $\operatorname{cosech} x = \dfrac{1}{\sinh x} = \dfrac{(1-t^2)}{2t}$.

$$\therefore \quad I = \int \frac{(1-t^2)}{2t}\frac{2\, dt}{(1-t^2)} = \int \frac{1}{t}\, dt = \log_e Ct$$

$$= \log_e C \tanh \tfrac{1}{2}x.$$

2·3. Quotients involving $\sin^2 x$ and $\cos^2 x$

This type of integral can be reduced to an algebraic form by dividing numerator and denominator by $\cos^2 x$.

EXAMPLES

(1) Evaluate $$I = \int_0^{\frac{1}{4}\pi} \frac{dx}{4 - 5\sin^2 x}.$$

Dividing numerator and denominator by $\cos^2 x$:

$$I = \int_0^{\frac{1}{4}\pi} \frac{\sec^2 x\, dx}{4\sec^2 x - 5\tan^2 x} \quad \left(\text{using } \frac{1}{\cos^2 x} = \sec^2 x\right)$$

$$= \int_0^{\frac{1}{4}\pi} \frac{\sec^2 x\, dx}{(4 + 4\tan^2 x) - 5\tan^2 x} = \int_0^{\frac{1}{4}\pi} \frac{\sec^2 x\, dx}{4 - \tan^2 x}.$$

Now put $u = \tan x$, then $du = \sec^2 x\, dx$.

When $x = 0$, $u = \tan 0 = 0$.

When $x = \frac{1}{4}\pi$, $u = \tan \frac{1}{4}\pi = 1$.

$$\therefore I = \int_0^1 \frac{du}{4 - u^2} = \frac{1}{4}\left[\log\left(\frac{2+u}{2-u}\right)\right]_0^1$$

$$= \tfrac{1}{4}(\log_e 3 - \log_e 1) = \underline{\tfrac{1}{4}\log_e 3.}$$

(2) Evaluate $\qquad I = \int \frac{2\sin^2 x\, dx}{1 + \cos^2 x}.$

Writing $1 - \cos^2 x$ for $\sin^2 x$:

$$I = 2\int \frac{1 - \cos^2 x}{1 + \cos^2 x}\, dx$$

$$= -2\int \frac{-1 + \cos^2 x}{1 + \cos^2 x}\, dx$$

$$= -2\int \frac{(1 + \cos^2 x) - 2}{1 + \cos^2 x}\, dx$$

$$= -2\int 1\, dx + 4\int \frac{1}{1 + \cos^2 x}\, dx$$

Alternatively this stage can be accomplished by dividing out

$$= -2x + 4\int \frac{\sec^2 x}{\sec^2 x + 1}\, dx$$

$$= -2x + 4\int \frac{\sec^2 x}{2 + \tan^2 x}\, dx$$

$$= -2x + 4\int \frac{du}{(2 + u^2)}, \quad \text{where } u = \tan x,$$

$$= -2x + \frac{4}{\sqrt{2}}\tan^{-1}\left(\frac{u}{\sqrt{2}}\right) + C$$

$$= \underline{-2x + 2\sqrt{2}\tan^{-1}\left(\frac{1}{\sqrt{2}}\tan x\right) + C.}$$

Integrals involving $\sinh^2 x$ and $\cosh^2 x$ can be dealt with in a similar manner.

EXERCISE 5

Evaluate the following integrals:

1. $\displaystyle\int \frac{dx}{1 + 2\cos x}$.

2. $\displaystyle\int \frac{dx}{3 + 5\cosh x}$.

3. $\displaystyle\int \frac{dx}{(x + 1)\sqrt{(x^2 + 1)}}$.

4. $\displaystyle\int_0^{\frac{1}{2}\pi} \frac{dx}{5 + 13\sin x}$.

5. $\displaystyle\int_0^{\frac{1}{4}\pi} \frac{dx}{1 + 3\cos^2 x}$.

6. $\displaystyle\int_{\frac{1}{5}}^{2} \frac{dx}{x\sqrt{(5x^2 + 4x - 1)}}$.

7. $\displaystyle\int_{\frac{1}{2}}^{\frac{2}{3}} \frac{dx}{x\sqrt{(3x - 2x^2 - 1)}}$. [L.U.]

8. $\displaystyle\int \frac{dx}{4 + 5\sinh^2 x}$.

9. $\displaystyle\int \frac{\cos\theta\, d\theta}{(5 + 4\cos\theta)}$. [L.U.]

10. $\displaystyle\int \frac{dx}{(x - 1)\sqrt{(4 + 3x - x^2)}}$. [L.U.]

11. $\displaystyle\int_0^{\frac{1}{2}\pi} \frac{(\cos x + 2\sin x)}{(\sin x + 2\cos x)}\,dx$.

12. $\displaystyle\int \frac{3\sin x.\cos x\, dx}{\cos^2 x + 2\sin^2 x}$.

13. $\displaystyle\int_1^2 \frac{dx}{x\sqrt{(4x + x^2)}}$.

14. $\displaystyle\int_0^{\frac{1}{2}\pi} \frac{dx}{4 - 2\cos^2 x}$.

15. $\displaystyle\int \frac{5\, dx}{4\cos^2 x + 3\sin^2 x}$.

16. $\displaystyle\int \frac{dx}{5 - 3\cosh^2 x}$.

ANSWERS

1. $\displaystyle\frac{1}{\sqrt{3}}\log_e C\left(\frac{\sqrt{3} + \tan\frac{1}{2}x}{\sqrt{3} - \tan\frac{1}{2}x}\right)$.

2. $\frac{1}{2}\tan^{-1}(\frac{1}{2}\tanh\frac{1}{2}x) + C$.

3. $\displaystyle -\frac{1}{\sqrt{2}}\sinh^{-1}\left(\frac{1 - x}{1 + x}\right) + C$.

4. $\frac{1}{12}\log_e 5$.

5. $\frac{1}{2}\tan^{-1}(\frac{1}{2}) \simeq 0{\cdot}232$.

6. $\frac{2}{3}\pi$.

7. $\frac{1}{2}\pi$.

8. $\frac{1}{2}\tan^{-1}(\frac{1}{2}\tanh x) + C$.

9. $\displaystyle\frac{\theta}{4} - \frac{5}{6}\tan^{-1}(\frac{1}{3}\tan\frac{1}{2}\theta) + C$.

[Write $\cos\theta$ as $\frac{1}{4}(5 + 4\cos\theta - 5)$ in numerator.]

10. $-\dfrac{1}{\sqrt{6}}\cosh^{-1}\left(\dfrac{x+11}{5x-5}\right)+C.$

11. $\frac{2}{5}\pi+\frac{3}{5}\log_e 2.$

[The derivative of $(\sin x+2\cos x)$ is $(\cos x-2\sin x)$. Write $(\cos x+2\sin x)\equiv a(\sin x+2\cos x)+b(\cos x-2\sin x)$ and find a and b by equating coefficients of $\sin x$ and $\cos x$.]

12. $\frac{3}{2}\log C\,(3-\cos 2x)$. [Work in double angles.]

13. $\frac{1}{2}(\sqrt{5}-\sqrt{3})$. **14.** $\dfrac{\sqrt{2}\pi}{8}.$

15. $\dfrac{5\sqrt{3}}{6}\tan^{-1}\left(\dfrac{\sqrt{3}}{2}\tan x\right)+C.$

16. $\dfrac{\sqrt{10}}{20}\log_e C\left(\dfrac{\sqrt{2}+\sqrt{5}\tanh x}{\sqrt{2}-\sqrt{5}\tanh x}\right).$

2·4. Integrals involving 'infinity'

Previously no detailed discussion has been given on integrals with infinite limits or on definite integrals in which the integrand itself becomes infinite within the range of integration.

So far, examples of this nature which have occurred have been simple ones where the result has been intuitively obvious. For example:

$$\int_0^\infty \frac{1}{1+x^2}\,dx=[\tan^{-1}x]_0^\infty=\tan^{-1}\infty-\tan^{-1}0$$

$$=\tfrac{1}{2}\pi,$$

$$\left[\log_e\frac{x}{x+1}\right]_1^\infty=\left[\log_e\frac{1}{1+1/x}\right]_1^\infty$$

$$=\log_e 1-\log_e\tfrac{1}{2}$$

$$=\log_e 2.$$

This section will deal with such integrals in a little more detail.

2·41. Infinite limits

Let $F'(x) = f(x)$.

Then $$\int_a^b f(x)\,dx = [F(x)]_a^b = F(b) - F(a).$$

If $F(b)$ approaches a finite value, l say, as $b \to \infty$, then $\int_a^\infty f(x)\,dx$ is said to have the value $l - F(a)$.

If $F(b)$ becomes immeasurably large as $b \to \infty$ the value of $\int_a^\infty f(x)\,dx$ is said to be infinite.

If $F(b)$ tends to no definite value as $b \to \infty$, then $\int_a^\infty f(x)\,dx$ cannot be evaluated.

An example worked in detail would be

$$\int_1^b \frac{1}{x^2}\,dx = \left[-\frac{1}{x}\right]_1^b = 1 - \frac{1}{b}.$$

As $b \to \infty$, $1/b \to 0$.

Hence $$\int_1^\infty \frac{1}{x^2}\,dx = 1.$$

Using this method for the evaluation at the infinite limit, the work may be set down thus:

$$\int_1^\infty \frac{1}{x^2}\,dx = \left[-\frac{1}{x}\right]_1^\infty = 1 - 0 = 1.$$

Again,

$$\int_2^b \frac{1}{x}\,dx = [\log_e x]_2^b = \log_e b - \log_e 2.$$

As $b \to \infty$, $\log_e b \to \infty$.

Hence $\int_2^\infty \frac{1}{x}\,dx$ is immeasurably large.

Similar reasoning applies if the other limit of integration is infinite.

2·42. Discontinuous integrand

Consider $$\int_0^3 \frac{dx}{\sqrt[3]{(x-2)}}.$$

When $x = 2$, which is *between* 0 and 3, $x - 2$ is zero and the integrand therefore infinite.

That is, the integrand has an infinite discontinuity at $x = 2$ (see vol. I, § 1·71).

To discuss the value of such an integral, it is first split into two parts:

$$\int_0^3 \frac{dx}{\sqrt[3]{(x-2)}} = \int_0^2 \frac{dx}{\sqrt[3]{(x-2)}} + \int_2^3 \frac{dx}{\sqrt[3]{(x-2)}}.$$

The value of x at which the integrand is infinite is made one of the new limits of integration.

If ϵ is a small positive quantity the value of the integral is taken as the limiting value of

$$\int_0^{2-\epsilon} \frac{dx}{\sqrt[3]{(x-2)}} + \int_{2+\epsilon}^3 \frac{dx}{\sqrt[3]{(x-2)}} \quad \text{as} \quad \epsilon \to 0,$$

if such a limit exists. If the limit does not exist, the integral has no value.

Now

$$\int_0^{2-\epsilon} \frac{dx}{\sqrt[3]{(x-2)}} = \tfrac{3}{2}\left[(x-2)^{\frac{2}{3}}\right]_0^{2-\epsilon}$$

$$= \tfrac{3}{2}\left[(-\epsilon)^{\frac{2}{3}} - (-2)^{\frac{2}{3}}\right]$$

$$= \tfrac{3}{2}\left(-2^{\frac{2}{3}} + \epsilon^{\frac{2}{3}}\right) \to -\frac{3 \,.\, 2^{\frac{2}{3}}}{2} \quad \text{as} \quad \epsilon \to 0. \quad (1)$$

Also

$$\int_{2+\epsilon}^3 \frac{dx}{\sqrt[3]{(x-2)}} = \tfrac{3}{2}\left[(x-2)^{\frac{2}{3}}\right]_{2+\epsilon}^3$$

$$= \tfrac{3}{2}\left(1^{\frac{2}{3}} - \epsilon^{\frac{2}{3}}\right) \to \tfrac{3}{2} \quad \text{as} \quad \epsilon \to 0. \qquad (2)$$

The value of $\displaystyle\int_0^3 \frac{dx}{\sqrt[3]{(x-2)}}$ is therefore $-\tfrac{3}{2} \,.\, 2^{\frac{2}{3}} + \tfrac{3}{2}$

$$= \underline{\tfrac{3}{2}(1 - \sqrt[3]{4})}.$$

As long as the above discussion is borne in mind, the working may be written

$$\int_0^3 \frac{dx}{\sqrt[3]{(x-2)}} = \int_0^2 \frac{dx}{\sqrt[3]{(x-2)}} + \int_2^3 \frac{dx}{\sqrt[3]{(x-2)}}$$

$$= \tfrac{3}{2}[(x-2)^{\frac{2}{3}}]_0^2 + \tfrac{3}{2}[(x-2)^{\frac{2}{3}}]_2^3$$

$$= \tfrac{3}{2}[0-(-2)^{\frac{2}{3}}] + \tfrac{3}{2}[1^{\frac{2}{3}} - 0]$$

$$= \tfrac{3}{2}(1 - \sqrt[3]{4}).$$

2·43. Measurement of area

(*a*) *The case of infinite limits.* Fig. 1 is a sketch of the graph of $y = \dfrac{1}{1+x^2}$. The area bounded by OX, OY, the curve and the ordinate $x = a$ is measured by

$$\int_0^a \frac{dx}{1+x^2} = [\tan^{-1} x]_0^a = \tan^{-1} a.$$

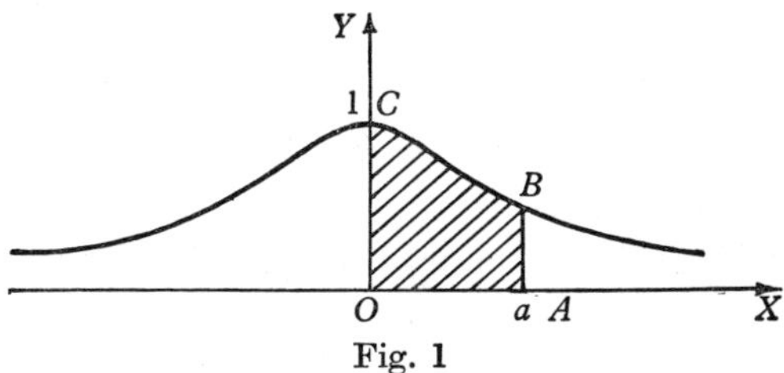

Fig. 1

When the point A moves off to infinity along the x-axis, $a \to \infty$ and $\tan^{-1} a$ becomes $\frac{1}{2}\pi$.

The area between OX, OY and that part of the curve for which x is positive is said to be measured by the integral $\displaystyle\int_0^\infty \frac{dx}{1+x^2}$ and has a value $\frac{1}{2}\pi$.

Other 'infinite' integrals can give rise to associated areas (or volumes) in a similar manner.

When the value of the integral is infinite (as in the second example in § 2·41) the area or volume concerned is immeasurably large and is said to be infinite.

(*b*) *The case of a discontinuous integrand.* Fig. 2 shows a sketch of the graph of $y = \dfrac{1}{\sqrt[3]{(x-2)}}$.

If the area between the curve and the x-axis is required from $x = 0$ to $x = 3$, it must be considered in two portions as already done in § 2·42.

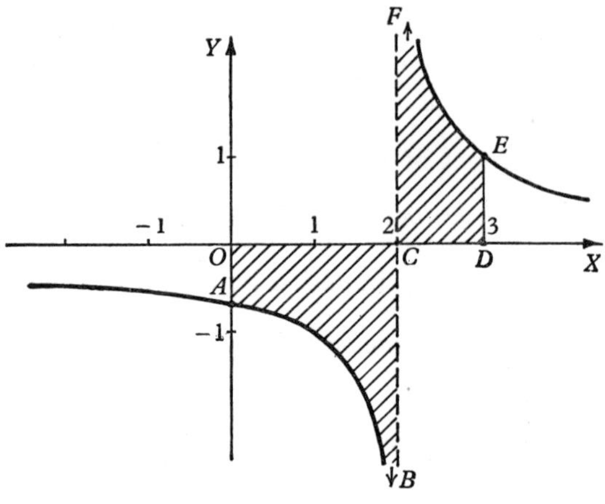

Fig. 2

Equation (1) of § 2·42 shows the area $OABC$ to be $\frac{3}{2}\sqrt[3]{4}$ square units. (The negative sign merely indicates that the area is below the x-axis.)

Equation (2) of § 2·42 shows the area $CDEF$ to be $\frac{3}{2}$ square units.

Thus the total area is $\frac{3}{2}(1 + \sqrt[3]{4})$ square units.

2·44. Further examples

(1) Find, if it exists, the value of $\displaystyle\int_0^3 \frac{dx}{\sqrt{(3-x)}}$.

$$\int_0^{3-\epsilon} \frac{dx}{\sqrt{(3-x)}} = -2[\sqrt{(3-x)}]_0^{3-\epsilon}$$

$$= -2[\sqrt{\epsilon} - \sqrt{3}].$$

As $\epsilon \to 0$, $\sqrt{\epsilon} \to 0$.

Therefore $$\int_0^3 \frac{dx}{\sqrt{(3-x)}} = \underline{2\sqrt{3}.}$$

(2) Find, if it exists, the value of $\int_1^\infty \frac{dx}{x(x^2+1)}$.

$$\frac{1}{x(x^2+1)} = \frac{1}{x} - \frac{x}{x^2+1} \quad \text{(by partial fractions).}$$

$$\therefore \int_1^a \frac{dx}{x(x^2+1)} = \int_1^a \frac{1}{x}\,dx - \int_1^a \frac{x\,dx}{(x^2+1)}$$

$$= [\log_e x]_1^a - \tfrac{1}{2}[\log_e (x^2+1)]_1^a$$

$$= \log_e a - \tfrac{1}{2}\log_e (a^2+1) + \tfrac{1}{2}\log_e 2$$

$$= \log_e \frac{a}{\sqrt{(a^2+1)}} + \tfrac{1}{2}\log_e 2$$

$$= \log_e \frac{1}{\sqrt{\left(1+\frac{1}{a^2}\right)}} + \tfrac{1}{2}\log_e 2.$$

As $a \to \infty$, $\frac{1}{\sqrt{\left(1+\frac{1}{a^2}\right)}} \to 1$,

therefore $\log_e \frac{1}{\sqrt{\left(1+\frac{1}{a^2}\right)}} \to \log_e 1 = 0$.

Thus
$$\underline{\int_1^\infty \frac{dx}{x(x^2+1)} = \tfrac{1}{2}\log_e 2.}$$

(3) Sketch the graph of $y^2 = \frac{2-x}{x}$ and find the total area between the curve and the y-axis.

Points for sketching the graph:

(i) It is symmetrical about the x-axis.

(ii) When $x < 0$, y^2 is negative and y is not real.

(iii) When $x > 2$, y^2 is negative and y is not real.

(iv) When $x = 0$, y is infinitely large.

(v) When $x = 2$, $y = 0$.

The curve is shown in fig. 3 below.

By symmetry the required area is $2\int_0^\infty x\,dy$.

Now
$$y^2 = \frac{2-x}{x} = \frac{2}{x} - 1.$$

$$\therefore x = \frac{2}{1+y^2}.$$

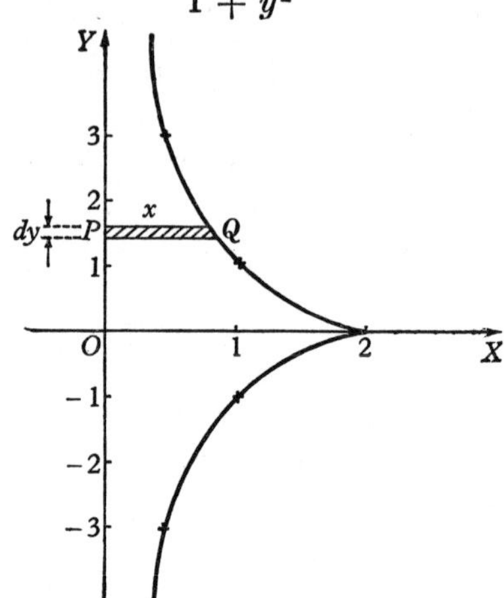

Fig. 3

Thus the area is
$$\begin{aligned} 2\int_0^\infty \frac{2}{1+y^2}\,dy &= 4[\tan^{-1} y]_0^\infty \\ &= 4[\tan^{-1}\infty - \tan^{-1} 0] \\ &= 4\,.\,\tfrac{1}{2}\pi \\ &= 2\pi. \end{aligned}$$

Exercise 6

Where possible, evaluate the following integrals:

1. $\displaystyle\int_2^\infty \frac{dx}{x^3}$.
2. $\displaystyle\int_1^\infty \frac{dx}{\sqrt{x}}$.
3. $\displaystyle\int_0^\infty \frac{dx}{x^2+16}$.
4. $\displaystyle\int_{-\infty}^{+\infty} \frac{4x}{x^4+1}\,dx$.
5. $\displaystyle\int_{-\infty}^{1} e^x\,dx$.
6. $\displaystyle\int_{-2}^{+2} \frac{dx}{x^3}$.
7. $\displaystyle\int_0^1 \frac{dx}{\sqrt{(1-x^2)}}$.
8. $\displaystyle\int_0^1 \frac{dx}{x(3+x)}$.

9. $\displaystyle\int_1^\infty \frac{dx}{x(2+x)}$. **10.** $\displaystyle\int_0^2 \log x\,dx$. [Take $\lim_{x\to 0}(x\log x) = 0$.]

11. Find the area between the x-axis, the line $x = 3$ and that part of the curve $y = 1/x^3$ for which $x > 3$.

12. Sketch the curve $y^2 = \dfrac{x-1}{3-x}$ and find the area between the curve and the line $x = 3$.

13. Show that the area contained between the curve $y = 1/x$ and the x-axis for all positive values of x greater than 1 is infinitely large; but the volume of revolution formed by rotating this 'area' about the x-axis can be determined accurately and evaluate it.

ANSWERS

1. $\frac{1}{8}$. **2.** Infinitely large. **3.** $\frac{1}{8}\pi$.

4. 0. **5.** $e = 2{\cdot}718$. **6.** Does not exist.

7. $\frac{1}{2}\pi$. **8.** Infinitely large.

9. $\frac{1}{2}\log_e 3$. **10.** $2\log_e 2 - 2$. **11.** $\frac{1}{18}$.

12. 2π. [Take strips parallel to x-axis.] **13.** π.

2·5. Reduction formulae

2·51. Introduction and method

Consider the integral $\int x^n \sin x\,dx$, where n is a positive integer. It may be denoted by the symbol

$$I_n = \int x^n \sin x\,dx. \qquad (1)$$

Thus, I_4 would be $\int x^4 \sin x\,dx$,

and I_{n-2} would be $\int x^{n-2} \sin x\,dx$.

Integrating by parts in equation (1) gives

$$I_n = -x^n \cos x - \int nx^{n-1}(-\cos x)\,dx$$

$$= -x^n \cos x + n\int x^{n-1} \cos x\,dx.$$

Integrating by parts a second time gives

$$I_n = -x^n \cos x + nx^{n-1} \sin x - n(n-1) \int x^{n-2} \sin x \, dx.$$

The integral on the right-hand side has the power of x *reduced* by two; otherwise it is the same form as the original integral. In symbolic notation

$$I_n = -x^n \cos x + nx^{n-1} \sin x - n(n-1) I_{n-2}. \qquad (2)$$

This is an example of a reduction formula. It can be applied successively until the final integral on the right-hand side contains only $x \sin x$ if n is odd or $x^0 \sin x$ if n is even. This simplifies the evaluation of the original integral, especially in the case of definite integrals.

EXAMPLE 1

Evaluate $$\int_0^{\frac{1}{2}\pi} x^4 \sin x \, dx.$$

Let $$I_4 = \int_0^{\frac{1}{2}\pi} x^4 \sin x \, dx.$$

From (2) above

$$I_n = [-x^n \cos x + nx^{n-1} \sin x]_0^{\frac{1}{2}\pi} - n(n-1)I_{n-2},$$

i.e. $$I_n = n(\tfrac{1}{2}\pi)^{n-1} - n(n-1)I_{n-2}. \qquad (1)$$

Thus $I_4 = 4(\frac{1}{2}\pi)^3 - 4.3.I_2$, putting $n = 4$.

But from (1), $I_2 = 2(\frac{1}{2}\pi) - 2.1I_0$, putting $n = 2$.

Therefore

$$I_4 = 4(\tfrac{1}{2}\pi)^3 - 4.3\{2(\tfrac{1}{2}\pi) - 2.1I_0\},$$

i.e. $$I_4 = \tfrac{1}{2}\pi^3 - 12\pi + 24I_0$$

$$= \tfrac{1}{2}\pi^3 - 12\pi + 24 \int_0^{\frac{1}{2}\pi} x^0 \sin x \, dx$$

$$= \tfrac{1}{2}\pi^3 - 12\pi + 24 \int_0^{\frac{1}{2}\pi} \sin x \, dx$$

$$= \tfrac{1}{2}\pi^3 - 12\pi + 24[-\cos x]_0^{\frac{1}{2}\pi}.$$

Thus $$\underline{I_4 = \tfrac{1}{2}\pi^3 - 12\pi + 24.}$$

EXAMPLE 2

Prove that

$$\int \frac{x^n\,dx}{\sqrt{(a^2+x^2)}} = \frac{x^{n-1}\sqrt{(a^2+x^2)}}{n} - \frac{(n-1)a^2}{n}\int \frac{x^{n-2}\,dx}{\sqrt{(a^2+x^2)}}.$$

Hence evaluate $\displaystyle\int_0^2 \frac{x^5\,dx}{\sqrt{(5+x^2)}}$. [L.U.]

Let $$I_n = \int \frac{x^n\,dx}{\sqrt{(a^2+x^2)}} = \int x^{n-1}\frac{x}{\sqrt{(a^2+x^2)}}\,dx. \qquad (1)$$

Now $$\frac{d}{dx}\sqrt{(a^2+x^2)} = \frac{x}{\sqrt{(a^2+x^2)}}.$$

Thus, integrating (1) by parts,

$$I_n = x^{n-1}\sqrt{(a^2+x^2)} - \int (n-1)x^{n-2}\sqrt{(a^2+x^2)}\,dx$$

$$= x^{n-1}\sqrt{(a^2+x^2)} - (n-1)\int \frac{x^{n-2}(a^2+x^2)}{\sqrt{(a^2+x^2)}}\,dx, \qquad (2)$$

$$= x^{n-1}\sqrt{(a^2+x^2)} - (n-1)\int \frac{a^2x^{n-2}+x^n}{\sqrt{(a^2+x^2)}}\,dx$$

$$= x^{n-1}\sqrt{(a^2+x^2)} - (n-1)a^2\int \frac{x^{n-2}}{\sqrt{(a^2+x^2)}}\,dx - (n-1)\int \frac{x^n}{\sqrt{(a^2+x^2)}}\,dx,$$

i.e. $I_n = x^{n-1}\sqrt{(a^2+x^2)} - (n-1)a^2I_{n-2} - (n-1)I_n.$

$\therefore\ I_n\{1+(n-1)\} = x^{n-1}\sqrt{(a^2+x^2)} - (n-1)a^2I_{n-2},$

giving $$I_n = \frac{x^{n-1}\sqrt{(a^2+x^2)}}{n} - \frac{(n-1)}{n}a^2I_{n-2}. \qquad (3)$$

Notice the technique used in equation (2) to obtain the original integral again on the right-hand side.

Let $$I_5 = \int_0^2 \frac{x^5\,dx}{\sqrt{(5+x^2)}}.$$

Using (3), with $n = 5$ and $a^2 = 5$,

$$I_5 = \left[\frac{x^4\sqrt{(5+x^2)}}{5}\right]_0^2 - \tfrac{4}{5}.5I_3.$$

$$\therefore\ I_5 = \tfrac{48}{5} - 4I_3. \tag{4}$$

Also from (3):

$$I_3 = \left[\frac{x^2\sqrt{(5+x^2)}}{3}\right]_0^2 - \tfrac{2}{3}.5I_1 \quad \text{(putting } n = 3\text{)}$$

$$= 4 - \tfrac{10}{3}I_1. \tag{5}$$

Thus $I_5 = \frac{48}{5} - 16 + \frac{40}{3}I_1$ (from (4) and (5)).

Now $I_1 = \int_0^2 \frac{x}{\sqrt{(5+x^2)}}\,dx = [\sqrt{(5+x^2)}]_0^2 = 3 - \sqrt{5}.$

Thus $I_5 = \frac{48}{5} - 16 + \frac{40}{3}.(3 - \sqrt{5}) = \underline{\frac{168}{5} - \frac{40}{3}\sqrt{5}}.$

Note. This example has been done by the method stated in the wording of the question, but in actual practice an integral such as $\int_0^2 \frac{x^5dx}{\sqrt{(5+x^2)}}$ is best done by a substitution such as $x = \sqrt{5}\sinh u$, or $u^2 = 5 + x^2$.

2·52. Reduction formulae for $\int \tan^n x\,dx$ and $\int \cot^n x\,dx$

Let $I_n = \int \tan^n x\,dx$, where n is a positive integer greater than one.

Now $\tan^n x = \tan^{n-2} x \tan^2 x = \tan^{n-2} x(\sec^2 x - 1).$

$$\therefore\ I_n = \int \tan^{n-2} x \sec^2 x\,dx - \int \tan^{n-2} x\,dx$$

$$= \int \tan^{n-2} x\,d(\tan x) - I_{n-2}.$$

$$\therefore\ \underline{I_n = \frac{1}{n-1}\tan^{n-1} x - I_{n-2}.} \tag{1}$$

By repeated application of this formula, I_n may be reduced to

(i) $I_1 = \int \tan x\,dx = \log_e \sec x$, if n is odd, or

(ii) $I_0 = \int 1.dx = x$, if n is even.

EXAMPLE

$$I_5 = \int \tan^5 x\,dx.$$

From (1),

$$\begin{aligned} I_5 &= \tfrac{1}{4}\tan^4 x - I_3, \quad \text{putting } n = 5 \\ &= \tfrac{1}{4}\tan^4 x - \{\tfrac{1}{2}\tan^2 x - I_1\}, \quad \text{putting } n = 3 \\ &= \tfrac{1}{4}\tan^4 x - \tfrac{1}{2}\tan^2 x + I_1 \\ &= \tfrac{1}{4}\tan^4 x - \tfrac{1}{2}\tan^2 x + \int \tan x\,dx \\ &= \tfrac{1}{4}\tan^4 x - \tfrac{1}{2}\tan^2 x + \log_e C\sec x. \end{aligned}$$

In a similar manner, using $\cot^2 x = \operatorname{cosec}^2 x - 1$, if $I_n = \int \cot^n x\,dx \; (n > 1)$, then

$$I_n = -\frac{\cot^{n-1} x}{(n-1)} - I_{n-2}. \qquad (2)$$

EXAMPLE

$$I_4 = \int \cot^4 x\,dx.$$

From (2),

$$\begin{aligned} I_4 &= -\frac{\cot^3 x}{3} - I_2, \quad \text{putting } n = 4, \\ &= -\frac{\cot^3 x}{3} - \{-\cot x - I_0\}, \quad \text{putting } n = 2, \\ &= -\frac{\cot^3 x}{3} + \cot x + \int 1.dx \\ &= x - \frac{\cot^3 x}{3} + \cot x + C. \end{aligned}$$

Exercise 7

1. If $I_n = \int x^n e^{2x}\,dx$, show that $I_n = \frac{x^n e^{2x}}{2} - \frac{n}{2} I_{n-1}$ and deduce that $I_n = \frac{x^n e^{2x}}{2} - \frac{n x^{n-1} e^{2x}}{2^2} + \frac{n(n-1)}{2^2} I_{n-2}$.

Hence evaluate $\int x^3 e^{2x}\,dx$.

2. If $I_n = \int_0^\infty x^n e^{-x}\,dx$, show that $I_n = nI_{n-1}$. Deduce that if n is a positive integer, the value of I_n is $n!$.

3. If
$$I_n = \int_0^a \frac{x^n\,dx}{\sqrt{(3a^2 + x^2)}},$$
show that
$$I_n = \frac{2a^n}{n} - \frac{3(n-1)a^2}{n} I_{n-2}.$$

Evaluate I_7. [L.U.]

4. If
$$I_n = \int (\log_e x)^n\,dx,$$
show that
$$I_n = x(\log_e x)^n - nI_{n-1}.$$

Use this formula to evaluate $\int (\log_e x)^4\,dx$.

5. Obtain the reduction formula
$$\int x^n \sin x\,dx = nx^{n-1} \sin x - x^n \cos x - n(n-1) \int x^{n-2} \sin x\,dx.$$

Hence evaluate

(i) $\int_0^\pi x^5 \sin x\,dx$, (ii) $\int_0^{\frac{1}{2}\pi} x^2 \sin x\,dx$. [L.U.]

6. Evaluate

(i) $\int_0^{\frac{1}{4}\pi} \tan^6 x\,dx$, (ii) $\int_{\frac{1}{4}\pi}^{\frac{1}{2}\pi} \cot^5 x\,dx$, (iii) $\int_0^{\frac{1}{4}\pi} \tan^7 x\,dx$.

ANSWERS

1. $e^{2x}(\frac{1}{2}x^3 - \frac{3}{4}x^2 + \frac{3}{4}x - \frac{3}{8}) + C.$

3. $\frac{a^7}{35}(432\sqrt{3} - 746).$

4. $x\{(\log x)^4 - 4(\log x)^3 + 12(\log x)^2 - 24\log x + 24\} + C.$

5. (i) $\pi^5 - 20\pi^3 + 120\pi$; (ii) $\pi - 2$.

6. (i) $\frac{13}{15} - \frac{1}{4}\pi$; (ii) $\frac{1}{2}\log 2 - \frac{1}{4}$; (iii) $\frac{5}{12} - \frac{1}{2}\log 2$.

2·53. $\int \sin^n x\,dx$ and $\int \cos^n x\,dx$. Wallis's formulae

Let $I_n = \int \sin^n x\,dx$, where n is a positive integer.

Then $I_n = \int \sin^{n-1} x . \sin x\,dx = -\int \sin^{n-1} x(-\sin x\,dx)$.

Integrating by parts, noting that $d(\cos x) = -\sin x\,dx$:

$$I_n = -\sin^{n-1} x \cos x + \int \{(n-1)\sin^{n-2} x \cos x\} \cos x\,dx$$

$$= -\sin^{n-1} x \cos x + (n-1)\int \sin^{n-2} x \cos^2 x\,dx$$

$$= -\sin^{n-1} x \cos x + (n-1)\int \sin^{n-2} x(1 - \sin^2 x)\,dx$$

$$= -\sin^{n-1} x \cos x + (n-1)\int \sin^{n-2} x\,dx - (n-1)\int \sin^n x\,dx.$$

Thus

$$I_n = -\sin^{n-1} x \cos x + (n-1)I_{n-2} - (n-1)I_n,$$

giving

$$nI_n = -\sin^{n-1} x \cos x + (n-1)I_{n-2}$$

and

$$I_n = -\frac{1}{n}\sin^{n-1} x \cos x + \frac{(n-1)}{n} I_{n-2}. \qquad (1)$$

In a similar manner, if $I_n = \int \cos^n x \, dx$,

$$I_n = \frac{1}{n} \cos^{n-1} x \sin x + \frac{(n-1)}{n} I_{n-2}. \tag{2}$$

If the integrals in (1) and (2) are taken between the limits 0 to $\frac{1}{2}\pi$ they both reduce to

$$I_n = \frac{(n-1)}{n} I_{n-2}, \tag{3}$$

as

$$\left[-\frac{1}{n} \sin^{n-1} x \cos x\right]_0^{\frac{1}{2}\pi} = 0 \quad \text{and} \quad \left[\frac{1}{n} \cos^{n-1} x \sin x\right]_0^{\frac{1}{2}\pi} = 0.$$

Putting $(n-2)$ for n in (3) gives

$$I_{n-2} = \frac{(n-3)}{(n-2)} I_{n-4}.$$

Thus $$I_n = \frac{(n-1)(n-3)}{n(n-2)} I_{n-4}.$$

Working in this way and successively reducing the integral on the right-hand side, the final results are:

(i) *n even:*

$$I_n = \frac{(n-1)(n-3)(n-5) \ldots 3.1}{n(n-2)(n-4) \ldots 4.2} I_0.$$

Now $$I_0 = \int_0^{\frac{1}{2}\pi} 1.dx = \tfrac{1}{2}\pi,$$

in cases of both $\cos x$ and $\sin x$.

$$\therefore \; I_n = \frac{(n-1)(n-3) \ldots 3.1}{n(n-2) \ldots 4.2} \frac{\pi}{2}. \tag{4}$$

(ii) *n odd:*

$$I_n = \frac{(n-1)(n-3) \ldots 4.2}{n(n-2) \ldots 5.3} I_1.$$

When the integral is

$$\int_0^{\frac{1}{2}\pi} \sin^n x \, dx, \quad I_1 = \int_0^{\frac{1}{2}\pi} \sin x \, dx = 1.$$

When the integral is

$$\int_0^{\frac{1}{2}\pi} \cos^n x\,dx, \quad I_1 = \int_0^{\frac{1}{2}\pi} \cos x\,dx = 1.$$

Thus, in both cases

$$I_n = \frac{(n-1)(n-3)\ldots 4.2}{n(n-2)\ldots 5.3}. \tag{5}$$

EXAMPLES

(1) $$\int_0^{\frac{1}{2}\pi} \sin^7 x\,dx$$

Using (5), $$I_7 = \frac{6.4.2}{7.5.3} = \frac{16}{35}.$$

(2) $$\int_0^{\frac{1}{2}\pi} \cos^6 x\,dx$$

Using (4), $$I_6 = \frac{5.3.1}{6.4.2}\frac{\pi}{2} = \tfrac{5}{32}\pi.$$

Summary and Abbreviations. $7\times5\times3\times1$ and $8\times6\times4\times2$, in which the factors decrease by 2 at each stage, are similar to factorial numbers except they decrease by 2 instead of 1.

For $7\times6\times5\times4\times3\times2\times1$ the symbol 7! is used. It is not surprising that $7\times5\times3\times1$ is often written as 7!!. $8\times6\times4\times2$ would be written 8!!.

The symbol !! denotes that the factors decrease by 2 at each step, ending with the factor 2 or 1.

With this notation (4) and (5) may be written

$$\int_0^{\frac{1}{2}\pi} \sin^n x\,dx = \int_0^{\frac{1}{2}\pi} \cos^n x\,dx = \frac{(n-1)!!}{n!!} \times \frac{\pi}{2} \text{ or } 1,$$ according as n is even or odd.

These are known as Wallis's formulae.

2·54. $\int_0^{\frac{1}{2}\pi} \sin^m x \cos^n x\,dx$ (m, n positive integers)

Integrals of this type, when m and n are reasonably small, have been dealt with in volume I (§ 5·53).

A reduction formula can be obtained as follows:

Let
$$I_{m,n} = \int_0^{\frac{1}{2}\pi} \sin^m x \cos^n x \, dx$$
$$= \int_0^{\frac{1}{2}\pi} \cos^{n-1} x \sin^m x \cos x \, dx$$
$$= \int_0^{\frac{1}{2}\pi} \cos^{n-1} x \sin^m x \, d(\sin x).$$

Integrating by parts:

$$I_{m,n} = \left[\cos^{n-1} x \frac{\sin^{m+1} x}{(m+1)}\right]_0^{\frac{1}{2}\pi}$$
$$- \int_0^{\frac{1}{2}\pi} \frac{\sin^{m+1} x}{(m+1)} (n-1) \cos^{n-2} x(-\sin x) \, dx$$
$$= 0 + \frac{(n-1)}{(m+1)} \int_0^{\frac{1}{2}\pi} \sin^{m+2} x \cos^{n-2} x \, dx$$
$$= \frac{(n-1)}{(m+1)} \int_0^{\frac{1}{2}\pi} \sin^m x(1 - \cos^2 x) \cos^{n-2} x \, dx$$
$$= \frac{(n-1)}{(m+1)} \int_0^{\frac{1}{2}\pi} \sin^m x \cos^{n-2} x \, dx$$
$$- \frac{(n-1)}{(m+1)} \int_0^{\frac{1}{2}\pi} \sin^m x \cos^n x \, dx.$$

Thus
$$I_{m,n} = \frac{(n-1)}{(m+1)} I_{m,n-2} - \frac{(n-1)}{(m+1)} I_{m,n}.$$

Solving for $I_{m,n}$:

$$I_{m,n} = \frac{(n-1)}{(m+n)} I_{m,n-2}. \tag{1}$$

Similarly, starting with

$$I_{m,n} = \int_0^{\frac{1}{2}\pi} \sin^{m-1} x \cos^n x \sin x \, dx$$
$$= \int_0^{\frac{1}{2}\pi} - \sin^{m-1} x \cos^n x \, d(\cos x),$$

it is easily shown that

$$I_{m,n} = \frac{(m-1)}{(m+n)} I_{m-2,n}. \tag{2}$$

It follows that the powers of sin x and cos x can be reduced by two in successive stages:

From (1), writing $(n-2)$ for n,

$$I_{m,n-2} = \frac{(n-3)}{(m+n-2)} I_{m,n-4}.$$

Substitution of this value into (1) then gives

$$I_{m,n} = \frac{(n-1)(n-3)}{(m+n)(m+n-2)} I_{m,n-4}. \tag{3}$$

Putting $(n-4)$ for n in (2) gives

$$I_{m,n-4} = \frac{(m-1)}{(m+n-4)} I_{m-2,n-4}.$$

Substituting this value into (3) gives

$$I_{m,n} = \frac{(n-1)(n-3)(m-1)}{(m+n)(m+n-2)(m+n-4)} I_{m-2,n-4}. \tag{4}$$

Continuing in this way, noticing that both numerator and denominator drop in twos and the number of factors in the numerator is always the same as the number of factors in the denominator, it is seen that the final result depends on the values of m and n.

(i) *m and n both even:*

$$I_{m,n} = \frac{(n-1)(n-3)\ldots 3.1.(m-1)(m-3)\ldots 3.1}{(m+n)(m+n-2)\ldots 4.2} I_{0,0}.$$

(ii) *m and n both odd:*

$$I_{m,n} = \frac{(n-1)(n-3)\ldots 4.2.(m-1)(m-3)\ldots 4.2}{(m+n)(m+n-2)\ldots 6.4} I_{1,1}.$$

(iii) *m odd, n even:*

$$I_{m,n} = \frac{(n-1)(n-3)\ldots 3.1.(m-1)(m-3)\ldots 4.2}{(m+n)(m+n-2)\ldots 5.3} I_{1,0}.$$

(iv) *m even, n odd:*

$$I_{m,n} = \frac{(n-1)(n-3)\ldots 4.2.(m-1)(m-3)\ldots 3.1}{(m+n)(m+n-2)\ldots 5.3} I_{0,1}.$$

Now

$$I_{0,0} = \int_0^{\frac{1}{2}\pi} 1\,dx = \tfrac{1}{2}\pi; \quad I_{1,1} = \int_0^{\frac{1}{2}\pi} \sin x \cos x\,dx = [\tfrac{1}{2}\sin^2 x]_0^{\frac{1}{2}\pi} = \tfrac{1}{2};$$

$$I_{1,0} = \int_0^{\frac{1}{2}\pi} \sin x\,dx = 1 \quad \text{and} \quad I_{0,1} = \int_0^{\frac{1}{2}\pi} \cos x\,dx = 1.$$

The four cases can thus be summarized as

$$I_{m,n} = \frac{(n-1)!!(m-1)!!}{(m+n)!!} \times \tfrac{1}{2}\pi \text{ or } 1, \quad \text{according as}$$

both m and n are even or one or both of m and n are odd.

EXAMPLES

(1) $$\int_0^{\frac{1}{2}\pi} \sin^5 x \cos^3 x\,dx = I_{5,3} = \frac{(4\,.\,2)\,.\,2}{8\,.\,6\,.\,4\,.\,2}\,.\,1 = \frac{1}{24}.$$

(2) $$\int_0^{\frac{1}{2}\pi} \sin^6 x \cos^4 x\,dx = I_{6,4} = \frac{(5\,.\,3\,.\,1)\,.\,(3\,.\,1)}{10\,.\,8\,.\,6\,.\,4\,.\,2}\cdot\frac{\pi}{2} = \frac{3\pi}{512}.$$

(3) $$\int_0^{\frac{1}{2}\pi} \sin^5 x \cos^4 x\,dx = I_{5,4} = \frac{(4\,.\,2)\,.\,(3\,.\,1)}{9\,.\,7\,.\,5\,.\,3}\,.\,1 = \frac{8}{315}.$$

EXERCISE 8

Evaluate:

1. $\int_0^{\frac{1}{2}\pi} \sin^3 x\,dx.$

2. $\int_0^{\frac{1}{2}\pi} \cos^7 x\,dx.$

3. $\int_0^{\frac{1}{2}\pi} \sin^2 x \cos^6 x\,dx.$

4. $\int_0^{\frac{1}{2}\pi} \sin^4 x \cos^5 x\,dx.$

5. $\int_0^{\frac{1}{2}\pi} \sin^9 x\,dx.$

6. $\int_0^{\frac{1}{2}\pi} \cos^{10} x\,dx.$

7. Evaluate $\int_0^3 (9 - x^2)^{\frac{3}{2}}\,dx$, using the substitution $x = 3\sin\theta$.

8. Evaluate $\int_0^a x^4(a^2 - x^2)^{\frac{5}{2}}\,dx$, using the substitution $x = a\sin\theta$.

9. Show that

$$n\int \cos^n x\,dx = \cos^{n-1} x \sin x + (n-1)\int \cos^{n-2} x\,dx.$$

Use the result to evaluate $\int_0^{\frac{1}{2}\pi} \cos^n x\,dx$ for the values $n = 3$ and $n = 4$. Check the value obtained when $n = 3$ by some other method of integration.

10. Evaluate

$$\text{(i)} \int_0^1 \frac{x^5\,dx}{\sqrt{(1-x^2)}}, \qquad \text{(ii)} \int_0^1 \frac{x^6\,dx}{\sqrt{(1-x^2)}}.$$

ANSWERS

1. $\frac{2}{3}$. **2.** $\frac{16}{35}$. **3.** $\frac{5}{256}\pi$.

4. $\frac{8}{315}$. **5.** $\frac{128}{315}$. **6.** $\frac{63}{512}\pi$.

7. $\frac{243}{16}\pi$. **8.** $\dfrac{3a^{10}}{512}\pi$.

9. $\dfrac{5}{12}\sqrt{2}$; $\dfrac{3\pi}{32}+\dfrac{1}{4}$. **10.** (i) $\frac{8}{15}$; (ii) $\frac{5}{32}\pi$.

2·6. Some properties of definite integrals

2·61. Introduction

There are several properties of definite integrals which are useful in simplifying the integral before evaluation.

Before proceeding, the student should notice that the value of a definite integral is a function of the limits of integration and *not* a function of the variable in the integrand.

For example:

$$\int_1^2 x^2\,dx = [\tfrac{1}{3}x^3]_1^2 = \tfrac{1}{3}(8-1) = \tfrac{7}{3},$$

$$\int_1^2 z^2\,dz = [\tfrac{1}{3}z^3]_1^2 = \tfrac{1}{3}(8-1) = \tfrac{7}{3}.$$

Similarly

$$\int_a^{2a} x^2\,dx = \tfrac{7}{3}a^3,$$

$$\int_a^{2a} z^2\,dz = \tfrac{7}{3}a^3.$$

It does not matter what letter, x, z, y, etc., is put as the variable in the integrand:

$$\int_a^{2a} f(x)\,dx = \int_a^{2a} f(z)\,dz = \int_a^{2a} f(y)\,dy.$$

In the following it is taken that

$$F'(x) = f(x) \quad \text{and thus} \quad \int f(x)\,dx = F(x) + C.$$

It follows at once that

$$\int_a^b f(x)\,dx = F(b) - F(a) = \int_b^a - f(x)\,dx \qquad (1)$$

and $\int_a^c f(x)\,dx = F(c) - F(a) = \{F(c) - F(b)\} + \{F(b) - F(a)\}$

$$= \int_a^b f(x)\,dx + \int_b^c f(x)\,dx. \qquad (2)$$

2·62. $\int_{-a}^{a} f(x)\,dx = 0$ or $2\int_0^a f(x)\,dx$, according as $f(x)$ is an odd or even function

$$\int_{-a}^{a} f(x)\,dx = \int_{-a}^{0} f(x)\,dx + \int_0^a f(x)\,dx, \text{ from (2) of § 2·61.} \qquad (1)$$

Substituting $z = -x$ into $\int_{-a}^{0} f(x)\,dx$ gives

$$\int_{-a}^{0} f(x)\,dx = \int_a^0 - f(-z)\,dz = \int_0^a f(-z)\,dz, \text{ from (1) of § 2·61}$$

$$= \int_0^a f(-x)\,dx. \qquad (2)$$

From (1) and (2) above

$$\int_{-a}^{a} f(x)\,dx = \int_0^a \{f(x) + f(-x)\}\,dx. \qquad (3)$$

If $f(x)$ is odd, $f(-x) = -f(x)$, therefore from (3) above

$$\underline{\int_{-a}^{a} f(x)\,dx = 0.}$$

If $f(x)$ is even, $f(-x) = f(x)$, therefore from (3) above

$$\underline{\int_{-a}^{a} f(x)\,dx = 2\int_0^a f(x)\,dx.}$$

Taking the definite integrals as representing areas under curves, these results are obvious from typical graphs of odd and even functions. Fig. 4 shows a sketch of the two cases.

The result follows if it is remembered that areas beneath the x-axis are taken as negative in sign.

Note. It would be advisable at this stage to read the discussion on odd and even functions given in volume I, § 9·5.

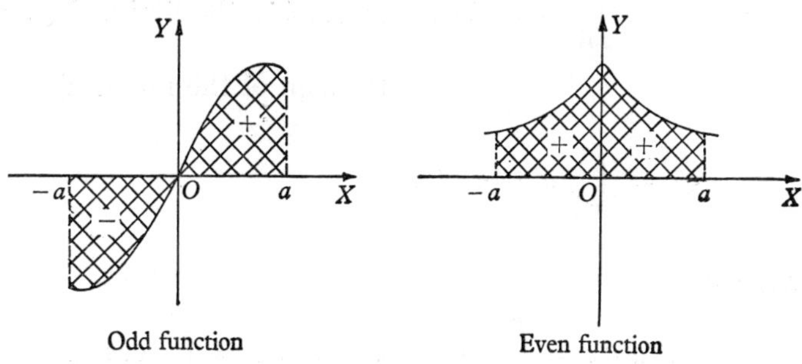

Fig. 4

EXAMPLES

(1) $\int_{-\frac{1}{2}\pi}^{\frac{1}{2}\pi} \sin^7 x\,dx = 0$, as $\sin^7 x$ is an odd function of x.

(2) $\int_{-\frac{1}{2}\pi}^{\frac{1}{2}\pi} \sin^6 x\,dx = 2\int_0^{\frac{1}{2}\pi} \sin^6 x\,dx,$ as $\sin^6 x$ is an even function of x

$$= 2\,\frac{5.3.1}{6.4.2}\,\frac{\pi}{2} = \frac{5\pi}{16}.$$

(3) $\int_{-\pi}^{\pi} \sin^5 x \cos^2 x\,dx = 0$, as $\sin^5 x\cos^2 x$ is an odd function.

(4) $\int_{-a}^{a} \frac{x^3}{a^2+x^2}\,dx = 0$, as $\frac{x^3}{a^2+x^2}$ is an odd function.

2·63. $\int_0^a f(x)\,dx = \int_0^a f(a-x)\,dx$

Substitute $x = a - z$ in $\int_0^a f(x)\,dx$.

Then $dx = -dz$. When $x = 0$, $z = a$ and when $x = a$, $z = 0$.

$$\therefore \int_0^a f(x)\,dx = \int_a^0 f(a-z)(-dz)$$

$$= -\int_a^0 f(a-z)\,dz$$

$$= \int_0^a f(a-z)\,dz \quad \text{(reversing the limits changes the sign of the integral)}$$

$$= \int_0^a f(a-x)\,dx.$$

EXAMPLES

(1) $$\int_0^{\frac{1}{2}\pi} \sin^n x\,dx = \int_0^{\frac{1}{2}\pi} \sin^n(\tfrac{1}{2}\pi - x)\,dx = \int_0^{\frac{1}{2}\pi} \cos^n x\,dx,$$

and in general

$$\int_0^{\frac{1}{2}\pi} f(\sin x)\,dx = \int_0^{\frac{1}{2}\pi} f(\cos x)\,dx,$$

where f is any function.

(2) $$\int_0^1 x(1-x)^7\,dx = \int_0^1 (1-x)\{1-(1-x)\}^7\,dx$$

$$= \int_0^1 (1-x)x^7\,dx$$

$$= [\tfrac{1}{8}x^8 - \tfrac{1}{9}x^9]_0^1 = \tfrac{1}{72}.$$

(3) $$\int_0^{\pi} \cos^7 x\,dx = \int_0^{\pi} \cos^7(\pi - x)\,dx = -\int_0^{\pi} \cos^7 x\,dx,$$

as $$\cos(\pi - x) = -\cos x.$$

$$\therefore 2\int_0^{\pi} \cos^7 x\,dx = 0,$$

i.e. $$\int_0^{\pi} \cos^7 x\,dx = 0.$$

2·64. $\int_0^{2a} f(x)\,dx = 0$ **or** $2\int_0^a f(x)\,dx$, **according as**

$$f(2a - x) = -f(x) \text{ or } +f(x)$$

$$\int_0^{2a} f(x)\,dx = \int_0^a f(x)\,dx + \int_a^{2a} f(x)\,dx. \tag{1}$$

Substitute $x = 2a - z$ in $\int_a^{2a} f(x)\,dx$.

Then $dx = -dz$. When $x = a, z = a$ and when $x = 2a, z = 0$.

Thus
$$\int_a^{2a} f(x)\,dx = \int_a^0 f(2a - z)(-dz)$$
$$= -\int_a^0 f(2a - z)\,dz$$
$$= \int_0^a f(2a - x)\,dx.$$

Putting this in (1) gives

$$\int_0^{2a} f(x)\,dx = \int_0^a f(x)\,dx + \int_0^a f(2a - x)\,dx$$
$$= \int_0^a \{f(x) + f(2a - x)\}\,dx.$$

The property given in the heading of this section is now obvious.

It is a useful property in dealing with trigonometric functions because

$$\sin(\pi - x) = \sin x, \qquad \cos(\pi - x) = -\cos x,$$
$$\sin(2\pi - x) = -\sin x, \quad \cos(2\pi - x) = \cos x.$$

Examples

(1)
$$\int_0^{\pi} \sin^7 x\,dx = 2\int_0^{\frac{1}{2}\pi} \sin^7 x\,dx$$
$$= 2\,\frac{6.4.2}{7.5.3} = \underline{\frac{32}{35}}.$$

(2) $\displaystyle\int_0^{\pi} \sin^6 x \cos^3 x \, dx = 0$, as $\sin^6 (\pi - x) \cos^3 (\pi - x)$

$$= - \sin^6 x \cos^3 x.$$

(3) $\displaystyle\int_0^{2\pi} \cos^5 x \, dx = 2 \int_0^{\pi} \cos^5 x \, dx$, as $\cos^5 (2\pi - x) = \cos^5 x$

$$= 0, \text{ as } \cos^5 (\pi - x) = - \cos^5 x.$$

(4) $\displaystyle\int_0^{2\pi} \sin^6 x \, dx = 2 \int_0^{\pi} \sin^6 x \, dx$, as $\sin^6 (2\pi - x) = \sin^6 x$

$$= 4 \int_0^{\frac{1}{2}\pi} \sin^6 x \, dx, \text{ as } \sin^6 (\pi - x) = \sin^6 x$$

$$= 4 \frac{5 . 3 . 1}{6 . 4 . 2} \times \tfrac{1}{2}\pi = \tfrac{5}{8}\pi.$$

Note. Even if the student does not remember how to *prove* these properties, he should familiarize himself with their *use.*

Exercise 9

Evaluate:

1. $\displaystyle\int_0^{\pi} \sin^9 x \, dx.$

2. $\displaystyle\int_0^{2} x^2(2 - x)^7 \, dx.$

3. $\displaystyle\int_{-\frac{1}{2}\pi}^{\frac{1}{2}\pi} \sin^2 x \cos^6 x \, dx.$

4. $\displaystyle\int_0^{2\pi} \sin^2 x \cos^7 x \, dx.$

5. $\displaystyle\int_{-2}^{+2} \frac{x^5}{9 + x^2} \, dx.$

6. Show that $I = \displaystyle\int_0^{\frac{1}{2}\pi} \frac{\sin x \, dx}{\sin x + \cos x} = \int_0^{\frac{1}{2}\pi} \frac{\cos x \, dx}{\sin x + \cos x}.$

Hence show that $2I = \displaystyle\int_0^{\frac{1}{2}\pi} 1 . dx$ and $I = \frac{1}{4}\pi.$

7. Evaluate $\displaystyle\int_{\pi}^{2\pi} \cos^9 x \, dx.$

8. Evaluate $\displaystyle\int_{-\frac{1}{2}\pi}^{\pi} \sin^7 x \, dx.$

9. Evaluate $\displaystyle\int_0^{3} x^3 \sqrt{(3 - x)} \, dx.$

10. Evaluate $\displaystyle\int_0^{\frac{1}{2}\pi} \frac{\cos x - \sin x}{\cos x + \sin x} \, dx.$

Answers

1. $\frac{256}{315}$.	**2.** $\frac{128}{45}$.	**3.** $\frac{5}{128}\pi$.
4. 0.	**5.** 0.	**7.** 0.
8. $\frac{16}{35}$.	**9.** $\frac{288}{35}\sqrt{3}$.	**10.** 0.

2·7. The mean-value theorem of integration

The (first) mean-value theorem of differentiation was given in volume I and later used in connexion with partial derivatives. The mean-value theorem of integration will now be given and later used in connexion with the differentiation of an integral (see § 3·3). The theorem will be deduced intuitively from geometrical considerations; a rigorous proof is beyond the scope of this book.

The student is already familiar with the mean value of a function over a given range. If the range is $x = a$ to $x = b$, the mean value, M, of $f(x)$ is given by

$$M = \frac{1}{(b-a)}\int_a^b f(x)\,dx,$$

i.e.

$$\underline{M.(b-a) = \int_a^b f(x)\,dx.} \qquad (1)$$

Referring to fig. 5, this means geometrically that M is the height of the rectangle, on base length $(b - a)$, which has an area equal to the area under the curve from $x = a$ to $x = b$.

Now if G and L are the greatest and least values of $f(x)$ between $x = a$ and $x = b$, the mean value M obviously lies between G and L. Assuming the function is continuous, $f(x)$ must take *all* values intermediate between L and G as x goes from a to b; therefore there must be some value of x, say $x = c$, between a and b for which $f(c) = M$.

As c lies between a and b it may be written

$$c = a + \theta(b-a), \quad \text{where} \quad 0 < \theta < 1.$$

Thus

$$M = f\{a + \theta(b-a)\},$$

and from (1)

$$\underline{\int_a^b f(x)\,dx = (b-a)\,f\{a+\theta(b-a)\}.} \qquad (2)$$

This is called the (first) mean-value theorem of integration.

In fig. 5 the area under the curve obviously lies between the values of the areas of the smallest and largest rectangles (shown dotted), i.e.

$$\underline{L(b-a) < \int_a^b f(x)\,dx < G(b-a).} \qquad (3)$$

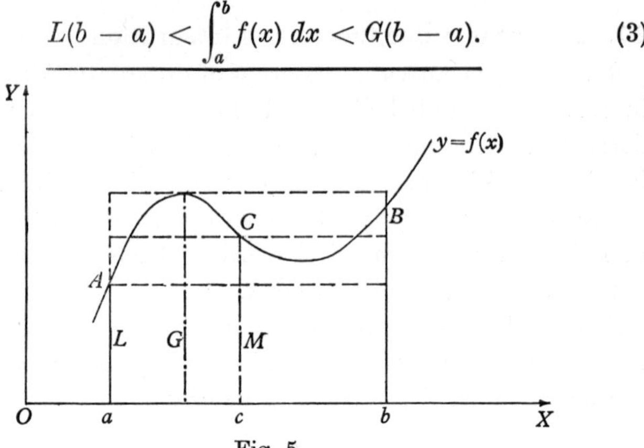

Fig. 5

This is sometimes used to find a rough approximation to the value of an integral. Limits can be found outside which the value of the integral cannot lie.

EXAMPLE

Estimate roughly $I = \int_0^{0\cdot2} e^{-x^2}\,dx$.

In the interval $x = 0$ to $0\cdot2$ the greatest value of e^{-x^2} is $e^0 = 1$, and the least value is $e^{-0\cdot04} \simeq 0\cdot9608$.

$$\therefore \; 0\cdot9608(0\cdot2 - 0) < I < 1(0\cdot2 - 0),$$

i.e.
$$0\cdot19216 < I < 0\cdot2.$$

The value therefore certainly lies between 0·192 and 0·2.

EXERCISE 10

1. Without evaluating the integrals accurately, show that

$$\int_0^{\frac{1}{2}\pi} \sin^8 x\,dx < \int_0^{\frac{1}{2}\pi} \sin^7 x\,dx,$$

and in general show that if $n > 0$,

$$0 < \int_0^{\frac{1}{2}\pi} \sin^{n+1} x\,dx < \int_0^{\frac{1}{2}\pi} \sin^n x\,dx.$$

2. Find limits between which $\int_{1\cdot5}^{1\cdot6} \log_e x^2\,dx$ must lie.

3. Find limits between which $\int_{1}^{1\cdot1} xe^{x^2}\,dx$ must lie.

Evaluate the integral accurately by using the substitution $u = x^2$ and find the value c of x such that

$$(1\cdot1 - 1)ce^{c^2} = \int_{1}^{1\cdot1} xe^{x^2}\,dx,$$

showing that c lies between 1 and 1·1.

Answers

1. [Discuss the integrals as areas under curves.]
2. Between 0·0811 and 0·094.
3. Between 0·272 and 0·369. $\frac{1}{2}(e^{1\cdot21} - e) \simeq 0\cdot318$. $c \simeq 1\cdot05$.

[Use Newton's method to find approximate solution to $0\cdot1ce^{c^2} = 0\cdot318$.]

Miscellaneous Exercises on Integration

These exercises include revision of the work done in volume I. In the harder cases hints are given in the Answers.

1. Evaluate:

(i) $\int \frac{(x-1)\,dx}{(x+1)(x^2+1)}$; (ii) $\int_1^2 x^2 \log_e x\,dx$;

(iii) $\int_0^{\frac{1}{2}\pi} \sin^2\theta \cos^4\theta\,d\theta$.

(iv) The density of a sphere of radius a at a distance r from the centre is $k(1 - 2r/3a)$. Prove that the mean density is half the density at the centre. [L.U.]

2. Prove that $\int_0^n x^2(n-x)^p\,dx = \int_0^n x^p(n-x)^2\,dx$ and find the common value of the two integrals. [L.U.]

3. Find the following indefinite integrals:

(i) $\int \tan^4\theta\,d\theta$; (ii) $\int \frac{x^3\,dx}{(a^2+x^2)^2}$.

Evaluate correct to three significant figures:

$$\int_{0\cdot5}^{2} x^3 \log_e x \, dx. \qquad \text{[L.U.]}$$

4. Evaluate:

(i) $\displaystyle\int_0^{\frac{1}{4}\pi} \sec^3 \theta \, d\theta$; (ii) $\displaystyle\int \frac{dx}{\cosh x}$;

(iii) $\displaystyle\int \frac{d\theta}{5 + 3\cos\theta}$. [L.U.]

5. Evaluate:

(i) $\displaystyle\int \frac{(x-1)\,dx}{\sqrt{(4+3x-x^2)}}$; (ii) $\displaystyle\int_2^5 \frac{2x^2-3x+4}{(x-1)^2}\,dx$;

(iii) $\displaystyle\int_0^\infty \frac{dx}{(x^3+8)}$. [L.U.]

6. Put $e^x = z$ and show that

$$\int_0^\infty \operatorname{sech} x \, dx = [2\tan^{-1}(e^x)]_0^\infty = \tfrac{1}{2}\pi. \qquad \text{[L.U.]}$$

7. By using the substitution $2 - x = \dfrac{1}{t}$, or otherwise, find

$$\int \frac{dx}{(2-x)\sqrt{(x^2-3x+1)}}. \qquad \text{[C]}$$

8. Evaluate:

(i) $\displaystyle\int \tan^3 x\,dx$; (ii) $\displaystyle\int \frac{x^4+x+3}{x^2-1}\,dx$.

9. Prove the rule for integration by parts.

Find $\displaystyle\int x\sin^2 x\,dx$, $\displaystyle\int \frac{\log_e x}{x^2}\,dx$. [C]

10. Evaluate:

(i) $\displaystyle\int_0^a x^3(a^2-x^2)^{\frac{1}{2}}\,dx$; (ii) $\displaystyle\int \frac{x^2+2}{x(x-1)^2}\,dx$;

(iii) $\displaystyle\int_0^2 \frac{x^2\,dx}{\sqrt{(4x^2+9)}}$. [C]

11. Prove that

$$\int_0^a f(x)\,dx = \int_0^a f(a-x)\,dx.$$

Evaluate $$\int_0^\pi \frac{x \sin x\,dx}{1+\cos^2 x}.$$ [C]

12. Evaluate $$\int_0^{\frac{1}{2}\pi} \frac{x \sin 2x\,dx}{1+\cos^2 2x}$$

and show that

$$\int_0^{\frac{1}{2}\pi} \frac{a \sin x + b \cos x}{\sin x + \cos x}\,dx = \frac{(a+b)}{4}\pi.$$ [L.U.]

13. Find

(i) $\int x(\log_e x)^2\,dx$; (ii) $\int \frac{dx}{\cos x - \cos^2 x}$.

Evaluate $$\int_0^{\frac{1}{2}a} x^2\sqrt{(a^2-x^2)}\,dx.$$ [C]

14. (i) Find

$$\int \frac{1+x}{1+\cos x}\,dx \quad \text{and} \quad \int \frac{(x+1)}{x^2(x^2+1)}\,dx.$$

(ii) Evaluate $$\int_0^1 x^2(1-x)^{\frac{1}{3}}\,dx.$$ [C]

15. Show that

$$\int_0^{\log_e 2} \frac{dx}{\sinh x + 5\cosh x} = \frac{1}{\sqrt{6}}\left(\tan^{-1}\sqrt{6} - \tan^{-1}\frac{\sqrt{6}}{2}\right).$$

16. If $I_n = \int \sec^n \theta\,d\theta$, show that, when $n \geqslant 2$,

$$(n-1)I_n = \sec^{n-2}\theta \tan\theta + (n-2)I_{n-2}.$$

Show that

$$8\int_0^{\frac{1}{4}\pi} \sec^5\theta\,d\theta = 7\sqrt{2} + 3\log_e(1+\sqrt{2}),$$

and evaluate

$$\int_0^a \frac{dx}{(2a^2-x^2)^3}.$$ [L.U.]

17. Using the fact that

$$\sin nx - \sin (n-2)x = 2 \cos (n-1)x \, . \sin x,$$

show that $\dfrac{\sin nx}{\sin x} = 2 \cos (n-1)x + \dfrac{\sin (n-2)x}{\sin x}$.

Hence prove that if $\displaystyle\int \frac{\sin nx}{\sin x}\, dx = I_n$, then

$$I_n = \frac{2 \sin (n-1)x}{(n-1)} + I_{n-2}.$$

Deduce the value of $\displaystyle\int_{\frac{1}{4}\pi}^{\frac{1}{2}\pi} \frac{\sin 4x}{\sin x}\, dx$.

18. A spherical surface of radius a with centre at O carries an electric charge of surface density σ, symmetrically distributed with respect to an axis OX. P is a point on OX such that $OP = x$. The potential at P is given by

$$\int_0^{\pi} \frac{a^2\sigma \sin \theta \, d\theta}{2\epsilon_0(a^2 + x^2 - 2ax \cos \theta)^{\frac{1}{2}}}.$$

Evaluate this integral, with ϵ_0 and σ constant, in the two cases $x > a$ and $x < a$. [I.E.E.]

Answers

1. (i) $\frac{1}{2} \log_e (x^2 + 1) - \log_e C(x + 1)$; (ii) $\dfrac{8}{3} \log_e 2 - \dfrac{7}{9}$;

(iii) $\dfrac{(3\pi + 4)}{192}$.

2. $n^{p+3}\left(\dfrac{1}{p+1} - \dfrac{2}{p+2} + \dfrac{1}{p+3}\right)$.

3. (i) $\frac{1}{3} \tan^3 \theta - \tan \theta + \theta + C$;

(ii) $\frac{1}{2} \log_e C(a^2 + x^2) + \dfrac{a^2}{2(a^2 + x^2)}$; (iii) 1·79.

4. (i) 1·148; (ii) $2 \tan^{-1} (\tanh \frac{1}{2}x) + C$;

(iii) $\frac{1}{2} \tan^{-1} (\frac{1}{2} \tan \frac{1}{2}\theta) + C$.

5. (i) $-\sqrt{(4+3x-x^2)}+\frac{1}{2}\sin^{-1}\left(\frac{2x-3}{5}\right)+C$;

(ii) $\dfrac{33}{4}+\log_e 4$; (iii) $\dfrac{\pi\sqrt{3}}{18}$.

[*Note.* $x^3+8=(x+2)(x^2-2x+4)$.]

7. $\sin^{-1}\dfrac{(4-x)}{\sqrt{5}(2-x)}+C$.

8. (i) $\frac{1}{2}\tan^2 x-\log_e C\sec x$;

(ii) $\frac{1}{3}x^3+x+\frac{1}{2}\log_e(x^2-1)+2\log_e C\left(\dfrac{x-1}{x+1}\right)$.

9. $\dfrac{x^2}{4}-\dfrac{x\sin 2x}{4}-\dfrac{\cos 2x}{8}+C$ [use $\sin^2 x=\frac{1}{2}(1-\cos 2x)$]

$-\dfrac{1}{x}\log_e x-\dfrac{1}{x}+C$.

10. (i) $\dfrac{2}{15}a^5$; (ii) $\log_e\dfrac{Cx^2}{(x-1)}-\dfrac{3}{(x-1)}$;

(iii) $\frac{5}{4}-\frac{9}{16}\log_e 3$. [Put $2x=3\sinh u$].

11. $\frac{1}{4}\pi^2$.

12. $\frac{1}{16}\pi^2$ [use $\int_0^{\frac{1}{2}\pi} f(x)\,dx=\int_0^{\frac{1}{2}\pi} f(\frac{1}{2}\pi-x)\,dx$].

13. (i) $\frac{1}{2}x^2[(\log_e x)^2-\log_e x+\frac{1}{2}]+C$;

(ii) $\log_e(\sec x+\tan x)-\cot\frac{1}{2}x+C$

$$\left[\text{use } \frac{1}{\cos x-\cos^2 x}=\frac{1}{\cos x}+\frac{1}{1-\cos x}=\sec x+\tfrac{1}{2}\operatorname{cosec}^2\tfrac{1}{2}x\right];$$

(iii) $\dfrac{a^4}{16}\left(\dfrac{\pi}{3}-\dfrac{\sqrt{3}}{4}\right)$.

14. (i) $(1 + x) \tan \frac{1}{2}x - 2 \log_e C \sec \frac{1}{2}x$

$$\left[\text{use } \frac{1}{1 + \cos x} = \tfrac{1}{2} \sec^2 \tfrac{1}{2}x \text{ and then integrate by parts}\right];$$

(ii) $\log_e x - \frac{1}{2} \log_e C\,(x^2 + 1) - \frac{1}{x} - \tan^{-1} x$;

(iii) $\frac{27}{140}$ $\left[\text{use } \int_0^1 f(x)\,dx = \int_0^1 f(1 - x)\,dx\right]$.

15. [It is best here to write

$$\sinh x = \tfrac{1}{2}(e^x - e^{-x}),\ \cosh x = \tfrac{1}{2}(e^x + e^{-x}).$$

Then put $u = e^x$ and the new limits are $u = 1$ to $u = 2$.]

16. [Write $I_n = \int \sec^{n-2} \theta \sec^2 \theta \, d\theta$ and integrate by parts, taking $\sec^2 \theta \, d\theta$ as $d(\tan \theta)$. In the resulting right-hand side integral write $\tan^2 \theta = (\sec^2 \theta - 1)$.

For last part put $x = \sqrt{2}a \sin \theta$ and use the first part.]

Ans. $[7\sqrt{2} + 3 \log_e (1 + \sqrt{2})]/32\sqrt{2}a^5$.

17. $\frac{4}{3}(1 - \sqrt{2})$.

18. $x > a$; $a^2\sigma/x\epsilon_0$. $x < a$; $a\sigma/\epsilon_0$.

[*Note.* It is instructive to verify that the integral given does represent the potential at P.]

CHAPTER 3

FURTHER PARTIAL DIFFERENTIATION

3·1. Use of differentials in forming partial derivatives

In finding partial derivatives of functions of several variables it is often convenient to use differentials.

Suppose that a functional relationship is given by $f(x, y) = 0$. The total differential is given by (see vol. I, p. 192)

$$df = 0 = \frac{\partial f}{\partial x}\,dx + \frac{\partial f}{\partial y}\,dy. \tag{1}$$

Thus
$$\frac{dy}{dx} = -\frac{\partial f}{\partial x}\Big/\frac{\partial f}{\partial y}. \tag{2}$$

The symbol dy/dx has been used, and not $\partial y/\partial x$, because y is a function of a *single* variable x.

Now let a functional relationship be given by $f(x, y, z) = 0$. Theoretically any one of the three quantities x, y and z can be expressed as a function of the other two.

The total differential is given by

$$df = 0 = \frac{\partial f}{\partial x}\,dx + \frac{\partial f}{\partial y}\,dy + \frac{\partial f}{\partial z}\,dz. \tag{3}$$

Suppose that the partial derivative of z with respect to x keeping y constant is needed. This may be written $(\partial z/\partial x)_y$. Now as y is kept constant, $dy = 0$, thus from (3)

$$\frac{\partial f}{\partial x}\,dx + \frac{\partial f}{\partial z}\,dz = 0,$$

from which
$$\left(\frac{\partial z}{\partial x}\right)_y = -\frac{\partial f}{\partial x}\Big/\frac{\partial f}{\partial z}. \tag{4}$$

Similarly, if
$$f(x, y, z, t) = 0,$$

then
$$df = 0 = \frac{\partial f}{\partial x}\,dx + \frac{\partial f}{\partial y}\,dy + \frac{\partial f}{\partial z}\,dz + \frac{\partial f}{\partial t}\,dt. \tag{5}$$

If $(\partial y/\partial x)_{z,t}$ is required, then as z and t are kept constant,

$$dz = dt = 0$$

and
$$\frac{\partial f}{\partial x}\,dx + \frac{\partial f}{\partial y}\,dy = 0,$$

giving
$$\frac{\partial y}{\partial x} = -\frac{\partial f}{\partial x}\bigg/\frac{\partial f}{\partial y}. \tag{6}$$

In a similar manner equation (5) can be used to obtain any of the possible partial derivatives which might be required. Keeping two of the variables x, y, z, t constant the number of possible first partial derivatives is twelve, viz.

$$\left(\frac{\partial y}{\partial x}\right)_{z,t},\quad \left(\frac{\partial z}{\partial x}\right)_{y,t},\quad \left(\frac{\partial z}{\partial y}\right)_{x,t},\quad \left(\frac{\partial t}{\partial x}\right)_{y,z},\quad \left(\frac{\partial t}{\partial y}\right)_{x,z},\quad \left(\frac{\partial t}{\partial z}\right)_{x,y},$$

$$\left(\frac{\partial x}{\partial y}\right)_{z,t},\quad \left(\frac{\partial x}{\partial z}\right)_{y,t},\quad \left(\frac{\partial y}{\partial z}\right)_{x,t},\quad \left(\frac{\partial x}{\partial t}\right)_{y,z},\quad \left(\frac{\partial y}{\partial t}\right)_{x,z},\quad \left(\frac{\partial z}{\partial t}\right)_{x,y}.$$

The example which follows was given as an exercise in volume I, but can now be done in a neater way.

Example

If any one of x, y, z is expressed in terms of the other two by means of the relation $2xyz - 5x + 2y - 3z = 1$, prove that

$$\left(\frac{\partial y}{\partial z}\right)_x \left(\frac{\partial z}{\partial x}\right)_y \left(\frac{\partial x}{\partial y}\right)_z = -1. \qquad \text{[L.U.]}$$

Let $f(x, y, z) = 2xyz - 5x + 2y - 3z - 1 = 0.$

Then

$$df = 0 = (2yz - 5)\,dx + (2xz + 2)\,dy + (2xy - 3)\,dz. \tag{1}$$

When x is constant, $dx = 0$ and

$$(2xz + 2)\,dy + (2xy - 3)\,dz = 0,$$

giving
$$\left(\frac{\partial y}{\partial z}\right)_x = -\frac{(2xy - 3)}{(2xz + 2)}. \tag{2}$$

Similarly, when y is constant, $dy = 0$, and (1) gives

$$\left(\frac{\partial z}{\partial x}\right)_y = -\frac{(2yz - 5)}{(2xy - 3)}. \tag{3}$$

Again, when z is constant $dz = 0$ and (1) gives

$$\left(\frac{\partial x}{\partial y}\right)_z = -\frac{(2xz + 2)}{(2yz - 5)}. \qquad (4)$$

On multiplication (2), (3) and (4) give

$$\left(\frac{\partial y}{\partial z}\right)_x \left(\frac{\partial z}{\partial x}\right)_y \left(\frac{\partial x}{\partial y}\right)_z = -1.$$

3·11. Interrelationships between variables

If several relationships between the variables are given, the partial derivatives will depend on which variables are taken as independent ones.

For example, consider the equations

$$x = r \sin\theta \cos\phi, \qquad (1)$$

$$y = r \sin\theta \sin\phi, \qquad (2)$$

$$z = r \cos\theta. \qquad (3)$$

(These are actually the Cartesian co-ordinates of a point in space expressed in terms of spherical polar co-ordinates r, θ and ϕ.)

Here there are potentially six variables x, y, z, r, θ and ϕ; but they are connected by the three equations (1), (2) and (3). Thus, if values are given to any three of the variables, the values of the other three are determined. For example, given r, θ and ϕ, x, y and z can be calculated; alternatively, given x, y and z, r, θ and ϕ can be calculated. This means that any three of the six may be chosen as independent variables, but the other three must then be treated as dependent variables. Thus $\partial\theta/\partial z$ could have *six* different interpretations depending on which two variables were taken with z as the three independent variables. In this connexion suffixes are useful. For example, $(\partial\theta/\partial z)_{x,y}$ would imply differentiation with respect to z keeping x and y constant and taking x, y, z as the independent variables. $(\partial\theta/\partial z)_{x,r}$, $(\partial\theta/\partial z)_{\phi,r}$, etc., have similar meanings. *All* the variables to be taken as independent variables are often written in the suffix, e.g. $(\partial\phi/\partial\theta)_{\theta,x,z}$.

The method of § 3·1 can be used to calculate any of the first partial derivatives.

Differentiating (1), (2) and (3) completely:

$$dx = \sin\theta\cos\phi\,dr + r\cos\theta\cos\phi\,d\theta - r\sin\theta\sin\phi\,d\phi, \quad (4)$$

$$dy = \sin\theta\sin\phi\,dr + r\cos\theta\sin\phi\,d\theta + r\sin\theta\cos\phi\,d\phi, \quad (5)$$

$$dz = \cos\theta\,dr - r\sin\theta\,d\theta. \quad (6)$$

Suppose that $(\partial\phi/\partial\theta)_{x,z}$ is required. As x, z are kept constant $dx = dz = 0$.

Equations (4) and (6) then give

$$\sin\theta\cos\phi\,dr + r\cos\theta\cos\phi\,d\theta - r\sin\theta\sin\phi\,d\phi = 0, \quad (7)$$

$$\cos\theta\,dr - r\sin\theta\,d\theta = 0. \quad (8)$$

Multiply (7) by $\cos\theta$ and (8) by $\sin\theta\cos\phi$ and subtract, thereby eliminating dr:

$$(r\cos^2\theta\cos\phi + r\sin^2\theta\cos\phi)\,d\theta - r\sin\theta\cos\theta\sin\phi\,d\phi = 0.$$

Thus $$\cos\phi\,d\theta - \sin\theta\cos\theta\sin\phi\,d\phi = 0,$$

giving $$\underline{\left(\frac{\partial\phi}{\partial\theta}\right)_{x,z}} = \frac{\cos\phi}{\sin\theta\cos\theta\sin\phi} = \underline{\frac{\cot\phi}{\sin\theta\cos\theta}}.$$

As a further example, consider the equation for an ideal gas $pv = RT$; here only two of p, v and T may be taken as independent variables. If ϕ is some function of the state of the gas, then $(\partial\phi/\partial v)_{v,T}$ would mean that v and T were to be taken as independent variables and T treated as constant. The same derivative might be found written as $(\partial\phi/\partial v)_{T\,\text{const.}}$ or merely as $(\partial\phi/\partial v)_T$. In most cases the meaning is clear from the context.

EXAMPLE

If $x = r\cos\theta$, $y = r\sin\theta$, find $(\partial\theta/\partial r)_x$.

Now $$dx = \cos\theta\,dr - r\sin\theta\,d\theta$$

and $$dy = \sin\theta\,dr + r\cos\theta\,d\theta.$$

When x is constant, $dx = 0$ and

$$\cos\theta\, dr - r\sin\theta\, d\theta = 0,$$

giving
$$\left(\frac{\partial\theta}{\partial r}\right)_x = \frac{\cos\theta}{r\sin\theta} = \frac{\cot\theta}{r}.$$

3·2. Change of variables

If $z = f(x, y)$, where x and y are functions of u and v, then z must also be some function of u and v.

It may often be necessary to change partial derivatives from terms of the variables x and y into terms of the variables u and v. This can be done as follows:

From volume I, § 12·4,

$$\delta z = \frac{\partial z}{\partial x}\delta x + \frac{\partial z}{\partial y}\delta y + \epsilon_1\delta x + \epsilon_2\delta y,$$

where $\epsilon_1 \to 0$ and $\epsilon_2 \to 0$ as $\delta x \to 0$ and $\delta y \to 0$.

Dividing by δu

$$\frac{\delta z}{\delta u} = \frac{\partial z}{\partial x}\frac{\delta x}{\delta u} + \frac{\partial z}{\partial y}\frac{\delta y}{\delta u} + \epsilon_1\frac{\delta x}{\delta u} + \epsilon_2\frac{\delta y}{\delta u}.$$

Keeping v constant during the change in u of δu, and allowing δu to approach zero, and therefore δx and $\delta y \to 0$:

$$\frac{\partial z}{\partial u} = \frac{\partial z}{\partial x}\frac{\partial x}{\partial u} + \frac{\partial z}{\partial y}\frac{\partial y}{\partial u}, \tag{1}$$

or in functional notation

$$\frac{\partial f}{\partial u} = \frac{\partial f}{\partial x}\frac{\partial x}{\partial u} + \frac{\partial f}{\partial y}\frac{\partial y}{\partial u}. \tag{2}$$

Similarly
$$\frac{\partial f}{\partial v} = \frac{\partial f}{\partial x}\frac{\partial x}{\partial v} + \frac{\partial f}{\partial y}\frac{\partial y}{\partial v}. \tag{3}$$

Two examples are given below to illustrate such transformations.

Example 1

If $f(x, y)$ is a function of x and y, and if $x = e^u\cosh v$. $y = e^u\sinh v$, prove that

(i) $\dfrac{\partial f}{\partial u} = x\dfrac{\partial f}{\partial x} + y\dfrac{\partial f}{\partial y}$, $\quad \dfrac{\partial f}{\partial v} = y\dfrac{\partial f}{\partial x} + x\dfrac{\partial f}{\partial y}$;

(ii) $\dfrac{\partial^2 f}{\partial u\,\partial v} - \dfrac{\partial f}{\partial v} = xy\left(\dfrac{\partial^2 f}{\partial x^2} + \dfrac{\partial^2 f}{\partial y^2}\right)$

$$+ (x^2 + y^2)\frac{\partial^2 f}{\partial x\,\partial y}. \qquad \text{[L.U.]}$$

$x = e^u \cosh v$, thus

$$\frac{\partial x}{\partial u} = e^u \cosh v = x \quad \text{and} \quad \frac{\partial x}{\partial v} = e^u \sinh v = y. \qquad (1)$$

$y = e^u \sinh v$, thus

$$\frac{\partial y}{\partial u} = e^u \sinh v = y \quad \text{and} \quad \frac{\partial y}{\partial v} = e^u \cosh v = x. \qquad (2)$$

(i) Now $\dfrac{\partial f}{\partial u} = \dfrac{\partial f}{\partial x}\dfrac{\partial x}{\partial u} + \dfrac{\partial f}{\partial y}\dfrac{\partial y}{\partial u}$

$$= x\frac{\partial f}{\partial x} + y\frac{\partial f}{\partial y}, \quad \text{from (1) and (2).} \qquad (3)$$

Also $\dfrac{\partial f}{\partial v} = \dfrac{\partial f}{\partial x}\dfrac{\partial x}{\partial v} + \dfrac{\partial f}{\partial y}\dfrac{\partial y}{\partial v}$

$$= y\frac{\partial f}{\partial x} + x\frac{\partial f}{\partial y}, \quad \text{from (1) and (2).} \qquad (4)$$

(ii) f is an arbitrary function of x and y, and it follows from equations (3) and (4) above that the operation of partial differentiation of a function with respect to u implies the operation $\left(x\dfrac{\partial}{\partial x} + y\dfrac{\partial}{\partial y}\right)$ in terms of x and y, i.e.

$$\frac{\partial}{\partial u} \equiv \left(x\frac{\partial}{\partial x} + y\frac{\partial}{\partial y}\right). \qquad (5)$$

Similarly

$$\frac{\partial}{\partial v} \equiv \left(y\frac{\partial}{\partial x} + x\frac{\partial}{\partial y}\right). \qquad (6)$$

(5) and (6) are valid, of course, only for this particular example in which $x = e^u \cosh v$ and $y = e^u \sinh v$. Equations of the types given in (3) and (4) would have to be obtained for other given relationships.

Now $\dfrac{\partial^2 f}{\partial u\,\partial v} = \dfrac{\partial}{\partial u}\left(\dfrac{\partial f}{\partial v}\right)$

$$= \left(x\frac{\partial}{\partial x} + y\frac{\partial}{\partial y}\right)\left(y\frac{\partial f}{\partial x} + x\frac{\partial f}{\partial y}\right),$$

from (4) and (5)

$$= x\frac{\partial}{\partial x}\left(y\frac{\partial f}{\partial x}\right) + x\frac{\partial}{\partial x}\left(x\frac{\partial f}{\partial y}\right) + y\frac{\partial}{\partial y}\left(y\frac{\partial f}{\partial x}\right) + y\frac{\partial}{\partial y}\left(x\frac{\partial f}{\partial y}\right)$$

$$= xy\frac{\partial^2 f}{\partial x^2} + \left(x\frac{\partial f}{\partial y} + x^2\frac{\partial^2 f}{\partial x\,\partial y}\right) + \left(y\frac{\partial f}{\partial x} + y^2\frac{\partial^2 f}{\partial y\,\partial x}\right) + xy\frac{\partial^2 f}{\partial y^2}.$$

$$\therefore\quad \frac{\partial^2 f}{\partial u\,\partial v} - \left(y\frac{\partial f}{\partial x} + x\frac{\partial f}{\partial y}\right) = xy\left(\frac{\partial^2 f}{\partial x^2} + \frac{\partial^2 f}{\partial y^2}\right) + (x^2 + y^2)\frac{\partial^2 f}{\partial x\,\partial y}.$$

Using (4), this gives

$$\frac{\partial^2 f}{\partial u\,\partial v} - \frac{\partial f}{\partial v} = xy\left(\frac{\partial^2 f}{\partial x^2} + \frac{\partial^2 f}{\partial y^2}\right) + (x^2 + y^2)\frac{\partial^2 f}{\partial x\,\partial y}.$$

EXAMPLE 2

If $u = f\left(\dfrac{z}{x}, \dfrac{x}{y}\right)$, prove that

$$x\frac{\partial u}{\partial x} + y\frac{\partial u}{\partial y} + z\frac{\partial u}{\partial z} = 0. \qquad \text{[L.U.]}$$

Here it is convenient to introduce a change of variables.

Let $$p = \frac{z}{x} \quad\text{and}\quad q = \frac{x}{y}, \qquad (1)$$

then $u = f(p, q)$.

From (1), $$\frac{\partial p}{\partial x} = -\frac{z}{x^2}, \quad \frac{\partial p}{\partial y} = 0, \quad \frac{\partial p}{\partial z} = \frac{1}{x} \qquad (2)$$

and $$\frac{\partial q}{\partial x} = \frac{1}{y}, \quad \frac{\partial q}{\partial y} = -\frac{x}{y^2}, \quad \frac{\partial q}{\partial z} = 0. \qquad (3)$$

Now

$$\frac{\partial u}{\partial x} = \frac{\partial u}{\partial p}\frac{\partial p}{\partial x} + \frac{\partial u}{\partial q}\frac{\partial q}{\partial x} = -\frac{z}{x^2}\frac{\partial u}{\partial p} + \frac{1}{y}\frac{\partial u}{\partial q}, \text{ from (2) and (3).}$$

Similarly $$\frac{\partial u}{\partial y} = \frac{\partial u}{\partial p}\frac{\partial p}{\partial y} + \frac{\partial u}{\partial q}\frac{\partial q}{\partial y} = -\frac{x}{y^2}\frac{\partial u}{\partial q},$$

and $$\frac{\partial u}{\partial z} = \frac{\partial u}{\partial p}\frac{\partial p}{\partial z} + \frac{\partial u}{\partial q}\frac{\partial q}{\partial z} = \frac{1}{x}\frac{\partial u}{\partial p}.$$

Thus $x\dfrac{\partial u}{\partial x} + y\dfrac{\partial u}{\partial y} + z\dfrac{\partial u}{\partial z}$

$$= x\left(-\frac{z}{x^2}\frac{\partial u}{\partial p} + \frac{1}{y}\frac{\partial u}{\partial q}\right) + y\left(-\frac{x}{y^2}\frac{\partial u}{\partial q}\right) + z\left(\frac{1}{x}\frac{\partial u}{\partial p}\right)$$

$$= \underline{0}.$$

Exercise 11

1. If $5xyz - 2xy + 3yz - 2zx - 7 = 0$, show that

$$\left(\frac{\partial x}{\partial y}\right)_z \left(\frac{\partial y}{\partial z}\right)_x \left(\frac{\partial z}{\partial x}\right)_y = -1.$$

2. If $z^2 = x^2 + y^2 + 1$, show that:

(i) $$\frac{\partial^2 z}{\partial x\,\partial y} = \frac{\partial^2 z}{\partial y\,\partial x},$$

(ii) $$xy\left(\frac{\partial^2 z}{\partial x^2} - \frac{\partial^2 z}{\partial y^2}\right) = (x^2 - y^2)\frac{\partial^2 z}{\partial x\,\partial y}.$$ [L.U.]

3. Given that $x = e^u \cos v$, $y = e^u \sin v$ and that z is a function of x and y, and hence also of u and v, find $\partial z/\partial u$ and $\partial z/\partial v$ in terms of x, y, $\partial z/\partial x$ and $\partial z/\partial y$. Hence find $\partial z/\partial x$ and $\partial z/\partial y$ in terms of u, v, $\partial z/\partial u$ and $\partial z/\partial v$.

4. If z is a function of x and y and these are functions of u and v defined by $x = uv$, $y = (u + v)/(u - v)$, show that:

(i) $2x \dfrac{\partial z}{\partial x} = u \dfrac{\partial z}{\partial u} + v \dfrac{\partial z}{\partial v}$;

(ii) $2y \dfrac{\partial z}{\partial y} = \frac{1}{2}(u^2 - v^2)\left(\dfrac{1}{u}\dfrac{\partial z}{\partial v} - \dfrac{1}{v}\dfrac{\partial z}{\partial u}\right)$. [L.U.]

5. If $u = x^2 - y^2$, $v = xy$ and $z = \phi(x, y)$, where ϕ is an arbitrary function, find $\dfrac{\partial^2 z}{\partial x^2} + \dfrac{\partial^2 z}{\partial y^2}$ in terms of u, v, $\dfrac{\partial^2 z}{\partial u^2}$ and $\dfrac{\partial^2 z}{\partial v^2}$.

6. If $u = f(r)$, where $r = \sqrt{(x^2 + y^2)}$, prove that

$$\frac{\partial^2 u}{\partial x^2} + \frac{\partial^2 u}{\partial y^2} = f''(r) + \frac{1}{r}f'(r).$$ [L.U.]

7. If $u = (x^2 - y^2) f(t)$, where $t = xy$, prove that

$$(x^2 - y^2)\{tf''(t) + 3f'(t)\} = \frac{\partial^2 u}{\partial x\, \partial y}.$$ [L.U.]

8. If the variables x, y, z, u are connected by the equations $u = \sin x \cosh y$, $\log_e (x + y) + 2y - 3 \log_e z = 4$, find $\partial u/\partial x$ when x and z are the independent variables. [L.U.]

9. If x, y, z are connected by $f(x, y, z) = 0$ and $x^2 + y^2 + z^2 =$ constant, prove that

$$\frac{dy}{dx} = -\frac{\left(z\dfrac{\partial f}{\partial x} - x\dfrac{\partial f}{\partial z}\right)}{\left(z\dfrac{\partial f}{\partial y} - y\dfrac{\partial f}{\partial z}\right)}.$$ [L.U.]

(*Note.* dy/dx is used as only one variable can be taken as independent.)

10. If $u^2 = x^2 - yv$, $v^2 = y^2 - xu$, what two meanings can be assigned to the symbol $\partial v/\partial y$? Find its value in each case. [L.U.]

11. The variables x, y, z are connected by the relation $x^2y + 3y^2z - 2z = 0$. If $u = xy^2z^3$, find the value of $\partial u/\partial x$ when $x = 1$, $y = 1$, $z = -1$, if x and y are taken as independent variables. [L.U.]

12. If $V = \frac{1}{2}(\phi_1 q_1 + \phi_2 q_2)$, where $\phi_1 = p_{11}q_1 + p_{12}q_2$, $\phi_2 = p_{21}q_1 + p_{22}q_2$ and $p_{12} = p_{21}$, prove that

$$\left(\frac{\partial V}{\partial x}\right)_{q\ \text{const.}} = -\left(\frac{\partial V}{\partial x}\right)_{\phi\ \text{const.}}$$

$$= \frac{1}{2}\left(\frac{\partial p_{11}}{\partial x}q_1^2 + 2\frac{\partial p_{12}}{\partial x}q_1 q_2 + \frac{\partial p_{22}}{\partial x}q_2^2\right).$$

(*Note.* This type of relationship occurs in the theory of electrostatic potential energy.)

Answers

3. $\dfrac{\partial z}{\partial u} = x\dfrac{\partial z}{\partial x} + y\dfrac{\partial z}{\partial y}$; $\dfrac{\partial z}{\partial v} = x\dfrac{\partial z}{\partial y} - y\dfrac{\partial z}{\partial x}$.

$$\frac{\partial z}{\partial x} = e^{-u}\left(\cos v\frac{\partial z}{\partial u} - \sin v\frac{\partial z}{\partial v}\right);$$

$$\frac{\partial z}{\partial y} = e^{-u}\left(\sin v\frac{\partial z}{\partial u} + \cos v\frac{\partial z}{\partial v}\right).$$

5. $\dfrac{\partial^2 z}{\partial x^2} + \dfrac{\partial^2 z}{\partial y^2} = \sqrt{(u^2 + 4v^2)}\left\{4\dfrac{\partial^2 z}{\partial u^2} + \dfrac{\partial^2 z}{\partial v^2}\right\}$.

8. $\dfrac{\{2(x+y)+1\}\cos x\cosh y - \sin x\sinh y}{2(x+y)+1}$.

9. $\left[\text{Use } df = 0 = \dfrac{\partial f}{\partial x}dx + \dfrac{\partial f}{\partial y}dy + \dfrac{\partial f}{\partial z}dz,\right.$

and $0 = x\,dx + y\,dy + z\,dz$. Eliminate $dz.\Big]$

10. $(\partial v/\partial y)_{x,y}$, taking x and y as independent variables. $(\partial v/\partial y)_{u,y}$, taking y and u as independent variables.

$$\left(\frac{\partial v}{\partial y}\right)_{x,y} = \frac{4uy + vx}{4uv - xy};\quad \left(\frac{\partial v}{\partial y}\right)_{y,u} = \frac{4xy - uv}{4vx + uy}.$$

11. -7. [Use $du = y^2z^3\,dx + 2xyz^3\,dy + 3xy^2z^2\,dz$ and $2xy\,dx + (x^2 + 6yz)\,dy + (3y^2 - 2)\,dz = 0$. Eliminate dz.]

3·3. Differentiation of a definite integral

3·31. Differentiation under the integral sign

Consider the definite integral

$$I = \int_0^2 (3x^2 + 2tx + t^2)\,dx, \tag{1}$$

where t is a parameter independent of x.

$$I = [x^3 + tx^2 + t^2x]_0^2 = 8 + 4t + 2t^2. \tag{2}$$

I is thus a function of t only, and from (2)

$$\frac{dI}{dt} = 4 + 4t. \tag{3}$$

The question arises, can the differentiation be done *before* the integration?

Trying this:

$$\frac{d}{dt}\left[\int_0^2 (3x^2 + 2tx + t^2)\,dx\right] = \int_0^2 \left\{\frac{\partial}{\partial t}(3x^2 + 2tx + t^2)\right\} dx$$

$$= \int_0^2 (2x + 2t)\,dx$$

$$= [x^2 + 2tx]_0^2 = 4 + 4t.$$

In this particular case the differentiation can be done before the integration, as long as it is remembered that the integrand is differentiated partially with respect to t.

This procedure is always possible, that is

$$\frac{d}{dt}\int_a^b f(x, t)\,dx = \int_a^b \frac{\partial f}{\partial t}\,dx,$$

provided that $f(x, t)$ possesses an integral with respect to x and that $\partial f/\partial t$ is a continuous function of both the variables x and t.

As t is independent of x, the integral is a function of t and may be written

$$I(t) = \int_a^b f(x, t)\, dx.$$

Differentiating from first principles:

$$I(t + \delta t) = \int_a^b f(x, t + \delta t)\, dx.$$

Thus,

$$I(t + \delta t) - I(t) = \int_a^b \{f(x, t + \delta t) - f(x, t)\}\, dx.$$

$$\therefore \quad \frac{\delta I}{\delta t} = \int_a^b \frac{\{f(x, t + \delta t) - f(x, t)\}}{\delta t}\, dx$$

$$= \int_a^b f_t(x, t + \theta\, .\, \delta t)\, dx \quad (0 < \theta < 1). \tag{4}$$

[$f(x, t + \delta t) - f(x, t) = \delta t\, .\, f_t(x, t + \theta\, .\, \delta t)$, see vol. I, § 12·4.]

Now it can be proved, under the conditions stated above, that

$$\lim_{\delta t \to 0} \int_a^b f_t(x, t + \theta \delta t)\, dx = \int_a^b \left\{\lim_{\delta t \to 0} f_t(x, t + \theta \delta t)\right\} dx.$$

Thus, taking the limit as $\delta t \to 0$ in (4):

$$\underline{\frac{dI}{dt}} = \int_a^b f_t\, (x,t)\, dx = \underline{\int_a^b \frac{\partial f}{\partial t}\, dx.}$$

That is, the derivative of the integral is the integral of the derivative of the integrand. This will be made use of later (see § 10·21 (e)).

EXAMPLE 1

Show that $\displaystyle\int_0^\infty \frac{1}{a^2 + x^2}\, dx = \frac{\pi}{2a}$.

Differentiate this relationship with respect to a and deduce the values of:

(i) $\displaystyle\int_0^\infty \frac{1}{(a^2+x^2)^2}\,dx$, (ii) $\displaystyle\int_0^\infty \frac{x^2}{(a^2+x^2)^2}\,dx$.

$$I = \int_0^\infty \frac{1}{a^2+x^2}\,dx = \frac{1}{a}\left[\tan^{-1}\frac{x}{a}\right]_0^\infty = \frac{1}{a}\left(\tfrac{1}{2}\pi - 0\right) = \underline{\frac{\pi}{2a}}. \tag{1}$$

(i) From (1):
$$\frac{dI}{da} = -\frac{\pi}{2a^2} = \int_0^\infty \left\{\frac{\partial}{\partial a}\left(\frac{1}{a^2+x^2}\right)\right\} dx$$
$$= \int_0^\infty \frac{-2a}{(a^2+x^2)^2}\,dx.$$

Thus

$$\frac{\pi}{2a^2} = 2a\int_0^\infty \frac{1}{(a^2+x^2)^2}\,dx,$$

giving

$$\underline{\int_0^\infty \frac{1}{(a^2+x^2)^2}\,dx = \frac{\pi}{4a^3}}. \tag{2}$$

(ii)
$$\int_0^\infty \frac{x^2}{(a^2+x^2)^2}\,dx = \int_0^\infty \frac{(a^2+x^2)-a^2}{(a^2+x^2)^2}\,dx$$
$$= \int_0^\infty \frac{1}{a^2+x^2}\,dx - \int_0^\infty \frac{a^2}{(a^2+x^2)^2}\,dx$$
$$= \frac{\pi}{2a} - a^2\frac{\pi}{4a^3}, \quad \text{from (1) and (2)}$$
$$= \underline{\frac{\pi}{4a}}.$$

EXAMPLE 2

Show that $\displaystyle\int_0^\infty x^n e^{-x}\,dx = n\int_0^\infty x^{n-1}e^{-x}\,dx$, and deduce that when n is a positive integer the value of the integral is $n!$. Find the value of a for which $\displaystyle\int_0^\infty (1-ax^2)^2 e^{-x}\,dx$ is a minimum. [L.U.]

The first part of this question (i.e. $I_n = \int_0^\infty x^n e^{-x}\,dx = n!$ when n is a positive integer) occurs as question 2 of Exercise 7; for the second part let

$$F(a) = \int_0^\infty (1 - ax^2)^2 e^{-x}\,dx.$$

Then

$$\frac{dF}{da} = \int_0^\infty \left\{\frac{\partial}{\partial a}(1 - ax^2)^2 e^{-x}\right\} dx$$

$$= \int_0^\infty - 2x^2(1 - ax^2)e^{-x}\,dx$$

$$= \int_0^\infty (- 2x^2e^{-x} + 2ax^4e^{-x})\,dx \qquad (1)$$

$$= - 2I_2 + 2aI_4$$

$$= - 2 \,.\, 2! + 2a \,.\, 4!$$

$$= - 4 + 48a. \qquad (2)$$

From (2),

$$\frac{d^2F}{da^2} = 48.$$

From (2), when $dF/da = 0$,

$$48a = 4, \quad \underline{a = \tfrac{1}{12}}.$$

As d^2F/da^2 is positive, $F(a)$ is a minimum when $a = \frac{1}{12}$.

Note. (i) In this case differentiation under the integral sign is probably not so short as evaluating $F(a)$ in terms of a. Thus:

$$F(a) = \int_0^\infty (1 - ax^2)^2 e^{-x}\,dx = \int_0^\infty (1 - 2ax^2 + a^2x^4)e^{-x}\,dx$$

$$= I_0 - 2aI_2 + a^2I_4$$

$$= 0! - 2a \,.\, 2! + a^2 \,.\, 4!$$

$$= 24a^2 - 4a + 1.$$

$$\therefore \quad \frac{dF}{da} = 48a - 4,$$

$$\frac{d^2F}{da^2} = + 48.$$

When $dF/da = 0$, $a = \frac{1}{12}$ and d^2F/da^2 is positive.

(ii) As a point of interest $\int_0^\infty x^n e^{-x}\,dx$, where n is a positive integer, can be evaluated without using integration by parts. Thus if

$$I = \int_0^\infty e^{-ax}\,dx = \left[-\frac{1}{a}e^{-ax}\right]_0^\infty = \frac{1}{a} \quad (a>0),$$

then

$$\frac{dI}{da} = \int_0^\infty \frac{\partial}{\partial a}(e^{-ax})\,dx = -\frac{1}{a^2},$$

giving

$$\int_0^\infty -xe^{-ax}\,dx = -\frac{1}{a^2}$$

or

$$\int_0^\infty xe^{-ax}\,dx = \frac{1}{a^2}.$$

Differentiating this relationship with respect to a gives

$$\int_0^\infty x^2e^{-ax}\,dx = \frac{2\,.\,1}{a^3} = \frac{2!}{a^3}.$$

Again differentiating with respect to a:

$$\int_0^\infty x^3e^{-ax}\,dx = \frac{3\,.\,2!}{a^4} = \frac{3!}{a^4}$$

and so on.

Differentiating n times would give

$$\int_0^\infty x^ne^{-ax}\,dx = \frac{n!}{a^{n+1}}.$$

Putting $a = 1$ now gives

$$\int_0^\infty x^ne^{-x}\,dx = n!.$$

3·32. Differentiation of a definite integral with respect to either limit

Let

$$I = \int_a^b 3x^2\,dx = [x^3]_a^b = b^3 - a^3. \qquad (1)$$

I is a function of the limits a and b.

From (1), $\partial I/\partial b = 3b^2$ and $\partial I/\partial a = -3a^2$.

There appears to be a close connexion between $\partial I/\partial b$, $\partial I/\partial a$ and the original integrand $3x^2$.

A rule is now proved for differentiating an integral with respect to either limit *without* evaluating the integral.

Let

$$I = \int_a^b f(x)\,dx, \tag{2}$$

where $f(x)$ is independent of a and b.

Then, if

$$F'(x) = f(x), \tag{3}$$

$$I = \int_a^b f(x)\,dx = [F(x)]_a^b = F(b) - F(a).$$

Thus

$$\frac{\partial I}{\partial b} = F'(b) = \underline{f(b)}, \tag{4}$$

and
$$\frac{\partial I}{\partial a} = -F'(a) = \underline{-f(a)}. \tag{5}$$

EXAMPLE

If
$$I = \int_a^b (2x^3 - 3x^2 + 2)\,dx,$$

then

$$\frac{\partial I}{\partial b} = 2b^3 - 3b^2 + 2$$

and

$$\frac{\partial I}{\partial a} = -(2a^3 - 3a^2 + 2).$$

3·33. Functions defined by definite integrals

Some important functions are defined as definite integrals. They are usually integrals which could not be evaluated in terms of ordinary functions; although many elementary functions *could* be expressed as definite integrals.

Thus, some text-books *define* $\log_e x$ to be $\int_1^x \frac{1}{t}\,dt$.

From § 3·32, equation (4), this is equivalent to saying that $\frac{d}{dx}(\log_e x)$ must be $\frac{1}{x}$.

Below are given some important functions which are usually defined as definite integrals. They cannot be expressed explicitly in terms of the usual well-known functions.

(*a*) *The error function.* This is important in statistics and in the theory of heat conduction. It is defined as

$$\operatorname{erf} x = \frac{2}{\sqrt{\pi}} \int_0^x e^{-t^2}\, dt.$$

('erf' is an abbreviation of error function.)

Tables of values of this function, correct to a given number of decimal places, are often given in text-books on statistics.

(*b*) *Elliptic integrals.* An example of this type of function is

$$\int_0^\theta \sqrt{(1 - k^2 \sin^2 \theta)}\, d\theta,$$

which is an elliptic integral of the second kind. The name comes from the fact that this type of integral is met with when attempting to calculate by integration the length of arc of an ellipse.

(*c*) *The gamma function.* This is defined as

$$\Gamma(\nu) = \int_0^\infty x^{\nu-1} e^{-x}\, dx.$$

It is, of course, a function of ν.

It is also evident from § 3·31, Example (2), that when ν is a positive integer n, then $\Gamma(n) = (n-1)!$.

In general $\quad \Gamma(\nu) = (\nu - 1)\,\Gamma(\nu - 1).$

The gamma function provides a generalization of factorial numbers for non-integral numbers.

(*d*) *Bessel functions.* A Bessel function of order n is given by

$$J_n(x) = \frac{1}{\pi} \int_0^\pi \cos(n\theta - x \sin\theta)\, d\theta. \tag{1}$$

It is a function of x and n.

It satisfies Bessel's equation of order n:

$$\frac{d^2y}{dx^2} + \frac{1}{x}\frac{dy}{dx} + \left(1 - \frac{n^2}{x^2}\right) y = 0,$$

an important type of differential equation.

$J_n(x)$ can also be expressed as an infinite series:

$$J_n(x) = \sum_{r=0}^{\infty} \frac{(-1)^r (\frac{1}{2}x)^{n+2r}}{r!(n+r)!}. \qquad (2)$$

The more important applications are concerned with the cases when $n = 0$ or a positive integer.

Bessel functions are important in the advanced study of high-frequency work in electrical conductors.

Exercise 12

1. Prove that $\displaystyle\int_0^1 \frac{dx}{a^2 - b^2x^2} = \frac{1}{2ab} \log_e \left(\frac{a+b}{a-b}\right)$,

where $a > b > 0$.

Differentiate this relationship

(i) with respect to a,
(ii) with respect to b.

2. Differentiate with respect to n:

$$\int_0^1 x^n\, dx = \frac{1}{n+1}, \quad \text{where} \quad n > -1.$$

Hence prove that:

$$\text{(i)} \quad \int_0^1 x^n \log_e x\, dx = \frac{-1}{(n+1)^2},$$

$$\text{(ii)} \quad \int_0^1 x^n (\log_e x)^2\, dx = \frac{2}{(n+1)^3}.$$

3. If $a > b$, by the substitution $t = \tan \frac{1}{2}x$, show that

$$\int_0^{\pi} \frac{dx}{a + b\cos x} = \frac{\pi}{\sqrt{(a^2 - b^2)}}.$$

By differentiating this relationship (i) with respect to a, (ii) with respect to b, find the value of two more definite integrals.

4. Prove that

$$\int_0^{\infty} e^{-ax} \sin bx\, dx = \frac{b}{a^2 + b^2} \quad (a > 0).$$

By differentiation with respect to a, prove that

$$\int_0^\infty xe^{-ax} \sin bx\, dx = \frac{2ab}{(a^2+b^2)^2}.$$

Also find $$\int_0^\infty xe^{-ax} \cos bx\, dx.$$

5. If $I = \int_a^b f(x, y)\, dx$, it can be shown, by methods similar to those of §§ 3·31 and 3·32, that

(i) $$\frac{dI}{dy} = \int_a^b \frac{\partial}{\partial y} f(x, y)\, dx,$$

if a and b are independent of x and y.

(ii) $$\frac{dI}{dy} = \int_a^b \frac{\partial}{\partial y} f(x, y)\, dx - f(a, y)\frac{da}{dy} + f(b, y)\frac{db}{dy},$$

if a and b are both functions of y.

(iii) $$\frac{\partial I}{\partial y} = \int_a^b \frac{\partial}{\partial y} f(x, y)\, dx,$$

if a and b are both functions of x.

In performing the integrations, all symbols other than x are treated as constants.

Verify these rules by integrating *first* and then differentiating in the cases:

(*a*) $\int_1^2 (x^2 - 3xy + y^2)\, dx$, (*b*) $\int_y^{2y} \left(x + \frac{y}{x}\right) dx$,

(*c*) $\int_x^{\frac{x}{2}} (x^2 - y^2)x\, dx.$

6. Show that

$$\int_0^a \frac{a}{\sqrt{(a^2+x^2)}}\, dx = a \log_e (1 + \sqrt{2}).$$

Using rule (ii) given in question 5 above, find the relationship formed when this result is differentiated with respect to a.

7. Bessel's function of order zero is given by equation (1), § 3·33, as

$$J_0(x) = \frac{1}{\pi}\int_0^{\pi} \cos\,(x \sin\theta)\, d\theta.$$

Prove, by differentiation under the integral sign, that this satisfies the equation

$$\frac{d^2y}{dx^2} + \frac{1}{x}\frac{dy}{dx} + y = 0.$$

8. Show that

$$\int_0^{2\pi} \frac{d\theta}{\sqrt{(a^2+b^2+2ab\cos\theta)}} = \int_0^{2\pi} \frac{d\theta}{\sqrt{(a^2+b^2-2ab\cos\theta)}},$$

and that either integral may be expressed as a multiple of the complete elliptic integral

$$K(k) = \int_0^{\frac{1}{2}\pi} \frac{d\phi}{\sqrt{(1-k^2\sin^2\phi)}}$$

with a suitable value of k.

If $a = 3$, $b = 1$, determine the value of the given integrals, using the following table of values of $K(k)$.

k	0	sin 30°	sin 60°	sin 90°
$K(k)$	1·5708	1·6857	2·1565	∞

[I.E.E.]

Answers

1. (i) $\displaystyle\int_0^1 \frac{dx}{(a^2-b^2x^2)^2} = \frac{1}{4a^3b}\log_e\left(\frac{a+b}{a-b}\right) + \frac{1}{2a^2(a^2-b^2)}$;

(ii) $\displaystyle\int_0^1 \frac{x^2\,dx}{(a^2-b^2x^2)^2} = \frac{1}{2b^2(a^2-b^2)} - \frac{1}{4ab^3}\log_e\left(\frac{a+b}{a-b}\right)$.

2. $\left[\text{Note that } \dfrac{\partial}{\partial n}(x^n) = x^n \log_e x.\right]$

3. (i) $\displaystyle\int_0^{\pi} \frac{dx}{(a + b \cos x)^2} = \frac{\pi a}{(a^2 - b^2)^{\frac{3}{2}}}$;

(ii) $\displaystyle\int_0^{\pi} \frac{\cos x \, dx}{(a + b \cos x)^2} = \frac{-\pi b}{(a^2 - b^2)^{\frac{3}{2}}}$.

4. $\displaystyle\int_0^{\infty} xe^{-ax} \cos bx \, dx = \frac{a^2 - b^2}{(a^2 + b^2)^2}$.

[Differentiate the original integral with respect to b.]

5. (*a*) $\dfrac{dI}{dy} = -\dfrac{9}{2} + 2y$.

(*b*) $\dfrac{dI}{dy} = 3y + log_e 2$.

(*c*) $\dfrac{\partial I}{\partial y} = \dfrac{3}{4} x^2 y$.

6. $\displaystyle\int_0^{a} \frac{x^2 \, dx}{(a^2 + x^2)^{3/2}} = \log_e (1 + \sqrt{2}) - \frac{1}{\sqrt{2}}$.

[Think of a as standing for y in rule (ii) of question 5.]

8. 2·1565.

CHAPTER 4

SOME APPLICATIONS OF PARTIAL DIFFERENTIATION

4·1. Thermionic valves

Valve constants are often defined as partial derivatives. Examples are given here for the case of a triode valve.

4·11. Characteristics of a triode

The anode current of a triode is a function of both the anode and grid voltages (or potentials). It is thus a function of two

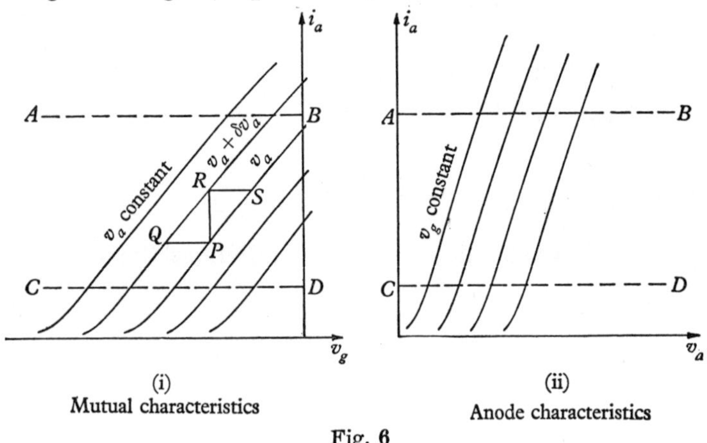

(i)
Mutual characteristics

(ii)
Anode characteristics

Fig. 6

variables. Unless surfaces in three dimensions are drawn, it is best to plot curves, first keeping the anode voltage constant and then the grid voltage constant. Such curves are known as the characteristics of the valve. A typical pair are shown in fig. 6.

When the anode voltage is kept constant the curves are called *mutual* characteristics; when the grid voltage is kept constant they are called *anode* characteristics.

In fig. 6, i_a is the anode current, v_a the anode voltage and v_g the grid voltage. The operating grid voltage is usually negative.

Suppose the triode is being operated under conditions represented by the point P in fig. 6 (i).

(*a*) Let the anode voltage be increased by a small amount δv_a, whilst simultaneously the grid voltage is changed by an amount δv_g (represented by PQ) so that the anode current, i_a, remains unchanged. Note that the change in grid voltage is a decrease. The valve now operates under conditions represented by the point Q. The ratio $\delta v_a/\delta v_g$ is inherently negative. The numerical value of this ratio is of practical importance. To ensure that the result is positive, it is best to deal with $-\delta v_a/\delta v_g$. The limiting value of this ratio as the changes become indefinitely small is known as the *amplification factor* of the triode and is denoted by μ.

Thus $$\mu = -\frac{\partial v_a}{\partial v_g}, \tag{1}$$

i_a being kept constant. μ, as can be seen, is a number; it has no dimensions.

(*b*) Now, starting from P, let the anode voltage increase by a small amount δv_a whilst the grid voltage is kept constant, so that the new operating point is at R. Let the change in anode current, an increase, be δi_a.

The limiting value of $\delta v_a/\delta i_a$ is known by any of the names *impedance*, *anode characteristic resistance* or *slope resistance* and is denoted by R_a or ρ.

Thus $$R_a = \frac{\partial v_a}{\partial i_a}, \tag{2}$$

v_g being kept constant.

It is usually measured in ohms or megohms.

(*c*) Finally, starting at P, let the grid voltage increase by a small amount δv_g, keeping v_a constant, so that the new operating point is at S. Let the change in the anode current, an increase, be δi_a (represented by PR). The limiting value of $\dfrac{\delta i_a}{\delta v_g}$ is known as the *mutual conductance* of the triode and is denoted by any of the symbols g, g_m, G_m, S, or κ. Here, g will be used.

Thus $$g = \frac{\partial i_a}{\partial v_g}, \tag{3}$$

v_a being kept constant.

g is measured in microsiemens or often in milliamps per volt.

In a similar manner, the anode characteristics could be used to discuss μ, R_a and g.

Summary

$$\mu = -\frac{\partial v_a}{\partial v_g}, \quad R_a = \frac{\partial v_a}{\partial i_a}, \quad g = \frac{\partial i_a}{\partial v_g}. \qquad (4)$$

4·12. Interrelationship between μ, R_a and g

Let the functional relationship between v_a, v_g and i_a be written as
$$F(v_a, v_g, i_a) = 0.$$

Then $$dF = 0 = \frac{\partial F}{\partial v_a} dv_a + \frac{\partial F}{\partial v_g} dv_g + \frac{\partial F}{\partial i_a} di_a. \qquad (1)$$

If v_a is constant, $dv_a = 0$ and (1) gives

$$g = \frac{\partial i_a}{\partial v_g} = -\left(\frac{\partial F}{\partial v_g}\right) \Big/ \left(\frac{\partial F}{\partial i_a}\right). \qquad (2)$$

If i_a is constant, $di_a = 0$ and (1) gives

$$\mu = -\left(\frac{\partial v_a}{\partial v_g}\right) = +\left(\frac{\partial F}{\partial v_g}\right) \Big/ \left(\frac{\partial F}{\partial v_a}\right). \qquad (3)$$

If v_g is constant, $dv_g = 0$ and (1) gives

$$R_a = \left(\frac{\partial v_a}{\partial i_a}\right) = -\left(\frac{\partial F}{\partial i_a}\right) \Big/ \left(\frac{\partial F}{\partial v_a}\right). \qquad (4)$$

From (2), (3) and (4):

$$\frac{\mu}{R_a} = -\frac{\left(\frac{\partial F}{\partial v_g}\right) \Big/ \left(\frac{\partial F}{\partial v_a}\right)}{\left(\frac{\partial F}{\partial i_a}\right) \Big/ \left(\frac{\partial F}{\partial v_a}\right)} = -\frac{\left(\frac{\partial F}{\partial v_g}\right)}{\left(\frac{\partial F}{\partial i_a}\right)} = g.$$

Thus $$\underline{g = \frac{\mu}{R_a}}. \qquad (5)$$

Note that this is only a special example of the use of differentials given in § 3·1.

The importance of the quantities μ, R_a and g lies in the fact that over a considerable range of the working characteristics they are approximately constant. This is the case in fig. 6 over the regions lying between the dotted lines AB and CD, where the curves are approximately straight lines.

4·13. Fundamental equations for the operation of a triode

When both the grid and anode voltages vary, the change in the anode current can be found from the formula for the total increment.

$$\delta i_a = \left(\frac{\partial i_a}{\partial v_a}\right)\delta v_a + \left(\frac{\partial i_a}{\partial v_g}\right)\delta v_g$$

$$= \frac{1}{R_a}\,\delta v_a + g\delta v_g\,.$$

Using equation (5) of § 4·12, this gives

$$\delta i_a = \frac{1}{R_a}\,(\delta v_a + gR_a\delta v_g),$$

$$\delta i_a = \frac{1}{R_a}\,(\delta v_a + \mu\delta v_g). \tag{1}$$

If the operating conditions do not depart from the straight parts of the characteristics, then μ and R_a remain constant. Equation (1) may then be written

$$di_a = \frac{1}{R_a}\,(dv_a + \mu dv_g), \tag{2}$$

where di_a, dv_g, dv_a denote changes, not necessarily small, from their original values.

The formula normally used to give the *total* current is

$$i_a = k(v_a + \mu v_g)^{\frac{3}{2}}, \tag{3}$$

where k is a constant for any particular tube.

4·2. Maxima and minima of a function of several variables

In this section the various rules and tests will be given for a function of two variables only. Similar results can be proved for a function of any finite number of variables.

4·21. Taylor's theorem for a function of two variables

Consider the function of x and y, $f(x + h, y + k)$.

Treated as a function of x only, that is, considering y as constant, § 1·11, (8) gives the expansion

$$f(x + h, y + k) = f(x, y + k) + h \frac{\partial}{\partial x} f(x, y + k)$$
$$+ \frac{h^2}{2!} \frac{\partial^2}{\partial x^2} f(x, y + k) + \dots, \qquad (1)$$

where the partial derivative signs must be used to show that y is treated as a constant.

Now taking the function $f(x, y + k)$, and treating this with x constant and y variable, then

$$f(x, y + k) = f(x, y) + k \frac{\partial}{\partial y} f(x, y) + \frac{k^2}{2!} \frac{\partial^2}{\partial y^2} f(x, y) + \dots. \qquad (2)$$

Similarly, treating $\frac{\partial}{\partial x} f(x, y + k)$ as a function with x constant and y variable:

$$\frac{\partial}{\partial x} f(x, y + k) = \left\{\frac{\partial}{\partial x} f(x, y)\right\} + k \frac{\partial}{\partial y} \left\{\frac{\partial}{\partial x} f(x, y)\right\} + \dots,$$

i.e. $$\frac{\partial}{\partial x} f(x, y + k) = \frac{\partial}{\partial x} f(x, y) + k \frac{\partial^2}{\partial y \partial x} f(x, y) + \dots. \qquad (3)$$

Similarly

$$\frac{\partial^2}{\partial x^2} f(x, y + k) = \left\{\frac{\partial^2}{\partial x^2} f(x, y)\right\} + k \frac{\partial}{\partial y} \left\{\frac{\partial^2}{\partial x^2} f(x, y)\right\} + \dots,$$

giving

$$\frac{\partial^2}{\partial x^2} f(x, y + k) = \frac{\partial^2}{\partial x^2} f(x, y) + k \frac{\partial^3}{\partial y \partial x^2} f(x, y) + \dots. \qquad (4)$$

Substituting for

$$f(x, y + k), \quad \frac{\partial}{\partial x} f(x, y + k) \quad \text{and} \quad \frac{\partial^2}{\partial x^2} f(x, y + k)$$

from (2), (3) and (4) into (1):

$$f(x+h, y+k) = \left\{f(x, y) + k\frac{\partial}{\partial y}f(x, y) + \frac{k^2}{2!}\frac{\partial^2}{\partial y^2}f(x, y) + \ldots\right\}$$
$$+ h\left\{\frac{\partial}{\partial x}f(x, y) + k\frac{\partial^2}{\partial y\partial x}f(x, y) + \ldots\right\}$$
$$+ \frac{h^2}{2!}\left\{\frac{\partial^2}{\partial x^2}f(x, y) + k\frac{\partial^3}{\partial y\partial x^2}f(x, y) + \ldots\right\}.$$

As far as the terms of 2nd degree in h and k, this gives

$$f(x+h, y+k) = f(x, y) + \left\{h\frac{\partial}{\partial x}f(x, y) + k\frac{\partial}{\partial y}f(x, y)\right\}$$
$$+ \frac{1}{2!}\left\{h^2\frac{\partial^2}{\partial x^2}f(x, y) + 2hk\frac{\partial^2}{\partial y\partial x}f(x, y) + k^2\frac{\partial^2}{\partial y^2}f(x, y)\right\} + \ldots. \quad (5)$$

Equation (5) is known as Taylor's theorem for a function of two variables. Note that, if h and k are small, neglecting the terms of 2nd degree in h and k gives

$$f(x+h, y+k) - f(x, y) \simeq h\frac{\partial}{\partial x}f(x, y) + k\frac{\partial}{\partial y}f(x, y).$$

If h and k are taken as small increments in x and y and $z = f(x, y)$, this gives the well-known formula

$$\delta z \simeq \frac{\partial z}{\partial x}\delta x + \frac{\partial z}{\partial y}\delta y \quad \text{(see vol. I, § 12·4).}$$

4·22. Maxima and minima tests

Let $z = f(x, y)$ be a continuous function of x and y, then maximum and minimum values of z are defined as follows: z has a maximum or minimum value when $x = x_1$, $y = y_1$ if $f(x_1, y_1)$ is always greater or smaller than $f(x_1 + h, y_1 + k)$, where h, k are small but non-zero, positive or negative. This means that $f(x+h, y+k) - f(x, y)$ must keep the same sign (negative for maximum, positive for minimum) as h and k vary over their range of small values.

Now by § 4·21, (5) above,

$$f(x+h, y+k) - f(x, y) = \left(h\frac{\partial z}{\partial x} + k\frac{\partial z}{\partial y}\right)$$
$$+ \frac{1}{2!}\left(h^2\frac{\partial^2 z}{\partial x^2} + 2hk\frac{\partial^2 z}{\partial y\,\partial x} + k^2\frac{\partial^2 z}{\partial y^2}\right) + \dots . \qquad (1)$$

When h and k are small enough the sign of the right-hand side will depend on the sign of $\left(h\frac{\partial z}{\partial x} + k\frac{\partial z}{\partial y}\right)$. This expression, however, changes sign if h and k change sign; thus it becomes a necessary condition for maximum or minimum that $h\frac{\partial z}{\partial x} + k\frac{\partial z}{\partial y} \equiv 0$ for varying values of h and k. This means that

$$\frac{\partial z}{\partial x} = 0 \quad \text{and} \quad \frac{\partial z}{\partial y} = 0. \qquad (2)$$

When this condition is satisfied, the sign of $f(x+h, y+k) - f(x, y)$ will depend on the second term of (1), viz.

$$h^2\frac{\partial^2 z}{\partial x^2} + 2hk\frac{\partial^2 z}{\partial x\,\partial y} + k^2\frac{\partial^2 z}{\partial y^2},$$

for small values of h and k.

Let $$a = \frac{\partial^2 z}{\partial x^2}, \quad b = \frac{\partial^2 z}{\partial x\,\partial y}, \quad c = \frac{\partial^2 z}{\partial y^2}, \qquad (3)$$

then z is a maximum or minimum according as $ah^2 + 2bhk + ck^2$ is negative or positive as h and k vary through small values.

Now $$ah^2 + 2bhk + ck^2 \equiv a\left(h^2 + 2\frac{b}{a}hk + \frac{c}{a}k^2\right)$$
$$= a\left\{\left(h + \frac{b}{a}k\right)^2 + \left(\frac{c}{a}k^2 - \frac{b^2}{a^2}k^2\right)\right\}$$
(on completing the square)
$$= a\left\{\left(h + \frac{b}{a}k\right)^2 + \frac{(ac - b^2)}{a^2}k^2\right\}.$$

This has the same sign as a for all values of h and k if, and only if,

$$ac > b^2, \tag{4}$$

implying that a and c have the same sign. (5)

When these conditions are satisfied, the expression is negative or positive according as a is negative or positive. That is, using (2), (3), (4) and (5), the function has a maximum or a minimum if

$$\frac{\partial z}{\partial x} = 0, \quad \frac{\partial z}{\partial y} = 0 \quad \text{and} \quad \frac{\partial^2 z}{\partial x^2}\frac{\partial^2 z}{\partial y^2} > \left(\frac{\partial^2 z}{\partial y \partial x}\right)^2, \tag{6}$$

and these conditions being satisfied, z has a maximum or minimum according to whether $\partial^2 z/\partial x^2$ (or $\partial^2 z/\partial y^2$) is negative or positive.

These tests are now summarized:

Maximum	Minimum
$\left.\begin{array}{c}\dfrac{\partial z}{\partial x} = 0 = \dfrac{\partial z}{\partial y} \\ \dfrac{\partial^2 z}{\partial x^2}\dfrac{\partial^2 z}{\partial y^2} > \left(\dfrac{\partial^2 z}{\partial y\, \partial x}\right)^2\end{array}\right\}$	$\left.\begin{array}{c}\dfrac{\partial z}{\partial x} = 0 = \dfrac{\partial z}{\partial y} \\ \dfrac{\partial^2 z}{\partial x^2}\dfrac{\partial^2 z}{\partial y^2} > \left(\dfrac{\partial^2 z}{\partial y\, \partial x}\right)^2\end{array}\right\}$
$\dfrac{\partial^2 z}{\partial x^2}\left(\text{or } \dfrac{\partial^2 z}{\partial y^2}\right)$ is negative.	$\dfrac{\partial^2 z}{\partial x^2}\left(\text{or } \dfrac{\partial^2 z}{\partial y^2}\right)$ is positive.

If $(\partial^2 z/\partial x^2)\ (\partial^2 z/\partial y^2) < (\partial^2 z/[\partial y \partial x])^2$, the stationary point is known as a saddle point. If $(\partial^2 z/\partial x^2)\ (\partial^2 z/\partial y^2) = (\partial^2 z/[\partial y \partial x])^2$, higher order terms of equation (1) must be examined.

For problems of a practical nature it is often taken as sufficient to apply the first test only and then argue from practical considerations whether the value obtained is a maximum or a minimum.

The tests can be extended to functions of more than two variables. For example, if $u = f(x, y, z)$, u has stationary values when $\partial u/\partial x = 0$, $\partial u/\partial y = 0$, $\partial u/\partial z = 0$.

4·23. Resistance networks

It can be proved that in any resistance network the current will so distribute itself that the total heat, H, generated is a minimum, i.e. $\partial H/\partial i_1 = 0$, $\partial H/\partial i_2 = 0$, etc., where $i_1, i_2, \ldots$ are the currents in the branches. This forms an alternative method to the Kirchhoff method for solving networks. However, it is not necessarily a shorter method.

4·24. Worked examples

(1) Fig. 7 shows a current i entering a network at A and leaving at C. Find the equivalent resistance of the network.

Let i_1 be the current flowing through the 1Ω resistance; then the current through the 2Ω resistance is $i - i_1$ (see point A).

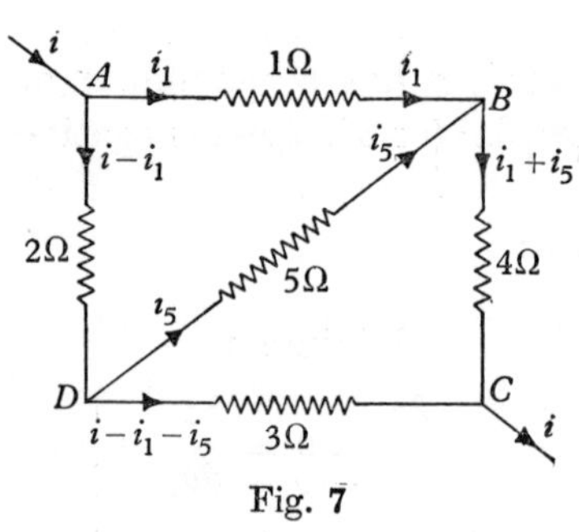

Fig. 7

Let i_5 be the current flowing through the 5Ω resistance; then the current through the 3Ω resistance is $i - i_1 - i_5$ (see point D), and the current through the 4Ω resistance is $i_1 + i_5$ (see point B).

The total heat, H, is given by

$$H = 1i_1^2 + 2(i - i_1)^2 + 5i_5^2 + 3(i - i_1 - i_5)^2 + 4(i_1 + i_5)^2, \quad (1)$$

where i is given and i_1, i_5 have to be found using the principle that H must be a minimum.

As H is a minimum,

$$\frac{\partial H}{\partial i_1} = 0 = \frac{\partial H}{\partial i_5}.$$

From equation (1):

$$2i_1 - 4(i - i_1) - 6(i - i_1 - i_5) + 8(i_1 + i_5) = 0, \quad (2)$$

and
$$10i_5 - 6(i - i_1 - i_5) + 8(i_1 + i_5) = 0. \quad (3)$$

These give

$$10i_1 + 7i_5 = 5i \quad (4)$$

and
$$7i_1 + 12i_5 = 3i. \quad (5)$$

Solving these equations for i_1 and i_5:

$$\underline{i_1 = \tfrac{39}{71}i, \quad i_5 = -\tfrac{5}{71}i.} \tag{6}$$

Now the voltage drop from A to C (taking link ABC) is

$$1i_1 + 4(i_1 + i_5) = 5i_1 + 4i_5.$$

From (6) this gives $\tfrac{175}{71}i$.

But if R is the equivalent resistance of the network, the voltage drop from A to C is $R\,.\,i$.

Thus $$R\,.\,i = \tfrac{175}{71}i,$$

i.e. $$\underline{R = \tfrac{175}{71}\Omega.}$$

Note. It is easily checked that equations (2) and (3) are really the back e.m.f. equations for the loops $ABCD$ and BCD, each multiplied by two.

(2) Find for what values of x and y the function

$$z = x^3 + y^3 - 3xy$$

is a maximum or a minimum.

$$\frac{\partial z}{\partial x} = 3x^2 - 3y, \quad \frac{\partial z}{\partial y} = 3y^2 - 3x.$$

Thus, when $\partial z/\partial x = 0 = \partial z/\partial y$,

$$x^2 - y = 0 \quad \text{and} \quad y^2 - x = 0.$$

Solving these equations:

$$(y^2)^2 - y = 0,$$

$$\therefore\ y(y^3 - 1) = 0.$$

$$\left.\begin{aligned} \therefore\ y = 0 \quad &\text{or} \quad y = 1, \\ x = 0 \quad &\text{or} \quad x = 1. \end{aligned}\right\} \tag{1}$$

$$\frac{\partial^2 z}{\partial x^2} = 6x, \quad \frac{\partial^2 z}{\partial y^2} = 6y \quad \text{and} \quad \frac{\partial^2 z}{\partial y\,\partial x} = -3. \tag{2}$$

(i) *For the values* $x = 0$, $y = 0$:

$$\frac{\partial^2 z}{\partial x^2} = 0 = \frac{\partial^2 z}{\partial y^2}.$$

Thus $$\frac{\partial^2 z}{\partial x^2}\frac{\partial^2 z}{\partial y^2} = 0 \quad \text{and} \quad \left(\frac{\partial^2 z}{\partial y\,\partial x}\right)^2 = 9.$$

$$\therefore \qquad \frac{\partial^2 z}{\partial x^2}\frac{\partial^2 z}{\partial y^2} < \left(\frac{\partial^2 z}{\partial y\,\partial x}\right)^2$$

and *when* $x = 0$, $y = 0$, z *has neither a maximum nor a minimum value.*

(ii) *For the values* $x = 1$, $y = 1$:

$$\frac{\partial^2 z}{\partial x^2} = 6 = \frac{\partial^2 z}{\partial y^2},$$

$$\therefore \qquad \frac{\partial^2 z}{\partial x^2}\frac{\partial^2 z}{\partial y^2} = 36 > \left(\frac{\partial^2 z}{\partial y\,\partial x}\right)^2.$$

Thus when $x = 1$, $y = 1$, z has a maximum or a minimum value.

As $\partial^2 z/\partial x^2$ (and $\partial^2 z/\partial y^2$) is positive, z *is a minimum when* $x = 1$, $y = 1$.

(3) Show that the rectangular closed box, of given surface area, which has maximum volume is a cube.

Let x, y, z be the length, breadth and height of the box.

Let S be the given surface area and V the volume.

Then $$S = 2(xy + yz + zx), \tag{1}$$

and $$V = xyz. \tag{2}$$

From (1), $z(x + y) = \frac{1}{2}S - xy$, giving

$$z = \frac{(S - 2xy)}{2(x + y)}. \tag{1)(a}$$

Thus $$V = \frac{xy(S - 2xy)}{2(x + y)}, \tag{3}$$

where S is a given constant.

$$\frac{\partial V}{\partial x} = \frac{1}{2}\left[\frac{(x + y)(Sy - 4xy^2) - xy(S - 2xy)}{(x + y)^2}\right]$$

$$= \frac{1}{2}\frac{(Sy^2 - 2x^2y^2 - 4xy^3)}{(x + y)^2}.$$

From the symmetry in x and y of equation (3):

$$\frac{\partial V}{\partial y} = \frac{1}{2}\frac{(Sx^2 - 2x^2y^2 - 4yx^3)}{(x + y)^2}.$$

When $\partial V/\partial x = 0 = \partial V/\partial y$,

$$Sy^2 - 2x^2y^2 - 4xy^3 = 0$$

and $$Sx^2 - 2x^2y^2 - 4yx^3 = 0.$$

One solution is obviously $x = 0 = y$, but this would give zero volume. Cancelling x^2 and y^2 in the equations:

$$\tfrac{1}{2}S = x^2 + 2xy \quad \text{and} \quad \tfrac{1}{2}S = y^2 + 2xy. \tag{4}$$

On subtraction these give $x^2 - y^2 = 0$.
Thus $x = y$ (ignoring the negative value as not practicable).
V is thus a maximum when $x = y$.

From (4), this gives $x = y = \sqrt{\dfrac{S}{6}}$.

From (1)(*a*),

$$z = \frac{\left(S - 2\dfrac{S}{6}\right)}{4\sqrt{\dfrac{S}{6}}} = \sqrt{\frac{S}{6}}.$$

Thus $x = y = z = \sqrt{\dfrac{S}{6}}$ make V a maximum.

4·25. Conditional maxima and minima

Example 3 above was a problem in which the function, $V = xyz$, contained variables which were not all independent; there was a relation between them, given by $2(xy + yz + zx) = S$ (a constant). Such problems are called conditional maxima and minima problems. The method used in the above example was to eliminate one of the variables, z, using the given condition. This procedure is often cumbersome. A neater way is to use the method of *undetermined multipliers*.

Let the function whose stationary values are required be

$$z = f(x, y), \tag{1}$$

and the given condition be

$$\phi(x, y) = 0. \tag{2}$$

Since z is stationary:

$$dz = \frac{\partial f}{\partial x}\,dx + \frac{\partial f}{\partial y}\,dy = 0.$$

As $\phi(x, y) = 0$

$$d\phi = \frac{\partial \phi}{\partial x}\,dx + \frac{\partial \phi}{\partial y}\,dy = 0.$$

For these equations to be consistent for the ratio $dx : dy$,

$$\frac{f_x}{\phi_x} = \frac{f_y}{\phi_y} = -\lambda \text{ say.}$$

Hence $$f_x + \lambda\phi_x = 0 \tag{3}$$

and $$f_y + \lambda\phi_y = 0. \tag{4}$$

Now consider the function

$$u = f(x, y) + \lambda\phi(x, y). \tag{5}$$

Equations (3) and (4) are equations that can be derived by putting $\partial u/\partial x = 0$ and $\partial u/\partial y = 0$, *treating λ as if it were a constant.*

Equations (2), (3) and (4) give three equations from which the values of x, y (and λ) at stationary points can be calculated.

This method can be extended to cover the general case of n variables and m $(<n)$ conditional equations. For example, if the stationary values of $V = f(x, y, z)$ are required, where x, y and z are subject to the relations

$$\phi(x, y, z) = 0 \quad \text{and} \quad \psi(x, y, z) = 0. \tag{6}$$

Form a function

$$u = f(x, y, z) + \lambda\phi(x, y, z) + \mu\psi(x, y, z)$$

and differentiate as if λ and μ were constants.

$$\left.\begin{aligned} f_x + \lambda\phi_x + \mu\psi_x &= 0, \\ f_y + \lambda\phi_y + \mu\psi_y &= 0, \\ f_z + \lambda\phi_z + \mu\psi_z &= 0. \end{aligned}\right\} \tag{7}$$

Equations (6) and (7) can then be solved to give values for x, y and z at stationary values of V.

Repeating Example 3 above using this method:

$$V = xyz,$$

where $$2(xy + yz + zx) = S \quad \text{(constant)}. \tag{8}$$

Forming $u = xyz + \lambda(2xy + 2yz + 2zx - S)$ and treating λ as if it were a constant:

$$yz + \lambda(2y + 2z) = 0,$$

$$zx + \lambda(2x + 2z) = 0,$$

$$xy + \lambda(2y + 2x) = 0.$$

By subtracting these equations from each other in turn, it is seen that

$$\text{either} \quad x = y = z = 0 \quad \text{or} \quad x = y = z = -4\lambda.$$

$x = y = z = 0$ would make V zero and require S to be zero.

Choosing the other values, equation (8) gives

$$x = y = z = \sqrt{\frac{S}{6}}.$$

Exercise 13

1. If $z = x^2 + y^3 - 6xy$, find the stationary values of z and state their nature.

2. Find the maximum value of x^3y^2z if $x + y + z = 6$.

3. Prove that the rectangular solid of maximum volume which can be inscribed in a sphere of given radius is a cube.

4. The sum of the volumes of a sphere and a right circular cylinder is constant. Find the relation between the radius of the sphere and the base-radius and height of the cylinder for the total surface area to be a minimum.

5. Show that $y = \sin ax \sin bt$ (a, b constants) has stationary points for any of the values $x = r\pi/2a$, $t = s\pi/2b$, where r and s are integers, either both even or both odd.

6. A resistance network consists of three branches in parallel, the branches being of resistance R_1, R_2, R_3 respectively. A total current i enters and leaves the network. Using the method of undetermined multipliers, find the distribution of the current in the separate branches.

7. A circuit is to be made up of three resistors in parallel to give an effective resistance of $\frac{1}{6}\Omega$. The first branch is to be made up of wire of resistance $\frac{1}{3}\Omega$ per metre, the second branch of wire of resistance $\frac{1}{4}\Omega$ per metre and the third of resistance $\frac{1}{5}\Omega$ per metre. Find the lengths of wire in each branch so that the sum of the lengths is a minimum.

8. A tetrahedron $ABCD$ consists of conducting wires such that the resistances of opposite edges are equal. The resistances of the edges AD, BD and CD are R/λ, R/μ and R, respectively. Determine the effective resistance of the tetrahedron when current enters at C and leaves at D. [C.U.]

9. Twelve equal straight wires are arranged as the edges of a cube. Current enters the network at a vertex A and leaves by the vertex B, AB being a space diagonal of the cube. Show that the resistance of the network is five-sixths of the resistance of one wire. Three of the wires are now removed, the three chosen having no points in common with each other and not containing A or B. Find the resistance of the network in this case. [C.U.]

10. Four concentric spherical shells are of radii a, x, y and b respectively, where $a < x < y < b$. The first and third are connected together by a fine wire passing through a small hole in the second. The second and fourth are connected together by a fine wire through a small hole in the third; the fourth shell is earthed. Treating the system as three capacitors in parallel, show that the total capacitance is

$$\frac{ax}{(x-a)} + \frac{yx}{(y-x)} + \frac{by}{(b-y)}.$$

Also find the values of x and y for minimum total capacitance, if x and y vary whilst a and b are fixed.

(For two concentric shells, one earthed, of radii r_1, r_2, the capacitance is $\dfrac{r_1 r_2}{(r_2 - r_1)}$, where $r_2 > r_1$.)

11. If $f(x,y) = (x^2 + 4xy + 2y^2 + 4x - 10)\,e^{-x-y}$, show that $f(x,y)$ has stationary values at two points for both of which $x = 2$. Find the corresponding y-values and show that $f(x,y)$ is a maximum for one value and is neither a maximum nor minimum for the other. [L.U.]

12. A factory has a rectangular floor plan and a flat roof and encloses a volume of 640,000 m^3. The front wall costs 3 times, and the roof 5/4 times as much per unit area as the other walls. Omitting the cost of the floor, find the length of frontage and the height for least cost.

ANSWERS

1. $x = 18$, $y = 6$ give z a minimum value of -108. $x = 0$, $y = 0$ give z a stationary value, neither max. nor min.

2. 108.

4. The radius of the sphere equals the base-radius of the cylinder and is half its height.

6. $$\dot{i}_1 = \frac{R_2R_3}{\mu}\dot{i}, \quad \dot{i}_2 = \frac{R_3R_1}{\mu}\dot{i}, \quad \dot{i}_3 = \frac{R_2R_1}{\mu}\dot{i},$$

where $\mu = R_1R_2 + R_2R_3 + R_3R_1$.

7. 1·723, 1·989 and 2·224 m.

8. $$\frac{(\lambda + \mu + 2)R}{2(\lambda + 1)(\mu + 1)}.$$

9. The resistance of the new network is the same as the resistance of one wire.

10. $$x = \frac{3ab}{(2b + a)}, \quad y = \frac{3ab}{(2a + b)}.$$

11. (2, 1) is a maximum. (2, $-$ 3) is neither maximum nor minimum.

12. Length 80 m; height 50 m.

Chapter 5

FURTHER COMPLEX VARIABLES

5·1. Locus problems and transformations

Let $z = x + jy$ be any complex number; then on an Argand diagram its position is fixed by the point P (or the vector $\overrightarrow{OP}$ if the complex number is to be represented by a vector),

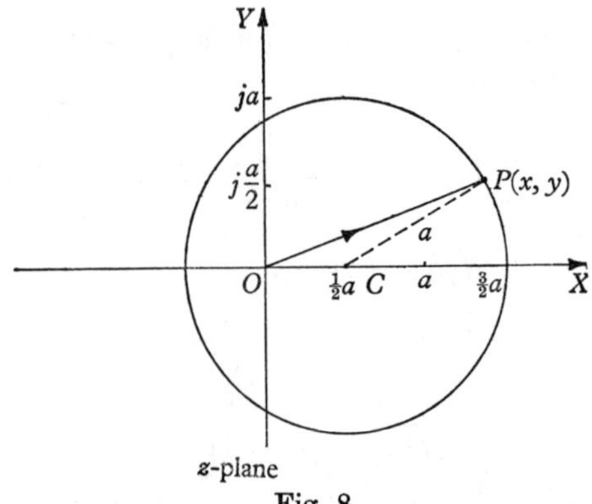

Fig. 8

the Cartesian co-ordinates of P being (x, y). If x and y vary to satisfy a functional relation $f(x, y) = 0$, P will describe a locus on the Argand diagram. This locus, or curve, is called the z-locus and the plane of the system of axes OX, OY, the z-plane.

For example, if $z = x + jy$ and (x, y) always satisfies the equation

$$(x - \tfrac{1}{2}a)^2 + y^2 = a^2, \qquad (1)$$

the locus of z in the z-plane is a circle, centre $(\frac{1}{2}a, 0)$, radius a (see fig. 8).

Examples of this nature have been met with already in volume I of this book.

Now suppose that another complex variable $w = u + jv$ is expressed as some function of z (i.e. $w = F(z)$); for example, take $w = 1/z$. Then for every pair of values of x, y there will correspond values of u and v. A locus of w can therefore be drawn. This is done on a second Argand diagram, which is usually called the w-plane, and the corresponding locus is called the w-locus. Its axes are labelled OU and OV.

Taking the example given above:

$$w = u + jv = \frac{1}{z} = \frac{1}{x + jy} = \frac{x - jy}{x^2 + y^2}.$$

Thus
$$u + jv = \frac{x}{x^2 + y^2} - j\frac{y}{x^2 + y^2},$$

giving
$$u = \frac{x}{x^2 + y^2} \quad \text{and} \quad v = \frac{-y}{x^2 + y^2}. \tag{2}$$

Squaring and adding these expressions:

$$u^2 + v^2 = \frac{x^2 + y^2}{(x^2 + y^2)^2} = \frac{1}{x^2 + y^2}. \tag{3}$$

But from (2), $x = u(x^2 + y^2)$ and $y = -v(x^2 + y^2)$, therefore from (3),

$$x = \frac{u}{u^2 + v^2}, \qquad y = \frac{-v}{u^2 + v^2}. \tag{4}$$

Now in the z-plane the point (x, y) describes the locus $(x - \frac{1}{2}a)^2 + y^2 = a^2$, thus the corresponding point (u, v) in the w-plane always satisfies the equation

$$\left(\frac{u}{u^2 + v^2} - \frac{a}{2}\right)^2 + \frac{v^2}{(u^2 + v^2)^2} = a^2,$$

i.e.
$$\frac{u^2 + v^2}{(u^2 + v^2)^2} - \frac{au}{(u^2 + v^2)} + \frac{a^2}{4} = a^2,$$

giving
$$1 - au = \tfrac{3}{4}a^2(u^2 + v^2),$$

i.e.
$$u^2 + v^2 + \frac{4u}{3a} = \frac{4}{3a^2}.$$

Completing the square of this expression:

$$\left(u+\frac{2}{3a}\right)^2+v^2=\frac{4}{3a^2}+\frac{4}{9a^2}=\frac{16}{9a^2}$$

or
$$\left(u+\frac{2}{3a}\right)^2+v^2=\left(\frac{4}{3a}\right)^2. \tag{5}$$

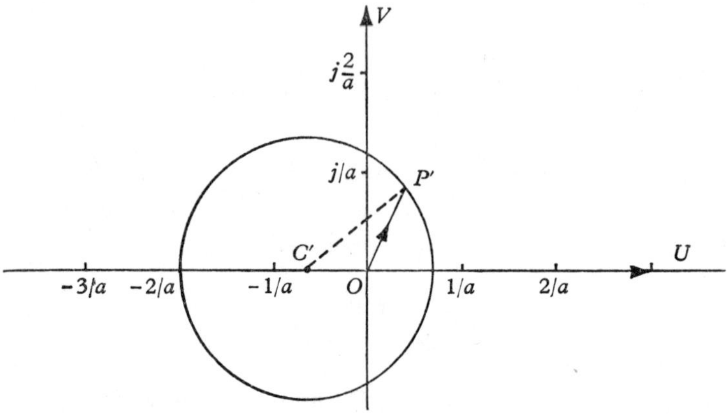

w-plane

Fig. 9

The w-locus is therefore a circle, centre $\left(\frac{-2}{3a}, 0\right)$ and radius $4/3a$ (see fig. 9).

The above process is called a *transformation.*

The given relationship $w = 1/z$ can be looked upon as transforming the circle of fig. 8 into the circle of fig. 9.

$w = 1/z$ is an example of a transformation that is called inversion plus reflexion. It is treated from a purely geometric point of view in the Appendix, and has important applications in electrical work.

In general, inversion transforms lines and circles into lines or circles.

The general process of finding the w-locus given the z-locus has been illustrated in the above example.

If the z-locus is known in the form $f(x, y) = 0$, u and v are determined in terms of x and y (see equation (2)); then x and y

are determined as functions of u and v (see equation (4)); finally, x and y are eliminated using the relationship $f(x, y) = 0$ (see equation (5)).

The two diagrams are usually put side by side so that they may be compared. This has been done in fig. 10.

To every point P on the z-locus there corresponds a point P' on the w-locus. As P traces out the z-locus in any given manner, it is instructive to find out the manner in which P' traces out the w-locus.

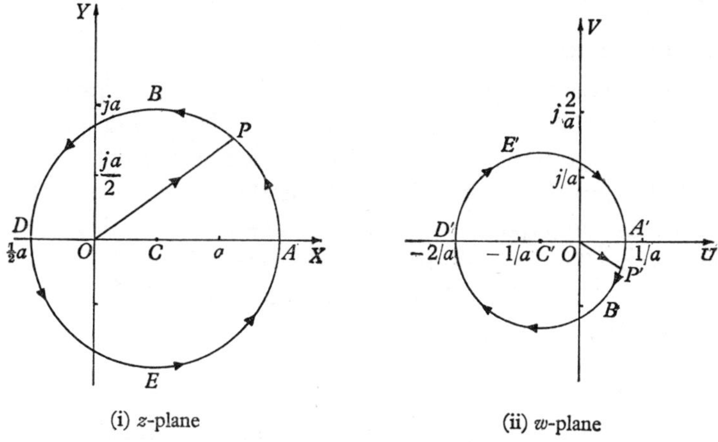

(i) z-plane (ii) w-plane

Fig. 10

Starting at A let P trace out the z-locus in an anticlockwise direction, taking the path $ABDEA$. Equation (2) above can be used to determine how the corresponding point P' will describe the w-locus. Normally a few corresponding positions are sufficient to give a clear picture.

At A, $x = \frac{3}{2}a$ and $y = 0$.

$\therefore$ From (2): $u = 2/3a$ and $v = 0$.

Thus $\qquad A, A'$ are corresponding points. $\qquad$ (6)

At B, $x = \frac{1}{2}a$, $y = a$.

$\therefore$ From (2): $\quad u = 2/5a \quad$ and $\quad v = -\,4/5a$.

Thus $\qquad B, B'$ are corresponding points. $\qquad$ (7)

At D, $x = -\,\frac{1}{2}a$, $y = 0$.

$\therefore$ From (2): $u = -2/a$ and $v = 0$.

Thus D, D' are corresponding points. (8)

Statements (6), (7) and (8) show that as P describes the semicircle ABD anticlockwise, P' describes the semicircle $A'B'D'$ clockwise.

Similarly, as P describes the semicircle DEA, P' describes the semicircle $D'E'A'$. Note the 'inversion' of one figure with respect to the other, and also the 'reflexion'.

5·2. Impedance and admittance loci

The representation of impedances as complex numbers is by now familiar to the student. Inversion plays a big part in such representations if the admittance is also needed. Thus, if z is the impedance of a circuit, $1/z$ is the admittance of the circuit. As the impedance z is varied a locus can be drawn in the z-plane. The symbol used for admittance is not w but Y. Thus $Y = 1/z$, and as z varies a locus can be drawn for Y in the Y-plane.

Example 1

A circuit consists of a fixed resistance R in parallel with a fixed inductance L. An alternating voltage of variable frequency $\omega/2\pi$ is to be applied to the circuit.

Draw the impedance and admittance loci as ω is varied.

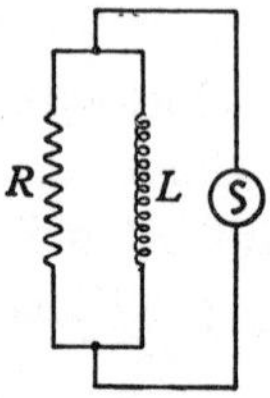

R L S

Fig. 11

Fig. 11 shows the circuit. As the elements are in parallel, the admittance is given by

$$Y = \frac{1}{z} = \frac{1}{R} + \frac{1}{j\omega L}. \tag{1}$$

Thus
$$Y = \frac{1}{R} - j\frac{1}{\omega L}.$$

Let $Y = u + jv$, then

$$u = \frac{1}{R} \quad \text{and} \quad v = -\frac{1}{\omega L}. \tag{2}$$

As R is fixed, the Y-locus as ω varies is the straight line $u = 1/R$. Now ω is essentially positive and the limits of variation are from 0 to $+\infty$; thus v can vary from $-\infty$ to zero. The Y-locus is shown in Fig. 12 (i).

From (1),
$$\frac{1}{z} = \frac{(R + j\omega L)}{j\omega LR}.$$

$$\therefore \qquad z = \frac{j\omega LR}{R + j\omega L} = \frac{j\omega LR(R - j\omega L)}{R^2 + \omega^2 L^2}.$$

If $z = x + jy$, then

$$x = \frac{\omega^2 L^2 R}{R^2 + \omega^2 L^2}, \qquad y = \frac{\omega L R^2}{R^2 + \omega^2 L^2}. \tag{3}$$

Squaring and adding these expressions:

$$x^2 + y^2 = \frac{\omega^2 L^2 R^2 (R^2 + \omega^2 L^2)}{(R^2 + \omega^2 L^2)^2} = \frac{\omega^2 L^2 R^2}{(R^2 + \omega^2 L^2)}.$$

Thus, from (3)

$$x^2 + y^2 = Rx.$$

Completing the square in x in this expression:

$$\left(x - \frac{R}{2}\right)^2 + y^2 = \frac{R^2}{4}. \tag{4}$$

As ω varies the locus of (x, y) is therefore a circle, centre $(\frac{1}{2}R, 0)$ and radius $\frac{1}{2}R$.

As ω, L, R are essentially positive, equation (3) shows that the actual impedance locus will only be that part of the circle which lies in the first quadrant.

When ω is zero, $x = 0 = y$ and when $\omega \to \infty$, $x \to R$, $y \to 0$. Fig. 12 (ii) shows the impedance locus.

The points P and P' are corresponding points for $\omega = R/L$. The vectors $\overrightarrow{OP}$, $\overrightarrow{OP'}$ represent the values of the impedance and admittance for this value of ω.

Note that when ω is zero the impedance is zero and the admittance is infinite; this is because the inductive branch is then effectively a short-circuit.

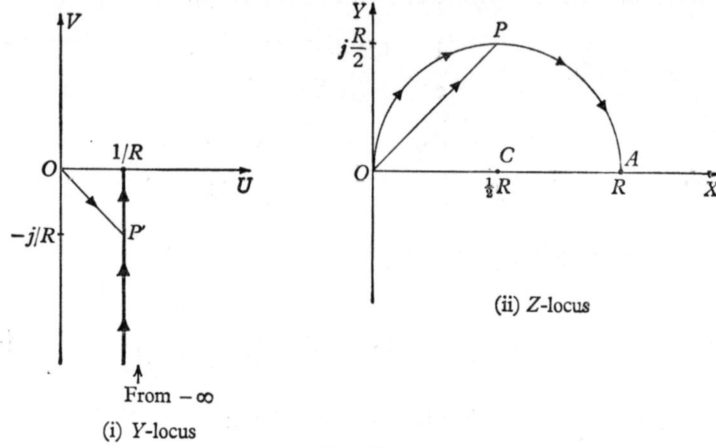

Fig. 12

EXAMPLE 2

A series L-C-R circuit has L and R fixed but C variable. An e.m.f. of fixed frequency $\omega/2\pi$ is to be applied. Draw the impedance and admittance loci as C varies.

Fig. 13 shows the circuit.

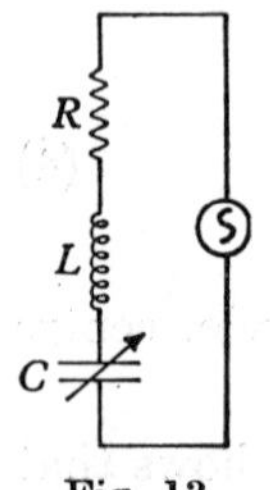

Fig. 13

As the elements are in series, the impedance z is given by

$$z = R + j\omega L - \frac{j}{\omega C}$$

$$= R + j\left(\omega L - \frac{1}{\omega C}\right). \quad (1)$$

Thus, if $z = x + jy$,

$$x = R, \quad y = \omega L - \frac{1}{\omega C}. \quad (2)$$

As C varies the (x, y) locus is the straight line $x = R$.

Now as C is essentially positive it may vary from 0 to $+\infty$; thus y varies from $-\infty$ to ωL (see (2)).

The impedance locus is shown in fig. 14 (i).

From (1), the admittance, Y, is given by

$$Y = \frac{1}{z} = \frac{1}{R + j\left(\omega L - \frac{1}{\omega C}\right)} = \frac{R - j\left(\omega L - \frac{1}{\omega C}\right)}{R^2 + \left(\omega L - \frac{1}{\omega C}\right)^2}.$$

If $Y = u + jv$, then

$$u = \frac{R}{R^2 + \left(\omega L - \frac{1}{\omega C}\right)^2}, \quad v = \frac{-\left(\omega L - \frac{1}{\omega C}\right)}{R^2 + \left(\omega L - \frac{1}{\omega C}\right)^2}. \quad (3)$$

Squaring and adding these expressions:

$$u^2 + v^2 = \frac{R^2 + \left(\omega L - \frac{1}{\omega C}\right)^2}{\left[R^2 + \left(\omega L - \frac{1}{\omega C}\right)^2\right]^2} = \frac{1}{R^2 + \left(\omega L - \frac{1}{\omega C}\right)^2}. \quad (4)$$

But from (3), $\qquad \dfrac{R}{u} = R^2 + \left(\omega L - \dfrac{1}{\omega C}\right)^2;$

thus from (4),

$$u^2 + v^2 = \frac{u}{R}.$$

Completing the square in u in this expression:

$$\left(u - \frac{1}{2R}\right)^2 + v^2 = \frac{1}{4R^2}.$$

This is a circle, centre $(1/2R, 0)$, radius $1/2R$.

From equation (3), as C varies from 0 to $1/\omega^2 L$ (where $\omega L = 1/\omega C$), u varies from 0 to $1/R$ and v from 0, through positive values, to 0 again; then as C varies from $1/\omega^2 L$ to $+\infty$, u varies from $1/R$ to $R/(R^2 + \omega^2 L^2)$ and v from 0 through negative values to $-\omega L/(R^2 + \omega^2 L^2)$.

The actual admittance locus is shown in fig. 14 (ii).

A, A' correspond to $C = 0$ (i.e. open-circuit).

B, B' correspond to $C = +\infty$ (i.e. shorted capacitor).

D, D' correspond to $C = 1/\omega^2 L$ (i.e. resonance).

Vectors drawn from the pole O to any points on the locus give the values for the impedance or admittance for the corresponding values of the capacitance C.

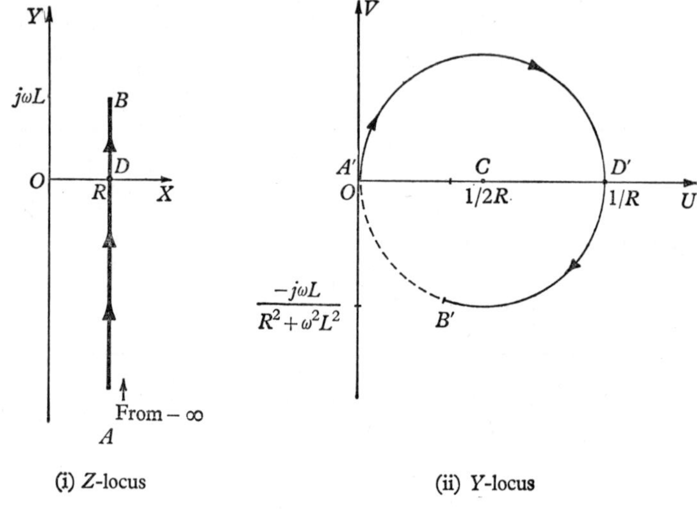

(i) Z-locus (ii) Y-locus

Fig. 14

As the current is given by the product of the voltage and the admittance, the admittance locus can also serve as a current locus merely by renumbering the scale (see Appendix).

Note. An alternative method of finding the admittance locus is to proceed as follows:

$$z = \frac{1}{Y} = \frac{1}{u+jv} = R + j\left(\omega L - \frac{1}{\omega C}\right), \quad \text{from (1) above.}$$

Thus
$$\frac{u-jv}{u^2+v^2} = R + j\left(\omega L - \frac{1}{\omega C}\right).$$

$$\therefore \quad \frac{u}{u^2+v^2} = R \quad \text{and} \quad \frac{v}{u^2+v^2} = -\left(\omega L - \frac{1}{\omega C}\right). \qquad (5)$$

The first of these equations gives at once

$$u^2 + v^2 = \frac{u}{R}, \quad \text{i.e.} \quad \left(u - \frac{1}{2R}\right)^2 + v^2 = \frac{1}{4R^2}.$$

Thus (u, v) lies on a circle.

Squaring and adding the equations in (5) gives

$$\frac{u^2 + v^2}{(u^2 + v^2)^2} = R^2 + \left(\omega L - \frac{1}{\omega C}\right)^2,$$

i.e.

$$\frac{1}{u^2 + v^2} = R^2 + \left(\omega L - \frac{1}{\omega C}\right)^2.$$

Substituting for $1/(u^2 + v^2)$ into equations (5) gives the equations for u and v separately as given in (3).

Note. In electrical technology the symbol Z is used for impedance and $R + jX$ as its $a + jb$ form. The admittance Y is also usually written in its complex form as $G + jB$ (see Appendix). In this section it has been more convenient to use $z = x + jy$ and $Y = u + jv$.

Exercise 14

1. If $z = x + jy$ and $w = 1/z$, find the locus of w $(= u + jv)$ when $x = a$ (constant) and y varies.

2. The point P represented in the Argand diagram by $z = (x + jy)$ lies on the line $6x + 8y = R$ where R is real. Q is the point represented by R^2/z. Show that the locus of Q is a circle and find its centre and radius. [L.U.]

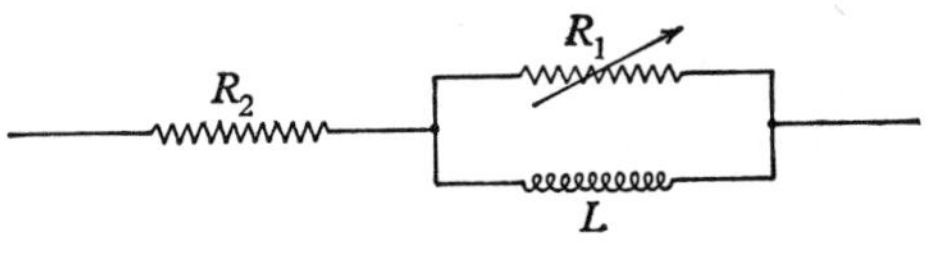

Fig. 15

3. The circuit shown in fig. 15 has resistances R_1, R_2 and a constant inductance L. The frequency of the applied e.m.f. is $\omega/2\pi$ (constant). If z is the total impedance of the circuit, show that $z = \left(R_2 + \dfrac{\omega^2 L^2 R_1}{\omega^2 L^2 + R_1^2}\right) + j\,\dfrac{\omega L R_1^2}{\omega^2 L^2 + R_1^2}$. Draw the impedance locus as R_1 varies and R_2 is kept constant.

4. In the circuit shown in fig. 16, C is a variable capacitance, R is a fixed resistance and L a fixed inductance. An alternating voltage of constant frequency $\omega/2\pi$ is applied. Find the impedance and admittance loci as C varies.

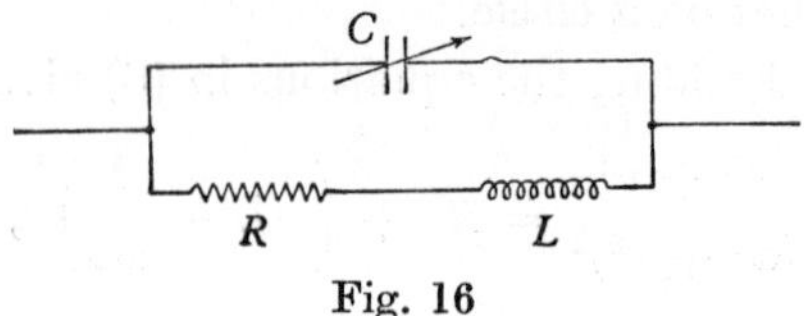

Fig. 16

5. A capacitance C shunted by a resistance R is put in series with an inductance L of resistance S. Show how to draw the vector locus of the impedance of the circuit as C is varied.

If $R = 800$ ohms, $L = 50$ millihenries, and $\omega = 5000$, find the values of C which make the circuit non-reactive. [L.U.]

6. A capacitor of 5 microfarads capacitance is hunted by a variable resistance R. Draw the vector locus of the impedance of the combination at 50 Hz as R varies from zero to infinity. Read off the impedance when the value of R is 500 ohms. [L.U.]

7. If $u + jv = (x + jy)^{\frac{1}{2}}$, where u, v are real and the point (x, y) describes the circle $(x - 1)^2 + y^2 = 1$, find the polar equation of the locus of the point (u, v). [L.U.]

8. If $x + jy = \dfrac{2j}{u + jv}$, where x, y, u, v are all real and $j = \sqrt{-1}$, express x, y in terms of u and v.

If the point (x, y) moves round a trapezium $ABCD$ whose vertices are the points $(1, 1)$, $(1, -1)$, $(2, -2)$, $(2, 2)$ in that order, find the equations of the path traced out by the point (u, v) and show it in a separate diagram. [L.U.]

ANSWERS

1. A circle with equation $(u - 1/2a)^2 + v^2 = 1/4a^2$, centre $(1/2a, 0)$ and radius $1/2a$. [As y goes from 0 to $+\infty$, w describes the semi-circle in the 4th quadrant; as y goes from 0 to $-\infty$, w describes the semi-circle in the 1st quadrant.]

2. Centre $(3R, -4R)$, radius $5R$. [This is the inversion plus reflexion of a straight line.]

3. That half of the circle $(x - R_2)^2 + (y - \frac{1}{2}\omega L)^2 = (\frac{1}{2}\omega L)^2$ for which $x \geqslant R_2$.

[*Note.* $\quad x - R_2 = \dfrac{R_1\omega^2L^2}{\omega^2L^2 + R_1^2}, \quad y = \dfrac{R_1^2\omega L}{\omega^2L^2 + R_1^2}.$

$$\therefore \qquad (x - R_2)^2 + y^2 = \frac{\omega^2L^2R_1^2}{\omega^2L^2 + R^2} = \omega Ly.]$$

4. Impedance locus:

part of the circle, centre $\left(\dfrac{R^2 + \omega^2L^2}{2R}, 0\right)$, radius $\dfrac{R^2 + \omega^2L^2}{2R}$.

The part required is the major arc between the point $(R, \omega L)$, when $C = 0$, and the point $(0, 0)$, when $C = \infty$.

Admittance locus:

that part of the line $u = \dfrac{R}{R^2 + \omega^2L^2}$ for which

$$v \geqslant \frac{-\omega L}{R^2 + \omega^2L^2}.$$

$\left[\text{Note that} \quad u = \dfrac{R}{R^2 + \omega^2L^2}, \quad v = \omega C - \dfrac{\omega L}{R^2 + \omega^2L^2}.\right]$

5. $x - S = \dfrac{R}{1 + \omega^2C^2R^2}, \quad y - \omega L = \dfrac{-\omega CR^2}{1 + \omega^2C^2R^2}.$

Locus is lower semicircle of circle $\{x - (S + \frac{1}{2}R)\}^2 + (y - \omega L)^2 = \frac{1}{4}R^2$. $C \simeq 0{\cdot}088\ \mu$F or $0{\cdot}712\ \mu$F.

6. Approx. 393Ω. Locus is the semicircle in the fourth quadrant of circle $x^2 + (y + 10^3/\pi)^2 = (10^3/\pi)^2$.

7. $r^2 = 2\cos 2\theta$.

8. Path is given by $x = \dfrac{2v}{u^2 + v^2}, \; y = \dfrac{2u}{u^2 + v^2}.$

The parts of the (u, v) locus, in order, are:

$u^2 + (v - 1)^2 = 1, \; u + v = 0, \; u^2 + (v - \frac{1}{2})^2 = \frac{1}{4}, \; u - v = 0.$

The locus begins at $(1, 1)$, goes along the upper half of the first circle to $(-1, 1)$; then down the line $u + v = 0$ to the second circle; then along the upper half of this circle to the point $(\frac{1}{2}, \frac{1}{2})$, and finally up the line $u - v = 0$ to the point $(1, 1)$.

5·3. Exponential values of trigonometric functions

From volume I, § 11·33

$$e^{jx} = \cos x + j \sin x$$

and $$e^{-jx} = \cos x - j \sin x.$$

Adding these expressions:

$$2 \cos x = e^{jx} + e^{-jx},$$

$$\therefore \quad \underline{\cos x = \tfrac{1}{2}(e^{jx} + e^{-jx}).} \tag{1}$$

Subtracting the expressions:

$$2j \sin x = e^{jx} - e^{-jx},$$

$$\therefore \sin x = \frac{1}{2j}(e^{jx} - e^{-jx}). \tag{2}$$

The similarity between formulae (1) and (2) and the definitions of sinh x and cosh x is obvious.

If cosh z and sinh z, when z may be complex, are defined by

$$\cosh z = \frac{e^z + e^{-z}}{2} \quad \text{and} \quad \sinh z = \frac{e^z - e^{-z}}{2},$$

then $$\cosh jx = \frac{e^{jx} + e^{-jx}}{2} = \cos x, \tag{3}$$

and $$\sinh jx = \frac{e^{jx} - e^{-jx}}{2} = j \sin x. \tag{4}$$

Writing jx for x in (3) and (4), remembering that $j^2 = -1$:

$$\cosh x = \cos jx \tag{5}$$

and $$\sinh(-x) = j \sin jx,$$

i.e. $$-\sinh x = j^2 \sinh x = j \sin jx.$$

$$\therefore \quad j \sinh x = \sin jx. \tag{6}$$

Summarizing these results:

$$\left.\begin{aligned} \cosh jx &= \cos x, \quad \cos jx = \cosh x, \\ \sinh jx &= j \sin x, \quad \sin jx = j \sinh x. \end{aligned}\right\} \tag{7}$$

These results explain the rule given in volume I, § 3·3, for converting relationships between trigonometric functions to corresponding relationships between hyperbolic functions.

It is now seen that a multiplying factor j occurs in the relationship between sine and sinh, but does not occur between cosine and cosh. Thus when there is a sine product (actual or implied), a factor $j^2 = -1$ will occur in the corresponding sinh product, e.g.

$$\underline{1} = \cos^2 jx + \sin^2 jx = \cosh^2 x + j^2 \sinh^2 x = \underline{\cosh^2 x - \sinh^2 x}.$$

By dividing the relationships in equations (7):

$$j \tan x = \tanh jx \quad \text{and} \quad j \tanh x = \tan jx. \tag{8}$$

5·4. Real and imaginary parts of trigonometric and hyperbolic functions of a complex variable

It is assumed that the usual addition and product formulae, etc., can be used to expand such expressions as $\sin(x + jy)$, etc.

Thus

$$\sin(x + jy) = \sin x \cos jy + \cos x \sin jy.$$

$$\therefore \sin(x + jy) = \sin x \cosh y + j \cos x \sinh y, \tag{1}$$

from § 5·3, equations (7).

$\sin(x + jy)$ has been evaluated as a complex number in its algebraic form. Its real part is $\sin x \cosh y$, and its imaginary part is $\cos x \sinh y$.

Similarly, $\cos(x + jy) = \cos x \cos jy - \sin x \sin jy$.

$$\therefore \cos(x + jy) = \cos x \cosh y - j \sin x \sinh y. \tag{2}$$

Analogous formulae can be easily found for other trigonometric functions. They are listed below, the verification of the others being left to the student:

$$\left.\begin{aligned} \sin(x \pm jy) &= \sin x \cosh y \pm j \cos x \sinh y, \\ \cos(x \pm jy) &= \cos x \cosh y \mp j \sin x \sinh y, \\ \tan(x \pm jy) &= \frac{\tan x \pm j \tanh y}{1 \mp j \tan x \tanh y}. \end{aligned}\right\} \tag{3}$$

Expansions for hyperbolic functions can be found in a similar manner, e.g.

$$\begin{aligned} \sinh(x + jy) &= \sinh x \cosh jy + \cosh x \sinh jy \\ &= \sinh x \cos y + j \cosh x \sin y. \end{aligned}$$

EXAMPLES

(1) If $\sin(x + jy) = r(\cos\theta + j\sin\theta)$, find the numerical values of r and θ when $x = y = 1$.

$$\sin(x + jy) = \sin x \cos jy + \cos x \sin jy$$
$$= \sin x \cosh y + j\cos x \sinh y.$$

$$\therefore\ r\cos\theta = \sin x \cosh y \quad \text{and} \quad r\sin\theta = \cos x \sinh y.$$

When $x = y = 1$:

$$r\cos\theta = \sin 1 \cosh 1 = 0{\cdot}8415 \times 1{\cdot}5431 \simeq 1{\cdot}299$$

and $$r\sin\theta = \cos 1 \sinh 1 = 0{\cdot}5403 \times 1{\cdot}1752 \simeq 0{\cdot}635.$$

$$\therefore\ r = \sqrt{(1{\cdot}298^2 + 0{\cdot}635^2)} \simeq 1{\cdot}445$$

and $$\theta = \tan^{-1}\frac{0{\cdot}635}{1{\cdot}298} \simeq 0{\cdot}455 \text{ radians (principal value).}$$

$$\underline{r = 1{\cdot}445, \quad \theta = 0{\cdot}455 \text{ rad.}}$$

(2) If $\sinh(x + jy) = j2$, find real positive values for x and y.

$$\sinh(x + jy) = j2 = \sinh x \cos y + j\cosh x \sin y.$$

Equating real and imaginary parts:

$$\sinh x \cos y = 0, \qquad (1)$$
$$\cosh x \sin y = 2. \qquad (2)$$

Equation (1) is satisfied when *either* $\sinh x = 0$ *or* $\cos y = 0$, i.e. either $x = 0$ or $y = \frac{1}{2}(2k + 1)\pi$, k any integer.

Taking $x = 0$. Equation (2) then gives

$$\cosh 0 \sin y = 2,$$
i.e. $$\sin y = 2.$$

This is not satisfied by any real value of y.

Taking $y = \frac{1}{2}(2k + 1)\pi$. Equation (2) now gives

$$\cosh x \sin \tfrac{1}{2}(2k + 1)\pi = 2.$$

Now if k is an *odd* number, $\sin\frac{1}{2}(2k + 1)\pi$ is -1, which makes $\cosh x = -2$. This is not satisfied by real x values. But if k is an even number or zero, $\sin\frac{1}{2}(2k + 1)\pi$ is $+1$, which makes $\cosh x = 2$ and $x = 1{\cdot}317$ (positive value).

Thus the general solution is

$$x = \cosh^{-1} 2 = 1{\cdot}317 \quad \text{(positive value)},$$

$$y = \tfrac{1}{2}(4n + 1)\pi \quad (n \text{ any integer}).$$

Taking the lowest positive value of y, when $n = 0$,

$$\left.\begin{aligned} &\underline{x = 1{\cdot}317,} \\ &\underline{y = \tfrac{1}{2}\pi = 1{\cdot}571,} \end{aligned}\right\}$$

is a pair of positive values of x and y satisfying the given equation.

Notes. (i) This type of equation often occurs in line transmission and filter circuit theory.

(ii) It is now clear that equations such as $\cos z = 10$ are perfectly feasible as long as z is taken as a complex number.

5·5. Logarithms of complex numbers

If $e^z = N$, where N, z may be complex numbers, then z is *defined* to be $\log_e N$.

Let $\qquad N = re^{j\theta} = re^{j(\theta + 2k\pi)},$

then $\qquad \log_e N = \log_e re^{j(\theta + 2k\pi)}$

$$= \log_e r + \log_e e^{j(\theta + 2k\pi)},$$

giving $\qquad \underline{\log_e N = \log_e r + j(\theta + 2k\pi).}$

It appears that the logarithm of a complex number has a single-valued real part ($\log_e r$), but a multivalued imaginary part.

Unless otherwise stated or required, the principal value, that value between $-\pi$ and $+\pi$, will be taken.

Note that if N is given in its $a + jb$ form, this must first be converted into its polar form.

EXAMPLE

Find $\log_e(-2)$ in its $a + jb$ form.

$$-2 = -2 + 0j = 2(-1 + 0j) = 2(\cos\pi + j\sin\pi).$$

In general polar form

$$-2 = 2e^{j(\pi + 2k\pi)}.$$

$$\text{Thus } \log_e (-2) = \log_e 2e^{j(\pi + 2k\pi)}$$
$$= \log_e 2 + j(\pi + 2k\pi).$$

The principal value is

$$\underline{\log_e 2 + j\pi = 0{\cdot}6931 + j3{\cdot}142.}$$

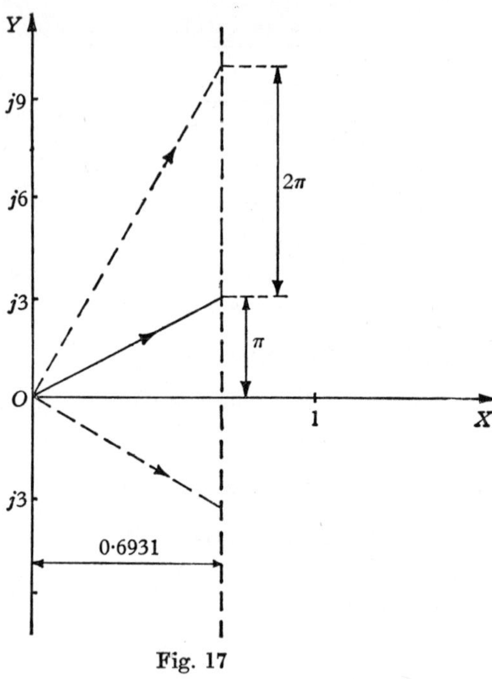

Fig. 17

In fig. 17 some of the multivalues of $\log_e (-2)$ are shown on an Argand diagram; the principal value is shown in full line.

Exercise 15

1. Derive formulae for $\cosh (x + jy)$, $\cosh (x - jy)$, $\tanh (x+jy)$ and $\tanh (x - jy)$ in the manner explained in § 5·4.

2. Evaluate in $a + jb$ form

(i) $\sin (2 + j3)$, (ii) $\log_e (2 + j) (3 - j2)$.

3. (i) Find a and b if $\log_e (a + jb) = 2 - j\dfrac{\pi}{4}$.

(ii) Find the principal value of $j \log_e j$.

4. Find real values for x and y if $\sin (x + jy) = 2$.

5. Find real values for x and y if $x + jy = \cosh^{-1}(-2)$.

6. If $z = x + jy = \tanh\left(u + j\dfrac{\pi}{4}\right)$, where u is real, find x and y in terms of u and show that for all values of u the point z lies on the circle $x^2 + y^2 = 1$ in the Argand diagram. [L.U.]

7. If $\tan \frac{1}{2}(x + jy) = u + jv$, find the value of u/v in terms of x and y.

8. If $\tan(A + jB) = x + jy$, prove that:

$$x^2 + y^2 + 2x \cot 2A - 1 = 0,$$

$$x^2 + y^2 - 2y \coth 2B + 1 = 0.$$ [L.U.]

9. If $\tan^{-1}(x + jy) = u + jv$, show that:

$$u = \tfrac{1}{2}\tan^{-1}\left(\frac{2x}{1 - x^2 - y^2}\right),$$

$$v = \tfrac{1}{2}\tanh^{-1}\left(\frac{2y}{1 + x^2 + y^2}\right).$$

Show that if the real part of $\tan^{-1}(x + jy)$ is $\frac{1}{8}\pi$ the representative point of the complex number $x + jy$ on the Argand diagram lies on the circle of radius $\sqrt{2}$ with its centre at the point $-1 + j0$. [L.U.]

Answers

1. $\cosh x \cos y + j \sinh x \sin y$, $\cosh x \cos y - j \sinh x \sin y$,

$$\frac{\tanh x + j \tan y}{1 + j \tanh x \tan y}, \quad \frac{\tanh x - j \tan y}{1 - j \tanh x \tan y}.$$

2. (i) $9{\cdot}15 - j4{\cdot}17$.

(ii) $\frac{1}{2}\log_e 65 - j0{\cdot}124 = 2{\cdot}087 - j0{\cdot}124$.

3. (i) $a = e^2/\sqrt{2}$, $b = -e^2/\sqrt{2}$;

(ii) $-\frac{1}{2}\pi = -1{\cdot}571$.

4. $x = \frac{1}{2}\pi \simeq 1{\cdot}571$,

$y = \cosh^{-1} 2 \simeq 1{\cdot}317$.

5. $x = \cosh^{-1} 2 \simeq 1{\cdot}317$, $y = \pi \simeq 3{\cdot}142$.

6. $x = \dfrac{2 \tanh u}{1 + \tanh^2 u} = \tanh 2u,$

$$y = \frac{1 - \tanh^2 u}{1 + \tanh^2 u} = \operatorname{sech} 2u.$$

(Note that $\operatorname{sech}^2 2u + \tanh^2 2u = 1$.)

7. $\dfrac{u}{v} = \dfrac{\tan \frac{1}{2}x(1 - \tanh^2 \frac{1}{2}y)}{\tanh \frac{1}{2}y(1 + \tan^2 \frac{1}{2}x)}.$

Using $1 - \tanh^2 \frac{1}{2}y = \operatorname{sech}^2 \frac{1}{2}y$ and $1 + \tan^2 \frac{1}{2}x = \sec^2 \frac{1}{2}x$, this reduces to

$$\frac{u}{v} = \frac{\tan \frac{1}{2}x \operatorname{sech}^2 \frac{1}{2}y}{\tanh \frac{1}{2}y \sec^2 \frac{1}{2}x} = \frac{\sin \frac{1}{2}x \cosh \frac{1}{2}y \cos^2 \frac{1}{2}x}{\cos \frac{1}{2}x \sinh \frac{1}{2}y \cosh^2 \frac{1}{2}y}$$

$$= \frac{\sin \frac{1}{2}x \cos \frac{1}{2}x}{\sinh \frac{1}{2}y \cosh \frac{1}{2}y} = \frac{\sin x}{\sinh y}.$$

8. A neat method for this problem is to use conjugates. If $\tan(A + jB) = x + jy$, then $\tan(A - jB) = x - jy$. Multiplying these expressions:

$$x^2 + y^2 = \tan(A + jB) \,.\, \tan(A - jB). \qquad (1)$$

Adding the expressions:

$$2x = \tan(A + jB) + \tan(A - jB). \qquad (2)$$

Subtracting the expressions:

$$2jy = \tan(A + jB) - \tan(A - jB). \qquad (3)$$

Now use

$$\tan[(A + jB) + (A - jB)] = \tan 2A$$

$$= \frac{\tan(A + jB) + \tan(A - jB)}{1 - \tan(A + jB) \,.\, \tan(A - jB)}.$$

Using (1) and (2), this gives

$$\tan 2A = \frac{2x}{1 - (x^2 + y^2)},$$

giving $$x^2 + y^2 + 2x \cot 2A - 1 = 0.$$

For the second part start with

$$\tan[(A + jB) - (A - jB)] = \tan j2B = j \tanh 2B$$

$$= \frac{\tan(A + jB) - \tan(A - jB)}{1 + \tan(A + jB) . \tan(A - jB)}.$$

9. [For first part proceed as in question 8 above.]

5·6. General multiple-angle formulae

The multiple-angle formulae $\sin 2x = 2 \sin x \cos x$, $\cos 3x = 4 \cos^3 x - 3 \cos x$, etc., are met with in an S 3 Course. With the aid of De Moivre's theorem these formulae can be generalized.

Taking n as a positive integer:

$$\cos n\theta + j \sin n\theta = (\cos \theta + j \sin \theta)^n.$$

Now expand the right-hand side by the binomial theorem, writing c for $\cos \theta$ and s for $\sin \theta$:

$$\cos n\theta + j \sin n\theta = c^n + nc^{n-1}(js) + \frac{n(n-1)}{2!} c^{n-2}(j^2s^2) + \dots + (js)^n.$$

Equating real and imaginary parts, noting that $j^2 = -1$, $j^3 = -j$, $j^4 = +1$, etc.,

$$\cos n\theta = c^n - \frac{n(n-1)}{2!} c^{n-2}s^2 + \frac{n(n-1)(n-2)(n-3)}{4!} c^{n-4}s^4 \dots \quad (1)$$

and

$$\sin n\theta = nc^{n-1}s - \frac{n(n-1)(n-2)}{3!} c^{n-3}s^3 \dots . \quad (2)$$

The final term in each expansion is that in which the power of c is 1 or 0.

Dividing (2) by (1) and remembering that

$$\frac{s}{c} = \frac{\sin \theta}{\cos \theta} = \tan \theta = t \text{ say,}$$

$$\tan n\theta = \frac{nt - \dfrac{n(n-1)(n-2)}{3!} t^3 + \dots}{1 - \dfrac{n(n-1)}{2!} t^2 + \dots}. \quad (3)$$

EXAMPLES

(1) $$\sin 4\theta = 4\cos^3\theta\sin\theta - \frac{4\,.\,3\,.\,2}{3!}\cos\theta\sin^3\theta$$
$$= 4\cos^3\theta\sin\theta - 4\cos\theta\sin^3\theta.$$

(2) $$\cos 5\theta = \cos^5\theta - \frac{5\,.\,4}{2!}\cos^3\theta\sin^2\theta + \frac{5\,.\,4\,.\,3\,.\,2}{4!}\cos\theta\sin^4\theta$$
$$= \cos^5\theta - 10\cos^3\theta\sin^2\theta + 5\cos\theta\sin^4\theta.$$

Note that as all the powers of $\sin\theta$ occurring in the expansion of $\cos n\theta$ are even, then $\cos n\theta$ can always be expanded in terms of powers of $\cos\theta$, using $\sin^2\theta = 1 - \cos^2\theta$. Thus

$$\cos 5\theta = \cos^5\theta - 10\cos^3\theta(1 - \cos^2\theta) + 5\cos\theta(1 - \cos^2\theta)^2$$
$$= 16\cos^5\theta - 20\cos^3\theta + 5\cos\theta.$$

The same thing cannot be done with $\sin n\theta$ but *can* be done with $\dfrac{\sin n\theta}{\sin\theta}$.

(3) $$\tan 4\theta = \frac{4\tan\theta - \dfrac{4\,.\,3\,.\,2}{3!}\tan^3\theta}{1 - \dfrac{4\,.\,3}{2!}\tan^2\theta + \dfrac{4\,.\,3\,.\,2\,.\,1}{4!}\tan^4\theta}$$
$$= \frac{4\tan\theta - 4\tan^3\theta}{1 - 6\tan^2\theta + \tan^4\theta}.$$

5·7. $\cos^n\theta$ and $\sin^n\theta$ in terms of cosines and sines of multiple angles

Let

$$z = \cos\theta + j\sin\theta,$$

then

$$\frac{1}{z} = z^{-1} = \cos(-\theta) + j\sin(-\theta) = \cos\theta - j\sin\theta.$$

Adding the expressions:

$$z + \frac{1}{z} = 2\cos\theta.$$

Subtracting the expressions: (1)

$$z - \frac{1}{z} = j2\sin\theta.$$

Again,

$$z^n = (\cos\theta + j\sin\theta)^n = \cos n\theta + j\sin n\theta$$

and $$\frac{1}{z^n} = z^{-n} = \cos n\theta - j\sin n\theta.$$

Adding these expressions:

$$z^n + \frac{1}{z^n} = 2\cos n\theta.$$

Subtracting the expressions: (2)

$$z^n - \frac{1}{z^n} = j2\sin n\theta.$$

Formulae (1) and (2) can be used as required in the expansion of $\cos^n\theta$ and $\sin^n\theta$ in terms of multiple angles.

EXAMPLES

(1) Expand $\cos^6\theta$ in terms of multiple angles.

From (1):

$$2^6\cos^6\theta = \left(z + \frac{1}{z}\right)^6$$

$$= z^6 + 6z^4 + 15z^2 + 20 + \frac{15}{z^2} + \frac{6}{z^4} + \frac{1}{z^6},$$

i.e.

$$64\cos^6\theta = \left(z^6 + \frac{1}{z^6}\right) + 6\left(z^4 + \frac{1}{z^4}\right) + 15\left(z^2 + \frac{1}{z^2}\right) + 20.$$

Using (2):

$$64\cos^6\theta = 2\cos 6\theta + 6(2\cos 4\theta) + 15(2\cos 2\theta) + 20.$$

$$\therefore \cos^6\theta = \tfrac{1}{32}(\cos 6\theta + 6\cos 4\theta + 15\cos 2\theta + 10).$$

(2) Expand $\sin^5\theta$ in terms of multiple angles.
From (1):

$$2^5 j^5 \sin^5\theta = \left(z - \frac{1}{z}\right)^5$$

$$= z^5 - 5z^3 + 10z - \frac{10}{z} + \frac{5}{z^3} - \frac{1}{z^5},$$

i.e. $$j32\sin^5\theta = \left(z^5 - \frac{1}{z^5}\right) - 5\left(z^3 - \frac{1}{z^3}\right) + 10\left(z - \frac{1}{z}\right).$$

Using (2):

$$j32 \sin^5 \theta = j2 \sin 5\theta - 5(j2 \sin 3\theta) + 10(j2 \sin \theta),$$

giving $$\sin^5 \theta = \frac{1}{16} (\sin 5\theta - 5 \sin 3\theta + 10 \sin \theta).$$

Notes. (i) Such results as these could be used for integrating $\cos^6 \theta$, $\sin^5 \theta$, etc., but reduction formulae are usually shorter.

(ii) These results can also be looked on as Fourier series for $\cos^6 \theta$, $\sin^5 \theta$, etc.

EXERCISE 16

1. Expand $\cos^5 \theta$ in a series of cosines of multiples of θ.

2. Expand $\sin 6x/\sin x$ as a power series in $\cos x$.

3. Expand $\tan 6x$ in terms of $\tan x$.

4. Expand $\sin^6 x$ in a series of cosines of multiples of x.

5. Writing

$$C = 1 + \cos x + \frac{1}{2!} \cos 2x + \ldots + \frac{1}{n!} \cos nx + \ldots,$$

and

$$S = \sin x + \frac{1}{2!} \sin 2x + \ldots + \frac{1}{n!} \sin nx + \ldots,$$

show that

$$C + jS = 1 + e^{jx} + \frac{e^{j2x}}{2!} + \ldots + \frac{1}{n!} e^{jnx} + \ldots.$$

Hence, comparing this series with the expansion of e^z, show that $C + jS = \exp (e^{jx})$.

Deduce that $$C = e^{\cos x} \cos (\sin x)$$

and $$S = e^{\cos x} \sin (\sin x).$$

6. Find the sum to infinity of the series

$$1 + \tfrac{1}{3} \cos x + \tfrac{1}{9} \cos 2\,x + \tfrac{1}{27} \cos 3\,x + \ldots,$$

using a method similar to that of question 5.

7. Prove that

$$\sec \theta \cos 7\theta = 1 - 24 \sin^2 \theta + 80 \sin^4 \theta - 64 \sin^6 \theta.$$

ANSWERS

1. $\frac{1}{16}(\cos 5\theta + 5 \cos 3\theta + 10 \cos \theta)$.
2. $32 \cos^5 x - 32 \cos^3 x + 6 \cos x$.
3. $$\frac{6 \tan x - 20 \tan^3 x + 6 \tan^5 x}{1 - 15 \tan^2 x + 15 \tan^4 x - \tan^6 x}.$$
4. $\frac{1}{32}(10 - 15 \cos 2x + 6 \cos 4x - \cos 6x)$.
6. $$\frac{3(3 - \cos x)}{2(5 - 3 \cos x)}.$$

[Note that

$$C + jS = 1 + \tfrac{1}{3}e^{jx} + \frac{1}{3^2} e^{j2x} \dots,$$

which is a geometrical progression whose sum to infinity is $\dfrac{1}{1 - e^{jx}/3}$. For C, find the real part of this expression.]

5·8. Symmetrical components

5·81. Introduction

If the voltages, currents and impedances are balanced, in a 3-phase network, it is a comparatively simple matter to reduce the system to an equivalent single-phase network. If they are unbalanced, the general analysis of a 3-phase network becomes complicated, as it is then necessary to consider the complete network, which will have almost three times as many branches as the equivalent single-phase system for a balanced network. In this case a more orderly method of analysis is to use the method of symmetrical components. Applying this method to networks in which the voltages or currents are unbalanced, or in which a limited amount of unbalance of impedance exists, it is possible to replace the normal equivalent network by three separate but very similar networks. The method involves resolving the unbalanced voltages and currents into 'symmetrical components'. The impedances of the various branches of the network to these components are then determined. In a network in which *all* the impedances are unbalanced, the method can be extended by resolving the impedances into symmetrical components.

However, as this does not effectively simplify the analysis it is not often used in practice.

The student is referred to text-books on Power System Analysis for a detailed treatment of networks. This section will be mainly concerned with the actual resolution of three 'unbalanced' vectors into symmetrical components.

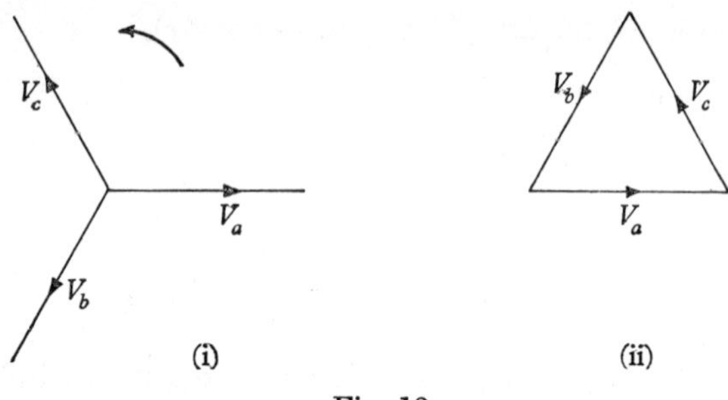

Fig. 18

The unit operator h. In a 3-phase balanced circuit the general vector diagram for voltages (or currents) takes the form of three vectors of equal magnitudes and with phase angles differing by $\frac{2}{3}\pi$ radians, or 120° (see fig. 18 (i)).

The arrangement in fig. 18 is called a *positive sequence* system; V_a leads V_b, which in turn leads V_c.

As complex numbers, the vectors may be represented, taking V_a as a reference, as

$$V_a = r, \qquad V_b = re^{j\,4\pi/3}, \qquad V_c = re^{j\,2\pi/3}, \tag{1}$$

where r is the magnitude of each vector.

Thus $$V_b = V_a e^{j\,4\pi/3}, \quad V_c = V_a e^{j\,2\pi/3}\,. \tag{2}$$

On a vector diagram $e^{j\,2\pi/3}$ has modulus one and argument (phase displacement) $\frac{2}{3}\pi$. Multiplication of any vector by $e^{j\,2\pi/3}$ therefore leaves its modulus unaltered but increases its phase angle by $\frac{2}{3}\pi$.

In work of this kind $e^{j\,2\pi/3}$ is often called the operator h (some books use a or λ).

If $h \equiv e^{j\,2\pi/3}$, then

$$h^2 = e^{j\,4\pi/3} \quad \text{and} \quad h^3 = e^{j\,6\pi/3} = e^{j\,2\pi} = 1\,. \tag{3}$$

Thus

$$h^3 = 1, \quad h^2 = \frac{h^3}{h} = \frac{1}{h} \quad \text{and} \quad h = \frac{h^3}{h^2} = \frac{1}{h^2}. \tag{4}$$

An important relationship can be found from the identity

$$h^3 - 1 \equiv (h-1)(1+h+h^2).$$

As $h^3 = 1$, then $\qquad (h-1)(1+h+h^2) = 0.$

As $h \neq 1$, this gives $\qquad \underline{1 + h + h^2 = 0.} \qquad (5)$

From (2) $\qquad V_a + V_b + V_c = V_a(1 + h^2 + h) = 0.$

This can easily be seen from the fact that the three vectors form a closed triangle (see fig. 18 (ii)). In any balanced circuit then,

$$V_a + V_b + V_c = 0\,. \tag{6}$$

5·82. Phase-sequence components

In a 3-phase unbalanced circuit the general vector diagram for voltages (or currents) can be represented by three arbitrary vectors as shown in fig. 19.

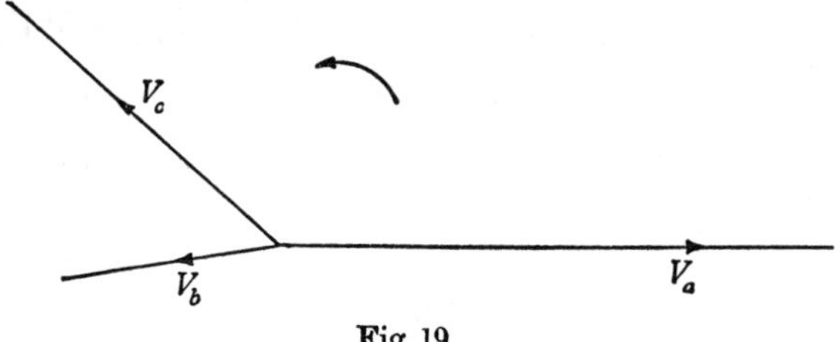

Fig. 19

These may be considered as the sum of three other systems of vectors as follows:

I. A system of three balanced vectors V_{a1}, V_{b1}, V_{c1}, in the same sequence as V_a, V_b, V_c.

This system is called the *positive phase-sequence* system. It is shown in fig. 20.

Here, $\qquad V_{b1} = h^2 V_{a1}$ and $V_{c1} = hV_{a1}$. $\qquad$ (1)

II. A system of three balanced vectors V_{a2}, V_{b2}, V_{c2}, of *reversed* sequence; that is, one in which V_{c2} leads V_{b2}, which in turn leads V_{a2}.

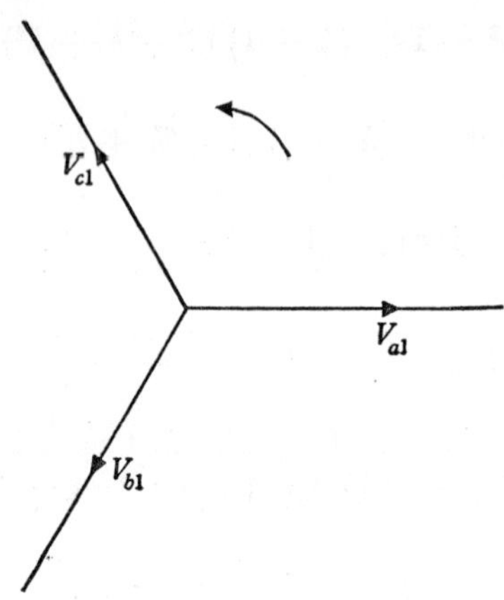

Fig. 20

This is called the *negative phase-sequence* system and is shown in fig. 21.

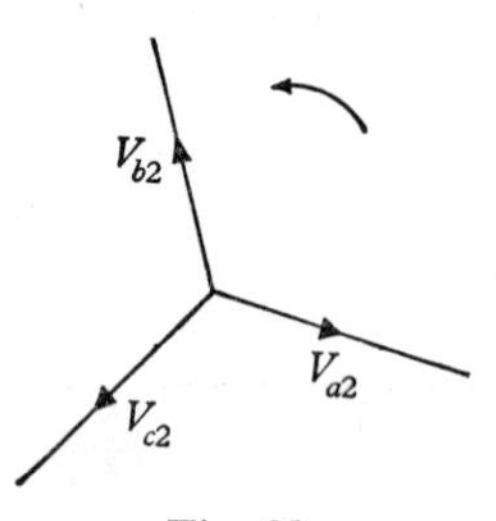

Fig. 21

Here

$$V_{b2} = hV_{a2} \quad \text{and} \quad V_{c2} = h^2 V_{a2}. \qquad (2)$$

III. A system of three equal vectors in phase, V_{a0}, V_{b0}, V_{c0}, as shown in fig. 22.

This is called the zero phase-sequence system.

Here $\qquad V_{a0} = V_{b0} = V_{c0}$. $\qquad$ (3)

It can be shown that any system of three vectors can be analysed into such a set of three symmetrical systems. Fig. 23 shows a diagram of the three systems combining to give the original three vectors.

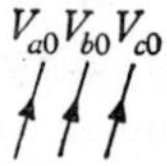

V_{a0} V_{b0} V_{c0}

Fig. 22

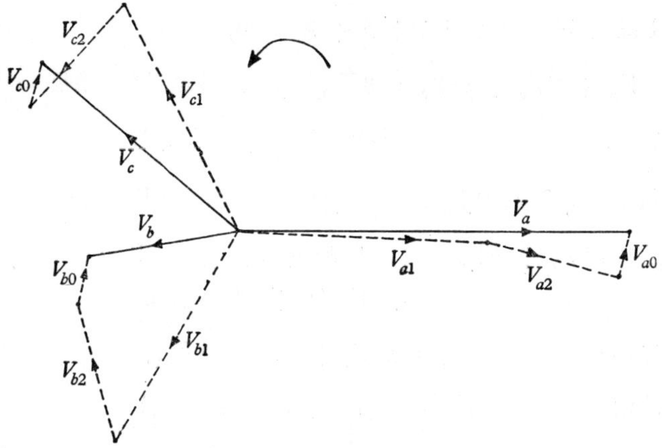

Fig. 23

5·83. Calculation of the symmetrical components

Taking the relationships

$$V_a = V_{a0} + V_{a1} + V_{a2},$$
$$V_b = V_{b0} + V_{b1} + V_{b2},$$
$$V_c = V_{c0} + V_{c1} + V_{c2},$$

and using equations (1), (2) and (3) of § 5·82 in order to express all the right-hand sides in terms of V_{a0}, V_{a1}, V_{a2}:

$$V_a = V_{a0} + V_{a1} + V_{a2}, \tag{1}$$
$$V_b = V_{a0} + h^2 V_{a1} + h V_{a2}, \tag{2}$$
$$V_c = V_{a0} + h V_{a1} + h^2 V_{a2}. \tag{3}$$

Making use of the fact that $1 + h + h^2 = 0$, the values of V_{a0}, V_{a1} and V_{a2} are now determined.

Adding (1), (2) and (3) gives

$$V_a + V_b + V_c = 3V_{a0} + (1 + h + h^2)V_{a1} + (1 + h + h^2)V_{a2}.$$

Thus $$\underline{V_{a0} = \tfrac{1}{3}(V_a + V_b + V_c).} \tag{4}$$

Multiplying (2) by h and (3) by h^2 and adding all three gives

$$V_a + hV_b + h^2V_c = V_{a0}(1 + h + h^2) + V_{a1}(1 + h^3 + h^3) + V_{a2}(1 + h^2 + h^4).$$

Using $h^3 = 1$ and $1 + h + h^2 = 0$,

$$V_a + hV_b + h^2V_c = 3V_{a1} \quad \text{(note that } h^4 = h^3h = h\text{)}.$$

Thus $$\underline{V_{a1} = \tfrac{1}{3}(V_a + hV_b + h^2V_c).} \tag{5}$$

Multiplying (2) by h^2 and (3) by h and adding all three gives

$$V_a + h^2V_b + hV_c = (1 + h^2 + h)V_{a0} + V_{a1}(1 + h^4 + h^2) + V_{a2}(1 + h^3 + h^3).$$

Thus $$V_a + h^2V_b + hV_c = 3V_{a2},$$

giving $$\underline{V_{a2} = \tfrac{1}{3}(V_a + h^2V_b + hV_c).} \tag{6}$$

Equations (4), (5) and (6) determine the values of V_{a0}, V_{a1}, V_{a2}. Knowing these, the other symmetrical components can easily be found from equations (1), (2) and (3) of § 5·82.

5·84. Numerical example

When working a numerical example use is made of the fact that

(i) $h = e^{j\,2\pi/3} = (\cos \tfrac{2}{3}\pi + j \sin \tfrac{2}{3}\pi) = -\tfrac{1}{2} + j\dfrac{\sqrt{3}}{2}$.

(ii) $h^2 = e^{j\,4\pi/3} = (\cos \tfrac{4}{3}\pi + j \sin \tfrac{4}{3}\pi) = -\tfrac{1}{2} - j\dfrac{\sqrt{3}}{2}$.

As an example, take

$$V_a = 76 + j2, \quad V_b = -30 - j5, \quad V_c = -40 + j30.$$

Using equation (4) above:

$$V_{a0} = \tfrac{1}{3}\{(76 + j2) + (-30 - j5) + (-40 + j30)\}$$
$$= \tfrac{1}{3}(6 + j27).$$

Thus $$\underline{V_{a0} = 2 + j9.} \tag{1}$$

From equation (5) above:

$$V_{a1} = \tfrac{1}{3}\{(76 + j2) + (-0{\cdot}5 + j0{\cdot}866)(-30 - j5) + (-0{\cdot}5 - j0{\cdot}866)(-40 + j30)\}$$

$$\simeq \tfrac{1}{3}(141{\cdot}3 - j1{\cdot}84),$$

i.e. $$\underline{V_{a1} = 47{\cdot}1 - j0{\cdot}61.} \tag{2}$$

From equation (6) above:

$$V_{a2} = \tfrac{1}{3}\{(76 + j2) + (-0\cdot5 - j0\cdot866)(-30 - j5) + (-0\cdot5 + j0\cdot866)(-40 + j30)\}$$

$$\simeq \tfrac{1}{3}(80\cdot69 - j19\cdot16),$$

i.e. $$\underline{V_{a2} = 26\cdot9 - j6\cdot39.} \quad (3)$$

Using § 5·82, equation (1):

$$\underline{V_{b1}} = h^2 V_{a1} = (-0\cdot5 - j0\cdot866)(47\cdot1 - j0\cdot61)$$

$$\simeq \underline{-24\cdot1 - j40\cdot5.}$$

$$\underline{V_{c1}} = hV_{a1} = (-0\cdot5 + j0\cdot866)(47\cdot1 - j0\cdot61)$$

$$\simeq \underline{-23 + j41\cdot1.}$$

Using § 5·82, equation (2):

$$\underline{V_{b2}} = hV_{a2} = (-0\cdot5 + j0\cdot866)(26\cdot9 - j6\cdot39)$$

$$\simeq \underline{-7\cdot92 + j26\cdot5.}$$

$$\underline{V_{c2}} = h^2 V_{a2} = (-0\cdot5 - j0\cdot866)(26\cdot9 - j6\cdot39)$$

$$\simeq \underline{-18\cdot98 - j20\cdot1.}$$

Using § 5·82, equation (3):

$$\underline{V_{a0} = V_{b0} = V_{c0} = 2 + j9.}$$

Figs. 19–23 are diagrams for this numerical example, and are approximately to scale.

5·85. Graphical construction of symmetrical components

Equations (4), (5) and (6) of § 5·83 can be used to give a method of graphical construction for the symmetrical components, given the original 'unbalanced' vectors. All three sets of symmetrical components can be constructed on the same diagram, but this would lead to very many construction lines. The construction will be easier to follow if each of the three sets are drawn in separate diagrams. Fig. 24 shows three given unbalanced vectors V_a, V_b and V_c, and also the positive phase-sequence system.

The dotted lines show hV_b and h^2V_c; these are easily

constructed using the fact that multiplication by h rotates a vector through 120° (anticlockwise).

From equation (5), § 5·83,

$$V_{a1} = \frac{V_a}{3} + \frac{hV_b}{3} + \frac{h^2V_c}{3}.$$

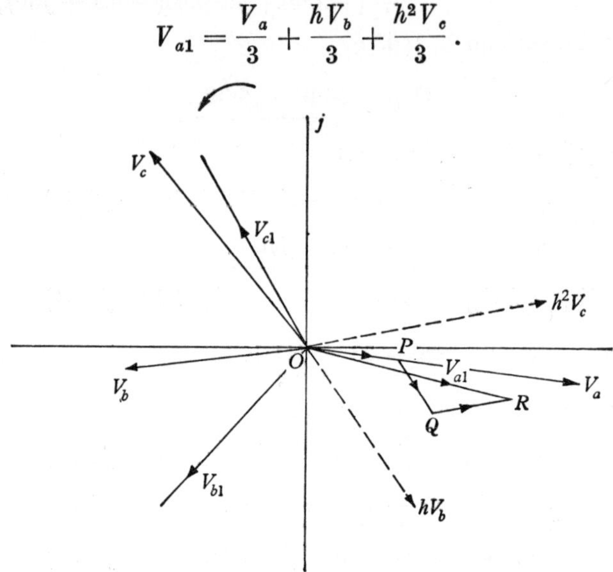

Positive phase-sequence system

Fig. 24

$\vec{OP}$ is easily constructed as one-third of V_a; $\vec{PQ}$ is drawn parallel to hV_b and of one-third the magnitude; $\vec{QR}$ is drawn parallel to h^2V_c and of one-third the magnitude.

Thus $\vec{OR} = \vec{OP} + \vec{PQ} + \vec{QR}$ represents V_{a1}.

V_{b1} and V_{c1} can now be drawn of the same length as V_{a1} and at the correct 120° intervals in the correct order.

Fig. 25 shows the original three vectors and the negative phase-sequence system. The construction for the latter is derived from equation (6), § 5·83.

$$V_{a2} = \frac{V_a}{3} + \frac{h^2V_b}{3} + \frac{hV_c}{3}.$$

$\vec{OP}$ is one-third of V_a, $\vec{PQ}$ one-third of h^2V_b and $\vec{QR}$ one-third of hV_c. V_{b2} and V_{c2} are drawn in the correct sequence, at 120° intervals.

Fig. 26 shows the original three vectors and the zero phase-sequence system. This construction is derived from equation (4), § 5·83, i.e. $V_{a0} = \frac{1}{3}(V_a + V_b + V_c)$.

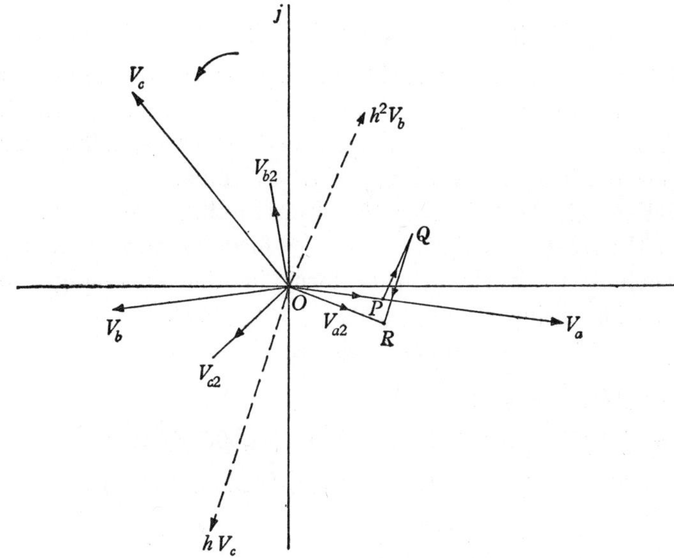

Negative phase-sequence system
Fig. 25

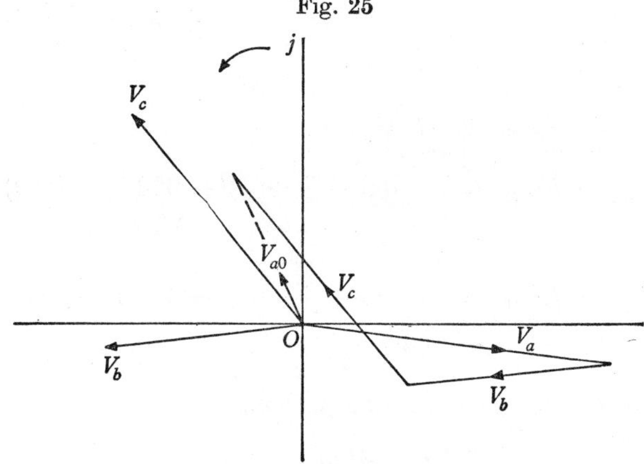

Zero phase-sequence system
Fig. 26

For this case the vectors V_a, V_b and V_c have been added vectorially and then their resultant divided by three. The system is three vectors $V_{a0} = V_{b0} = V_{c0}$.

5·86. Worked example

The three line currents in a 3-phase system under short-circuit conditions are given by

$$I_R = 1000 \angle 60°; \quad I_Y = 1500 \angle 120°; \quad I_B = 2000 \angle 270°.$$

Determine the values of the three symmetrical components and draw vector diagrams showing them to scale. [L.U.]

R, Y, B are letters often used for the phases, taken from the usual wiring colours red, yellow and blue.

R-Y-B will be taken as the positive-phase sequence.

As the vectors are given in $r\angle\theta$ form in this case, it is convenient to work in this form in the first stages of calculation, using $h = 1\angle 120°$, $h^2 = 1\angle 240°$.

(i) *Positive phase-sequence system.*

$$\begin{aligned} I_{R1} &= \tfrac{1}{3}(I_R + hI_Y + h^2I_B) \\ &= \tfrac{1}{3}(1000 \angle 60° + 1500 \angle 240° + 2000 \angle 510°) \\ &= \tfrac{1}{3}(1000 \angle 60° + 1500 \angle 240° + 2000 \angle 150°) \\ &= \frac{100}{3}\left\{10\left(\frac{1}{2} + j\frac{\sqrt{3}}{2}\right) + 15\left(-\frac{1}{2} - j\frac{\sqrt{3}}{2}\right)\right. \\ &\qquad\qquad \left. + 20\left(-\frac{\sqrt{3}}{2} + j\frac{1}{2}\right)\right\} \\ &= \frac{100}{3}(-19\cdot82 + j5\cdot67). \end{aligned}$$

$$\therefore \quad \underline{I_{R1} = -660\cdot7 + j189.}$$

$$\begin{aligned} I_{Y1} = h^2I_{R1} &= (-0\cdot5 - j0\cdot866)(-660\cdot7 + j189) \\ &= \underline{494\cdot0 + j477\cdot7,} \\ I_{B1} = hI_{R1} &= (-0\cdot5 + j0\cdot866)(-660\cdot7 + j189) \\ &= \underline{166\cdot7 - j666\cdot7.} \end{aligned}$$

(ii) *Negative phase-sequence system.*

$$\begin{aligned} I_{R2} &= \tfrac{1}{3}(I_R + h^2I_Y + hI_B) \\ &= \tfrac{1}{3}(1000 \angle 60° + 1500 \angle 360° + 2000 \angle 390°) \\ &= \tfrac{100}{3}\{10(0\cdot5 + j0\cdot866) + 15 + 20(0\cdot866 + j0\cdot5)\} \\ &= \tfrac{100}{3}(37\cdot32 + j18\cdot66). \end{aligned}$$

$$\underline{I_{R2} = 1244 + j622 = 622(2 + j).}$$

$$I_{Y2} = hI_{R2} = 622(-0{\cdot}5 + j0{\cdot}866)(2 + j)$$
$$= \underline{-1161 + j766{\cdot}3.}$$
$$I_{B2} = h^2 I_{R2} = 622(-0{\cdot}5 - j0{\cdot}866)(2 + j)$$
$$= \underline{-83{\cdot}33 - j1388.}$$

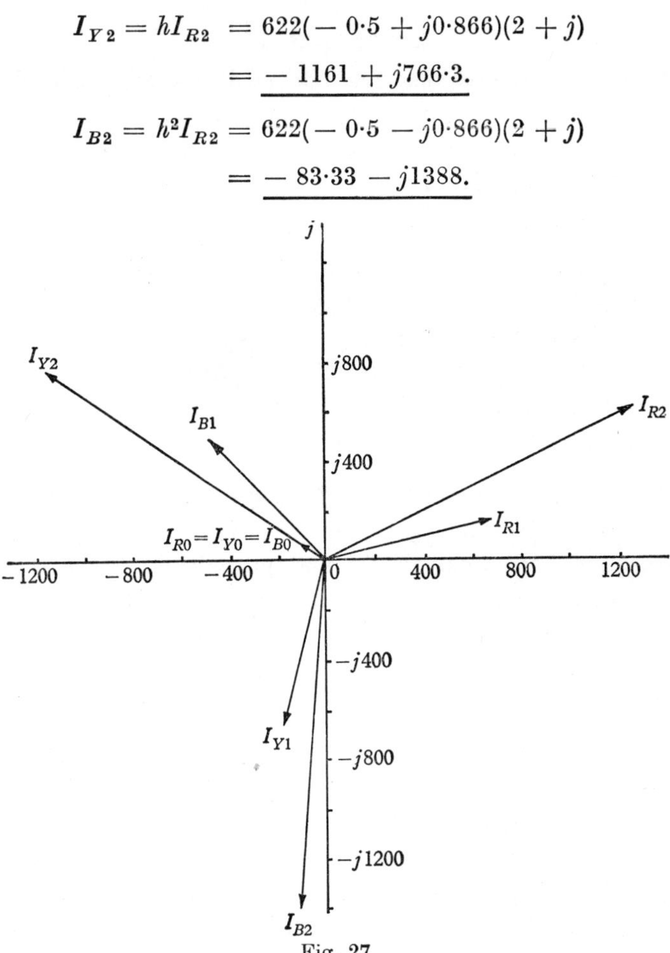

Fig. 27

(iii) *Zero phase-sequence system.*

$$I_{R0} = I_{Y0} = I_{B0} = \tfrac{1}{3}(I_R + I_Y + I_B)$$
$$= \tfrac{1}{3}(1000 \angle 60^\circ + 1500 \angle 120^\circ + 2000 \angle 270^\circ)$$
$$= \tfrac{100}{3}\{(5 + j8{\cdot}66) + (-7{\cdot}5 + j12{\cdot}99) - j20\}$$
$$= \tfrac{100}{3}(-2{\cdot}5 + j1{\cdot}65)$$
$$= \underline{-83{\cdot}3 + j55.}$$

Fig. 27 shows the symmetrical components on vector diagrams drawn to scale.

Exercise 17

1. Derive expressions for the positive- and zero-phase sequence components of the currents in an unbalanced three-phase system and calculate the components for the following case: $I_R = 20 - j10$, $I_Y = 100 + j0$, $I_B = 0 + j70$. [L.U.]

2. Explain the essential features in the representation of an unsymmetrical 3-phase system of voltages or currents by means of symmetrical components. In a 3-phase system the three line currents are: $I_R = 30 + j50$; $I_Y = 15 - j45$; $I_B = -40 + j70$ amp. Determine graphically or analytically the values of the positive, negative and zero sequence components. [L.U.]

3. The link currents in a 3-phase system are $\underset{\cdot}{I}_1 = 10 + j5$, $\underset{\cdot}{I}_2 = 15 - j15$ and $\underset{\cdot}{I}_3 = -20 + j10$ amp. respectively. Determine both analytically and graphically the symbolic expressions for the three symmetrical components of $\underset{\cdot}{I}_1$, $\underset{\cdot}{I}_2$ and $\underset{\cdot}{I}_3$.

Under what conditions must the zero-sequence component be absent? [L.U.]

4. Explain the principle of symmetrical component analysis of unsymmetrical 3-phase systems, giving the solution by vector algebra and by a graphical method.

Under fault conditions the three line currents in a 3-phase system are: $I_R = 1000 \overline{\diagdown} 0^\circ$; $I_Y = 2000 \overline{\diagdown} 90^\circ$; $I_B = 3000 \overline{\diagdown} 210^\circ$. Determine the values of the symmetrical components and draw vector diagrams showing them to scale. [L.U.]

5. The phase voltages of a 3-phase 4-wire system are V_{AN}, V_{BN} and V_{CN}, where $V_{AN} = 200 + j0$ volts, $V_{BN} = 0 - j200$ volts, $V_{CN} = -100 + j100$ volts. Show that these three voltages can be replaced by symmetrical components of positive, negative, and zero sequence, and calculate the magnitude of each component.

Sketch vector diagrams representing the positive, negative and zero sequence components for each of the three phases. [L.U.]

Answers

1. $I_{R1} = 10{\cdot}21 + j13{\cdot}87$; $I_{Y1} = 6{\cdot}9 - j15{\cdot}77$;
$I_{B1} = -17{\cdot}11 + j1{\cdot}91$;
$I_{R0} = I_{Y0} = I_{B0} = 40 + j20$.

2. $I_{R1} = 47{\cdot}36 + j28{\cdot}38$; $I_{Y1} = 0{\cdot}89 - j55{\cdot}31$;
$I_{B1} = -48{\cdot}26 + j26{\cdot}83$; $I_{R2} = -19{\cdot}03 - j3{\cdot}38$;
$I_{Y2} = 12{\cdot}44 - j14{\cdot}79$; $I_{B2} = 6{\cdot}59 + j18{\cdot}17$;
$I_{R0} = I_{Y0} = I_{B0} = 1{\cdot}67 + j25$.

3. $\dot{I}_{11} = 11{\cdot}38 + j12{\cdot}6$; $\dot{I}_{21} = 5{\cdot}22 - j16{\cdot}16$;
$\dot{I}_{31} = -16{\cdot}6 + j3{\cdot}56$; $\dot{I}_{12} = -3{\cdot}05 - j7{\cdot}6$;
$\dot{I}_{22} = 8{\cdot}11 + j1{\cdot}16$; $\dot{I}_{32} = -5{\cdot}06 + j6{\cdot}44$;
$\dot{I}_{10} = \dot{I}_{20} = \dot{I}_{30} = 1{\cdot}67 + j0$.

If $\dot{I}_1 + \dot{I}_2 + \dot{I}_3 \equiv 0$, i.e. the vector sum of the link currents is zero. In this case there is no need for the fourth 'neutral' wire.

4. $I_{R1} = 1776{\cdot}7 + j833{\cdot}3$; $I_{Y1} = 166{\cdot}7 - j1954{\cdot}7$;
$I_{B1} = -1609{\cdot}9 + j1122$; $I_{R2} = -244 - j666{\cdot}7$;
$I_{Y2} = 699{\cdot}3 + j122$; $I_{B2} = -455{\cdot}3 + j544{\cdot}7$;
$I_{R0} = I_{Y0} = I_{B0} = -532{\cdot}7 - j166{\cdot}7$.

5. (*Note.* The double-suffix notation is coming into frequent use. The letters A, B, C refer to the three phases and N refers to the 'neutral' point.)

$V_{A1} = 169{\cdot}93 + j45{\cdot}53$; $V_{B1} = -45{\cdot}53 - j169{\cdot}93$;
$V_{C1} = -124{\cdot}4 + j124{\cdot}4$; $V_{A2} = -3{\cdot}27 - j12{\cdot}2$;
$V_{B2} = 12{\cdot}2 + j3{\cdot}27$; $V_{C2} = -8{\cdot}93 + j8{\cdot}93$;
$V_{A0} = V_{B0} = V_{C0} = \frac{100}{3} - j\frac{100}{3}$.

The magnitude of the positive phase-sequence component is 176 volts. The magnitude of the negative phase-sequence component is 12·63 volts. The magnitude of the zero phase-sequence component is 47·14 volts. [Note that the symmetrical components could be written:

(1) 176 $\angle 15°$, $\angle 255°$, $\angle 135°$;
(2) 12·63 $\angle 255°$, $\angle 135°$, $\angle 15°$;
(3) 47·14 $\angle 315°$.]

CHAPTER 6

FURTHER FIRST-ORDER DIFFERENTIAL EQUATIONS

6·1. Equations reducible to separable variables

6·11. Homogeneous type

$x^3 - 3x^2y + 5xy^2 - 7y^3$ is said to be a homogeneous expression in x and y. The test is that the total degree in x and y must be the same for all terms; in this case it is three.

$x^3 - 3x^2 + 2y$ is not homogeneous.

Let the first-order equation to be solved be written in the form $dy/dx = P/Q$; then if P and Q are both homogeneous and of the same degree, the equation can be reduced to separable variables.

EXAMPLE

Solve
$$\frac{dy}{dx} = \frac{y^2 - xy}{2x^2}.$$

Here $P \equiv y^2 - xy$, $Q \equiv 2x^2$.

These are homogeneous and of the second degree.

The method is to put $y = vx$, where v is not a constant but a function of x.

Then $\dfrac{dy}{dx} = v1 + x\dfrac{dv}{dx}$ and the original equation reduces to

$$v + x\frac{dv}{dx} = \frac{v^2x^2 - vx^2}{2x^2},$$

i.e.
$$v + x\frac{dv}{dx} = \frac{v^2 - v}{2}.$$

$$\therefore \quad x\frac{dv}{dx} = \frac{v^2 - v}{2} - v = \frac{v^2 - 3v}{2},$$

giving
$$\frac{dx}{x} = \frac{2dv}{v^2 - 3v} = \frac{2}{v(v-3)}dv.$$

Using partial fractions:

$$\int \frac{dx}{x} = \int \left[\frac{-2}{3v} + \frac{2}{3(v-3)}\right] dv.$$

$$\therefore \log_e Cx = -\tfrac{2}{3}\log_e v + \tfrac{2}{3}\log_e (v-3)$$

$$= \tfrac{2}{3}\log_e \frac{v-3}{v}.$$

$$\therefore\ 3\log_e Cx = 2\log_e\left(\frac{v-3}{v}\right),\quad C \text{ arbitrary},$$

$$\log_e Ax^3 = \log_e\left(\frac{v-3}{v}\right)^2,\quad A \text{ arbitrary } (= C^3).$$

$$\therefore\ Ax^3 = \left(\frac{v-3}{v}\right)^2 = \left(\frac{y/x-3}{y/x}\right)^2 = \frac{(y-3x)^2}{y^2}.$$

Thus $\underline{(y-3x)^2 = Ax^3y^2.}$

6·12. Type: $\dfrac{dy}{dx} = \dfrac{ax+by+c}{px+qy+r}$

This type of equation is not homogeneous but can be made so by a simple device.

It will be noticed that both numerator and denominator are linear expressions. The device varies according as the 'lines' would be parallel (i.e. $a/p = b/q$) or not parallel (i.e. $a/p \neq b/q$).

(i) If $a/p = b/q$.

In this case the substitution $z = ax + by$ (or $z = px + qy$) is used.

Example

Solve
$$\frac{dy}{dx} = \frac{4x-2y+6}{2x-y+5}.$$

Put $z = 2x - y$.

Then $\dfrac{dz}{dx} = 2 - \dfrac{dy}{dx}$ and thus $\dfrac{dy}{dx} = 2 - \dfrac{dz}{dx}$.

The original equation now becomes

$$2 - \frac{dz}{dx} = \frac{2z+6}{z+5}.$$

$$\therefore \frac{dz}{dx} = 2 - \frac{(2z + 6)}{z + 5} = \frac{4}{z + 5},$$

giving $$4\,dx = (z + 5)\,dz.$$

Integrating this expression:

$$4x = \frac{z^2}{2} + 5z + C,$$

i.e. $$4x = \frac{(2x - y)^2}{2} + 5(2x - y) + C.$$

(ii) If $$\frac{a}{p} \neq \frac{b}{q}.$$

In this case the substitutions $u = ax + by + c$, $v = px + qy + r$ reduce the equation to one which is homogeneous.

EXAMPLE

Solve $$\frac{dy}{dx} = \frac{3x + y - 5}{x - 3y + 5}.$$

Put $u = 3x + y - 5$ and $v = x - 3y + 5$. (1)

Then $du = 3\,dx + dy$ and $dv = dx - 3\,dy$

Thus $$\frac{du}{dv} = \frac{3\,dx + dy}{dx - 3\,dy} = \frac{3 + \dfrac{dy}{dx}}{1 - 3\dfrac{dy}{dx}}.$$

But $dy/dx = u/v$, therefore

$$\frac{du}{dv} = \frac{3 + \dfrac{u}{v}}{1 - 3\dfrac{u}{v}},$$ a homogeneous equation.

Now put $$u = zv. \qquad (2)$$

Then $$\frac{du}{dv} = z + v\frac{dz}{dv}.$$

Thus $$z + v\frac{dz}{dv} = \frac{3 + z}{1 - 3z},$$

giving $$v\frac{dz}{dv} = \frac{3 + z}{1 - 3z} - z = \frac{3 + 3z^2}{1 - 3z}.$$

Thus $$\frac{dv}{v} = \frac{(1 - 3z)\,dz}{3(1 + z^2)} = \frac{dz}{3(1 + z^2)} - \frac{z\,dz}{1 + z^2}.$$

Integrating this expression:

$$\log_e Cv = \tfrac{1}{3}\tan^{-1} z - \tfrac{1}{2}\log_e (1 + z^2), \quad C \text{ arbitrary.}$$

i.e. $$\log_e Cv + \log_e \sqrt{(1 + z^2)} = \tfrac{1}{3}\tan^{-1} z.$$

From (2), $$1 + z^2 = \frac{u^2 + v^2}{v^2},$$

thus $$\log_e Cv + \log_e \frac{\sqrt{(u^2 + v^2)}}{v} = \tfrac{1}{3}\tan^{-1}\frac{u}{v},$$

i.e. $$\log_e C\sqrt{(u^2 + v^2)} = \tfrac{1}{3}\tan^{-1}\frac{u}{v}.$$

From (1):

$$\underline{\log_e C\sqrt{\{(3x + y - 5)^2 + (x - 3y + 5)^2\}}}$$
$$\underline{= \tfrac{1}{3}\tan^{-1}\frac{(3x + y - 5)}{(x - 3y + 5)}}.$$

Note. The homogeneous type of § 6·11 occurs sometimes in cases of resisted motion, but the type of § 6·12 seldom occurs in engineering.

Exercise 18

Solve the following equations:

1. $\dfrac{dy}{dx} = x + y$, by putting $v = x + y$.

2. $(x^2 + xy)\,dy = (xy - y^2)\,dx$.

3. $(x^2 + y^2)\dfrac{dy}{dx} = 2xy$.

4. $\dfrac{dy}{dx} = \dfrac{2x - y - 3}{x - y - 1}$.

5. $(x - 3y + 5)\,dx = (3 - x - y)\,dy$.

6. A body of unit mass moves in a straight line against a resistance of magnitude n^2x, where x is the distance travelled

from a fixed point. If there is a further resistance to motion of magnitude $2nv$, where v is the velocity, show that

$$v\frac{dv}{dx} = -(n^2x + 2nv).$$

Solve this equation to give a formula connecting v and x, given that $v = 4$ when $x = 0$.

7. Solve the equation $\dfrac{dy}{dx} = \dfrac{2x - 4y + 5}{x - 2y + 1}$.

ANSWERS

1. $y = Ce^x - x - 1$.

2. $\dfrac{x}{y} = \log_e Cxy$ or $y = \dfrac{A}{x} e^{x/y}$.

3. $x^2 - y^2 + Cy = 0$.

4. $2(x - 2)^2 - 2(x - 2)(y - 1) + (y - 1)^2 = C$

or $$2x^2 + y^2 - 2xy - 6x + 2y = C.$$

[The student may have noted that the given equation is exact (this can be seen by cross-multiplying and collecting up, giving $x\,dy + y\,dx - y\,dy - dy - 2x\,dx + 3\,dx = 0$). This gives a much quicker method of solution.]

5. $\log_e C(x - y + 1) = \dfrac{(3 - x - y)}{(x - y + 1)}$.

6. $v + nx = 4e^{\frac{-nx}{v+nx}}$.

7. $4x - 2y = \log_e C(x - 2y + 3)^2$.

6·2. Equations reducible to linear equations of the first order

Equations of the type $dy/dx + Py = Qy^n$, where P, Q are functions of x, and n is an integer, are reducible to first-order linear equations. The type is often called *Bernoulli's* equation.

The reduction is effected by writing the equation in the form $\dfrac{1}{y^n}\dfrac{dy}{dx} + P\dfrac{1}{y^{n-1}} = Q$ and then substituting $z = \dfrac{1}{y^{n-1}}$.

EXAMPLE

Solve the equation $x\dfrac{dy}{dx} + y = x^2y^3.$

$$\frac{dy}{dx} + \frac{1}{x}y = xy^3,$$

$$\therefore \quad \frac{1}{y^3}\frac{dy}{dx} + \frac{1}{x}\frac{1}{y^2} = x. \tag{1}$$

Put $$z = 1/y^2, \tag{2}$$

then $\dfrac{dz}{dx} = -\dfrac{2}{y^3}\dfrac{dy}{dx}$, giving $\dfrac{1}{y^3}\dfrac{dy}{dx} = -\dfrac{1}{2}\dfrac{dz}{dx}.$

Thus (1) becomes

$$-\frac{1}{2}\frac{dz}{dx} + \frac{1}{x}z = x,$$

i.e. $$\frac{dz}{dx} - \frac{2}{x}z = -2x.$$

This is a first-order linear equation.

The integrating factor (see vol. I, § 13·24) is

$$\exp\left(\int -\frac{2}{x}\,dx\right) = \exp(-2\log_e x) = \exp\left(\log_e \frac{1}{x^2}\right) = \frac{1}{x^2}.$$

$$\therefore \quad z\frac{1}{x^2} = \int -2x\left(\frac{1}{x^2}\right)dx = -2\log_e Cx, \quad C \text{ arbitrary},$$

i.e. $$\frac{z}{x^2} = \log_e \frac{A}{x^2}, \quad A \text{ arbitrary}.$$

From (2), $$z = \frac{1}{y^2} = x^2 \log_e \frac{A}{x^2}.$$

EXERCISE 19

Solve the following equations:

1. $x\dfrac{dy}{dx} + y = xy^3$

2. $x\dfrac{dy}{dx} + y = -\dfrac{x^2}{y}.$

3. $$\frac{dy}{dx}\cos x - y \sin x = -y^2.$$

4. $$\frac{dy}{dx} + \frac{1}{x}y = \frac{y^2}{x}\log_e x.$$

5. A moving body is opposed by a force n^2x and a resistance $k(dx/dt)^2$, where x is its distance from a fixed point and dx/dt is its velocity at any time t. Taking the body to be of unit mass, show that the equation of motion is given by

$$\frac{d^2x}{dt^2} + k\left(\frac{dx}{dt}\right)^2 + n^2x = 0.$$

Show that the substitution $(dx/dt)^2 = u$, makes the equation linear and deduce that the velocity $v(= dx/dt)$ is connected with the distance x by the relation

$$v^2 = Ae^{-2kx} - \frac{n^2x}{k} + \frac{n^2}{2k^2}.$$

(Assume that the body is moving away from the fixed point and x is measured in this direction.)

ANSWERS

1. $y^2 = \dfrac{1}{x(2 + Cx)}$.

2. $y^2 = \dfrac{C}{x^2} - \dfrac{x^2}{2}$.

3. $y = \dfrac{1}{\sin x + C\cos x}$.

4. $y = \dfrac{1}{1 + Cx + \log_e x}$.

5. [As an equation between v and x is required, in the equation write $v(dv/dx)$ for d^2x/dt^2 and v for dx/dt. Then substitute $u = v^2$, using $\dfrac{du}{dx} = 2v\dfrac{dv}{dx}$.]

6·3. Isoclinals

It has already been stressed in volume I of this book that differential equations which can be solved exactly form only a small proportion of those met with in practice. Machines, called Differential Analysers, have been built to give graphical and numerical approximations to solutions of equations which cannot be solved by exact methods. Graphical and numerical methods have also been evolved which, with care and patience, will give reasonably good approximations to any particular solution of an equation.

Graphically speaking, the general solution to the first-order equation $dy/dx = f(x, y)$ is a family of curves, each curve corresponding to a particular value of the arbitrary constant which would arise in the general solution.

Such curves are called the *characteristics* of the equation.

For example, if $dy/dx = x$, the general solution is $y = \frac{1}{2}x^2 + C$, and the characteristics would be a family of parabolas,

$$(y = \tfrac{1}{2}x^2 - 1, \quad y = \tfrac{1}{2}x^2, \quad y = \tfrac{1}{2}x^2 + \tfrac{1}{2} \quad \text{etc.}).$$

To avoid discussion of special, awkward, cases it will be assumed that in the equation $dy/dx = f(x, y)$, $f(x, y)$ has a determinate finite value for every pair of finite values given to x and y. (This would not be true in the special case of functions such as y/x, where $x = 0$, $y = 0$ would make y/x indeterminate.) It will also be assumed that in the cases dealt with there is only *one* curve, of the family corresponding to the general solution, which passes through each and every given point in the plane.

6·31. The method of isoclinals

Consider the equation $dy/dx = y - 1/x$. Giving the slope, dy/dx, a definite numerical value, say 1, the equation becomes $1 = y - 1/x$ giving $x(y - 1) = 1$. Thus all the points on the characteristics at which the slope is 1, lie on the curve $x(y - 1) = 1$, a hyperbola. Such a curve is called an *isoclinal* or *isocline*; this means 'same slope'.

By giving various numerical values to dy/dx numerous isoclinals can be drawn showing the slopes of the characteristics or integral curves at many points. In this way a reasonably clear picture can be obtained of the shape of the actual characteristics.

The equation
$$\frac{dy}{dx} = y - \frac{1}{x} \tag{1}$$

will now be considered in detail.

Let $$p = \frac{dy}{dx}, \tag{2}$$

then $p = y - 1/x$, giving

$$x(y - p) = 1. \tag{3}$$

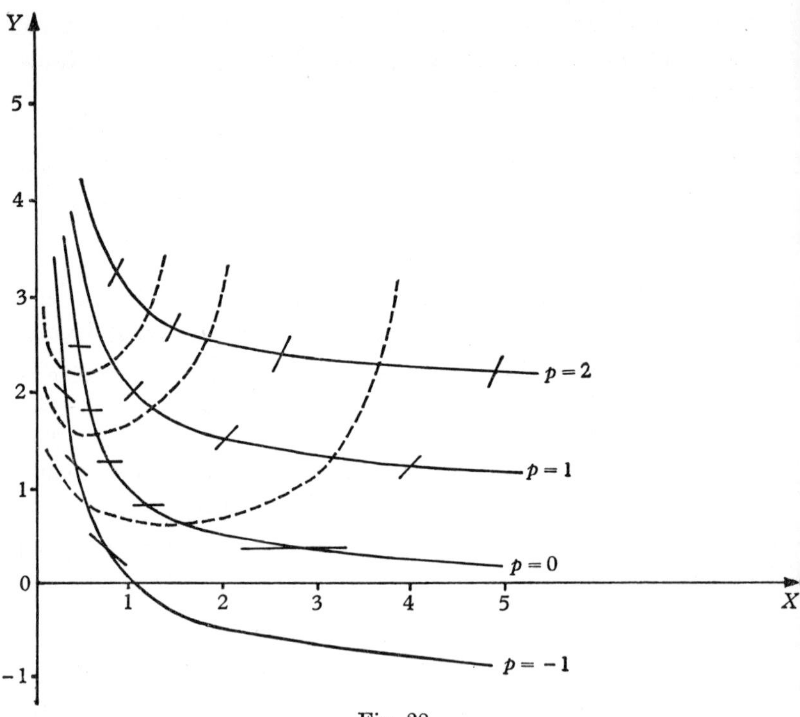

Fig. 28

Fig. 28 shows the isoclinals corresponding to values of the slope, $p = -1$, 0, 1 and 2. For convenience the isoclinals have mainly been drawn for positive values of x and y.

Taking the isoclinal $p = 0$, at any point on it the slope of the characteristic through that point is zero. At various points on this isoclinal short lines are drawn of zero slope.

Similarly on the isoclinal $p = 1$, short lines are drawn of slope 1; and so on for all the isoclinals.

The resulting system of these short 'slope' lines gives a reasonable picture of the shape the characteristic curves must take. Three characteristics have been drawn in dotted line. They correspond to three special values of the arbitrary constant in the general solution. The general solution is of course a whole family of such curves.

The isoclinal $p = 0$ must be the locus of all the turning points on the characteristics, since the slope of the characteristics is zero. It is seen from fig. 28 that all these turning points are in fact minimum points. This can easily be proved analytically in this case.

As $\dfrac{dy}{dx} = y - \dfrac{1}{x}$, then $\dfrac{d^2y}{dx^2} = \dfrac{dy}{dx} + \dfrac{1}{x^2}$.

Therefore at all turning points, where $\dfrac{dy}{dx} = 0$, $\dfrac{d^2y}{dx^2} = \dfrac{1}{x^2}$,

which, being a square, is positive.

This investigation of the sign of d^2y/dx^2 is often carried out before the characteristics are drawn, as it helps in their drawing. In some books the line $p = 0$ in this case would be called 'the locus of minima'.

Investigating the sign of d^2y/dx^2 can also be used to determine the concavity of the curve (concave upwards when d^2y/dx^2 is $+$, concave downwards when d^2y/dx^2 is $-$), and also to find the locus of points of inflexion (where $d^2y/dx^2 = 0$) if any.

In this case there are no points of inflexion for $\dfrac{dy}{dx} \geqslant 0$ as then $\dfrac{d^2y}{dx^2}$ is always positive. The characteristics must also be concave upwards for $\dfrac{dy}{dx} \geqslant 0$.

Exercise 20

1. Sketch the isoclinals for the equation $dy/dx = x(1 - y)$. Using the fact that $d^2y/dx^2 = 1 - y - x(dy/dx)$, show that the isoclinal for $dy/dx = 0$ is the locus of minima when $y < 1$ and the locus of maxima when $y > 1$.

Sketch the characteristics and verify your results by solving the equation completely as a first-order linear equation.

2. Sketch the isoclinals for the equation $dy/dx = y^2x$, and show that the y-axis is the locus of minimum points of the characteristics.

Sketch the characteristics and verify your results by solving the equation completely as a first-order equation with separable variables.

3. Indicate by a sketch the form of the isoclinals for the equation $dy/dx = x + y^2$. Show that the parabola $y^2 = -x$ is the locus of all minimum points of the characteristics.

Sketch the characteristic which passes through the origin.

Answers

1. The isoclinals have the equation $y = 1 - p/x$, where p is a parameter. They form a system of hyperbolas.

The complete solution of the equation is $y = 1 + Ce^{-\frac{1}{2}x^2}$.

2. The complete solution of the equation is $y = -\dfrac{2}{C + x^2}$.

3. The form of the required characteristic in the first quadrant is fairly obvious. For negative x values this characteristic keeps in the second quadrant and is roughly shown in fig. 29. The dotted parabola shown is the isoclinal for $dy/dx = 0$.

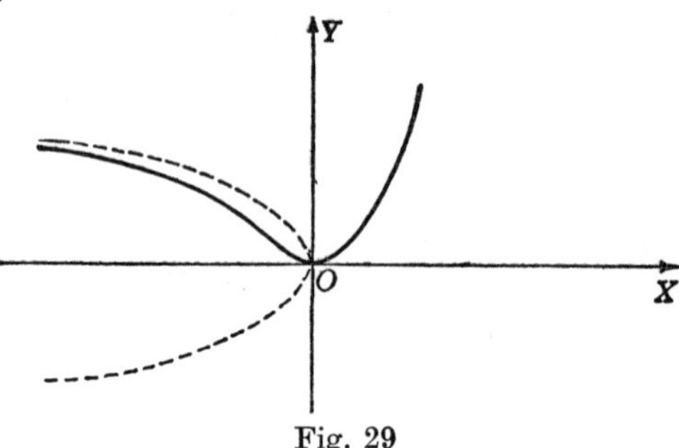

Fig. 29

6·4. Non-linear resistances and inductances

6·41. Non-linear resistances

In elementary electrical theory it is assumed that the voltage drop across a resistance is directly proportional to the current; that is, there is a *linear* relationship between current and voltage drop of the form $v = ki$, where k is constant (usually called the resistance R).

In various practical cases, for example in semi-conductors and electron tubes, the relationship is non-linear. The voltage drop and current are related in functional form by $i = f(v)$. The law most used in practice is $i = kv^n$, where k and n are constants.

EXAMPLE

A constant voltage E is applied at time $t = 0$ to an inductance L which is in series with a non-linear resistance for which $i = kv^{3/2}$, v being the voltage drop across the resistance and i the current. The initial current is zero. Show that at time T the voltage drop, V, across the resistance is given by

$$T = 3kL\left\{\sqrt{E}\tanh^{-1}\left(\sqrt{\frac{V}{E}}\right) - \sqrt{V}\right\}.$$

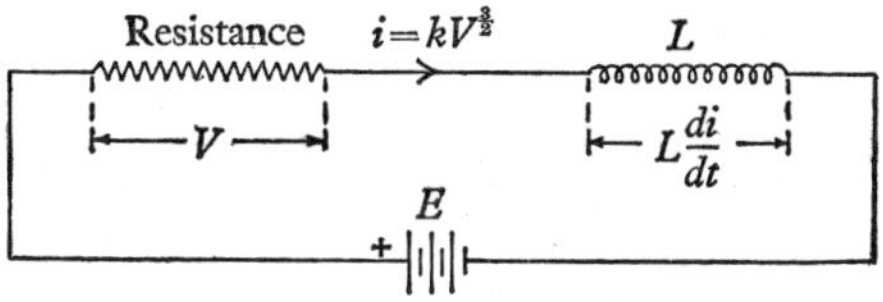

Fig. 30

Fig. 30 shows a diagram of the circuit with the voltage drops. Equating the driving voltage to the sum of the passive voltages:

$$v + L\frac{di}{dt} = E.$$

Now $$i = kv^{3/2}.$$

$$\therefore\ v + L\frac{d}{dt}(kv^{3/2}) = E,$$

i.e. $$v + \frac{3}{2}kLv^{1/2}\frac{dv}{dt} = E.$$

This is a first-order equation with separable variables and gives

$$v^{1/2}\frac{dv}{dt} = \frac{2(E - v)}{3kL},$$

or $$\frac{v^{1/2}\,dv}{(E - v)} = \frac{2}{3kL}\,dt. \tag{1}$$

To integrate, put $$u^2 = v, \tag{2}$$

then $$dv = 2u\,du \quad \text{and} \quad v^{1/2} = u.$$

Equation (1) now becomes

$$\frac{2u^2\,du}{E - u^2} = \frac{2}{3kL}\,dt,$$

i.e. $$\frac{u^2\,du}{E - u^2} = \frac{1}{3kL}\,dt.$$

On dividing out the left-hand side:

$$\left(\frac{E}{E - u^2} - 1\right)du = \frac{1}{3kL}\,dt.$$

Integrating this expression:

$$E\,\frac{1}{\sqrt{E}}\tanh^{-1}\left(\frac{u}{\sqrt{E}}\right) - u = \frac{t}{3kL} + C,$$

i.e. $$\sqrt{E}\tanh^{-1}\left(\sqrt{\frac{v}{E}}\right) - \sqrt{v} = \frac{t}{3kL} + C.$$

When $t = 0$, $i = 0$, therefore $v = 0$.

Thus $$0 = C.$$

Therefore, if the voltage drop is V at time T,

$$\sqrt{E}\tanh^{-1}\left(\sqrt{\frac{V}{E}}\right) - \sqrt{V} = \frac{T}{3kL},$$

i.e. $$\underline{T = 3kL\left\{\sqrt{E}\tanh^{-1}\left(\sqrt{\frac{V}{E}}\right) - \sqrt{V}\right\}}.$$

6·42. Non-linear inductances

It was stated in volume I of this book that iron-cored inductances cannot be treated in the usual manner, in which the voltage drop is $L\,(di/dt)$ and the flux, Φ, is Li.

For iron-cored inductances the flux is connected with the current flowing by a relation of the form $i = f(\Phi)$; for example, $i = a\Phi + b\Phi^3$, where a and b are constants. In this case the equation for a circuit in which a linear resistance R is in series with such a non-linear inductance would be

$$\frac{d\Phi}{dt} + R(a\Phi + b\Phi^3) = E,$$

where E is the applied voltage and the voltage drop across the inductance is given by $v = d\Phi/dt$.

Miscellaneous Exercises on Chapter 6

These exercises include revision questions on all first-order types done in volume I.

1. Solve the equation $x\dfrac{dy}{dx} + 2y = x^3$, given that $y = 0$ when $x = 1$. [L.U.]

2. Given that $y = 0$ when $x = 1$ solve the differential equation

$$(1 + x^2)\frac{dy}{dx} = x(1 - y^2).$$ [L.U.]

3. Solve the differential equations with the conditions given

(i) $dy/dx = (1 - x)\sin^2 y, \quad y = \frac{1}{4}\pi, \quad x = 0;$

(ii) $(x + y)\,dx + (x - y)\,dy = 0, \quad y = 0, \quad x = 0;$

(iii) $x^2\dfrac{dy}{dx} - xy = 1, \quad y = 2, \quad x = 1.$ [L.U.]

4. Solve the differential equations:

(i) $(x + y)^2\dfrac{dy}{dx} + 4 = 0;$ (ii) $(3x - 4y)\dfrac{dy}{dx} = x + 3y;$

(iii) $\dfrac{dy}{dx} - \dfrac{y}{x} = \log_e x.$ [L.U.]

5. Solve the differential equations:

(i) $2x^2\dfrac{dy}{dx} = (x + y)^2;$ (ii) $x^3\dfrac{dy}{dx} = y(x^2 + y);$

(iii) $(x - y)^2\dfrac{dy}{dx} = 1.$ [L.U.]

6. Find the general solutions of the differential equations

(i) $x^2 \dfrac{dy}{dx} + xy = y^2 \log_e x$; (ii) $(y^2 - 2x) \dfrac{dy}{dx} = y$;

(iii) $(x + 1) \dfrac{dy}{dx} = x + y + 1$. [L.U.]

7. A constant voltage E is applied at $t = 0$ to a capacitance C in series with a non-linear resistance for which $i = kv^3$, where i is the current and v the voltage drop across the resistance. If the capacitor is initially unchanged, show that the voltage drop, V, across the resistance at time t is given by

$$2kE^2V^2t = C(E^2 - V^2).$$

8. A non-linear inductance, in which the flux Φ is related to the current i by $i = a\Phi + b\Phi^2$, is in series with a linear resistance R. If the flux is Φ_0 at time $t = 0$, show that its value at time t is given by

$$\frac{\Phi}{a + b\Phi} = \frac{\Phi_0}{(a + b\Phi_0)} e^{-Rat}.$$

9. Define an isocline, and determine the isoclines of the equation $dy/dx = (2 - x)/(2 + y)$ showing that they form a family of straight lines, each perpendicular to the gradient to which it corresponds. Deduce that the integral curves are circles, and confirm by solving the equation. [I.E.E.]

10. Sketch the isoclines of the equation $dy/dx = x^2 - y^2$. Determine the locus of maxima and the locus of minima of the integral curves, and sketch the general form, for $x > -1$, of the integral curve through the point $(-1, 1)$. [I.E.E.]

ANSWERS

1. $y = \dfrac{1}{5}\left(x^3 - \dfrac{1}{x^2}\right)$.

2. $\dfrac{1 + y}{1 - y} = \frac{1}{2}(1 + x^2)$ or $y = \dfrac{x^2 - 1}{x^2 + 3}$.

3. (i) $\cot y = 1 - x + \frac{1}{2}x^2$;

(ii) $x^2 + 2xy - y^2 = 0$.

[Note that the equation is exact.]

(iii) $y = \dfrac{5x}{2} - \dfrac{1}{2x}$.

4. (i) $C(x + y + 2) = (x + y - 2)e^y$.

[Put $v = x + y$.]

(ii) $3 \tan^{-1} \dfrac{2y}{x} = \log_e C(x^2 + 4y^2)$.

(iii) $y = \frac{1}{2}x(\log_e x)^2 + Cx$.

5. (i) $y = x \tan (\frac{1}{2} \log_e Cx)$.

(ii) $y = \dfrac{x^2}{Cx + 1}$;

(iii) $x - y - 1 = Ce^{2y}(x - y + 1)$.

6. (i) $y = \dfrac{4x}{(Cx^2 + 2 \log_e x + 1)}$;

(ii) $y^4 = 4xy^2 + C$.

[The equation is exact if multiplied through by y.]

(iii) $y = (x + 1) \log_e C(x + 1)$.

7. The equation is $\quad v + q/C = E$.

Differentiating and using $i = dq/dt = kv^3$:

$$\frac{dv}{dt} + \frac{k}{C} v^3 = 0.$$

[Note that at $t = 0$, $v = E$.]

8. [The equation is $d\Phi/dt + R(a\Phi + b\Phi^2) = 0$, as no mention is made of any applied e.m.f.]

9. Isoclines are $y = -\dfrac{1}{p}x + 2\left(\dfrac{1}{p} - 1\right)$; solution is $x^2 + y^2 - 4x + 4y = C$.

10. Locus of maxima: the parts of the lines $y = \pm x$ for which $x < 0$. Locus of minima: the parts of the lines $y = \pm x$ for which $x > 0$.

CHAPTER 7

FURTHER SECOND-ORDER DIFFERENTIAL EQUATIONS

7·1. Introduction to the use of the operator D

The operation of differentiation, that is, d/dx, is symbolized by the letter D.

Thus
$$D \equiv \frac{d}{dx}.$$

D^2 is defined to mean that the function is to be differentiated twice. Thus

$$D^2(x^3) \equiv \frac{d^2}{dx^2}(x^3) = 6x.$$

In general D^n signifies that the function is to be differentiated n times, e.g.

$$D^4(\sin 2x) \equiv \frac{d^4}{dx^4}(\sin 2x) = 16 \sin 2x.$$

With this meaning $D^2\{D^3(\sin x)\}$ signifies that after differentiation of $\sin x$ three times the result is to be differentiated twice more. Since $\sin x$ will then have been differentiated five times it is clear that

$$D^2\{D^3(\sin x)\} \equiv D^5(\sin x).$$

This is an illustration of the rule that if y is any function of x, then

$$D^m\{D^n(y)\} \equiv D^{m+n}(y). \quad (1)$$

In fact, it can be proved in general that the operator D can be manipulated in combination with *constants and integral powers of itself* according to the usual algebraic rules.

Thus:

(i) $(D + a)y = (a + D)y$, where a is a constant, (2)

e.g. $(D - 3)x^2 = 2x - 3x^2 = (-3 + D)x^2$.

(ii) $$D(ay) = aD(y), \tag{3}$$

e.g. $$D(3 \sin x) = 3 \cos x = 3D(\sin x).$$

(iii) $$(D + a)y = Dy + ay, \tag{4}$$

e.g. $$(D + 2)e^x = e^x + 2e^x = De^x + 2e^x.$$

(iv) $$\begin{aligned}(D + a)(D + b)y &= (D + b)(D + a)y \\ &= \{D^2 + (a + b)D + ab\}y, \qquad (5)\end{aligned}$$

e.g. $$\begin{aligned}(D + 2)(D + 1)x^3 &= (D + 2)(3x^2 + x^3) \\ &= (6x + 3x^2) + (6x^2 + 2x^3) \\ &= 2x^3 + 9x^2 + 6x.\end{aligned}$$

And $(D^2 + 3D + 2)x^3 = 6x + 9x^2 + 2x^3$.

7·11. Application to linear equations

The second-order differential equation $\dfrac{d^2y}{dx^2} + a\dfrac{dy}{dx} + by = 0$ can now be written

$$D^2y + aDy + by = 0,$$

or more concisely

$$(D^2 + aD + b)y = 0.$$

In solving this equation by the standard method $D^2 + aD + b = 0$ can be used as the auxiliary equation to save time, if desired.

For example:

$$(D^2 - 3D + 2)y = 0.$$

Auxiliary equation: $D^2 - 3D + 2 = (D - 2)(D - 1) = 0$.

General solution : $\underline{y = Ae^{2x} + Be^x}$.

A more formal justification for the use and manipulation of the operator D in solving second-order linear equations with constant coefficients is given below.

7·111. Solution of $(D^2 + aD + b)y = 0$.

(*a*) *If* $(D^2 + aD + b)$ *has factors* $(D - h)(D - k)$. The equation may be written $(D - h)(D - k)y = 0$, meaning that y is first operated on by $(D - k)$ thereby obtaining $Dy - ky$; then this is operated on by $(D - h)$ giving

$$\begin{aligned}(D - h)(Dy - ky) &= D^2y - (h + k)Dy + hky \\ &= D^2y + aDy + by,\end{aligned}$$

as $(D - h)$, $(D - k)$ are the factors of $(D^2 + aD + b)$.

Hence, any value of y which satisfies $(D - k)y = 0$ also makes $(D - h)(D - k)y = 0$, as $(D - h)(0) = 0$, and so satisfies the original equation.

Solving $(D - k)y = 0$, i.e.

$$\frac{dy}{dx} - ky = 0,$$

$$\frac{1}{y}\,dy = k\,dx.$$

$$\therefore \log_e Cy = kx, \quad C \text{ arbitrary},$$

giving $$\underline{y = Ae^{kx}, \quad A \text{ arbitrary.}}$$

Similarly, any value of y which satisfies $(D - h)y = 0$ satisfies the original equation. This gives

$$\underline{y = Be^{hx}}, \quad B \text{ arbitrary, as a solution.}$$

Thus,

$$\underline{y = Ae^{kx} + Be^{hx} \text{ is the general solution}}$$

(see vol. I, § 13·321).

(*b*) *If* $D^2 + aD + b$ *has factors* $\{D - (p + jq)\}\{D - (p - jq)\}$. The same reasoning as in (*a*) above gives the general solution, as

$$\begin{aligned}y &= Ae^{(p + jq)x} + Be^{(p - jq)x}, \quad A \text{ and } B \text{ arbitrary}, \\ &= e^{px}(Ae^{jqx} + Be^{-jqx}) \\ &= e^{px}\{A(\cos qx + j \sin qx) + B(\cos qx - j \sin qx)\} \\ &= e^{px}\{(A + B) \cos qx + j(A - B) \sin qx\} \\ &= \underline{e^{px}(E \cos qx + F \sin qx)}, \quad E \text{ and } F \text{ arbitrary}, \\ &\text{or } \underline{Re^{px} \sin (qx + \alpha)}, \quad R \text{ and } \alpha \text{ arbitrary.}\end{aligned}$$

It should be noted that although $F = j(A - B)$, it does not follow that F must be an imaginary number.

If A and B are conjugate complex numbers, for example, both $A + B$ and $j(A - B)$ will be real numbers, e.g.

$$A = 3 + j2, \quad B = 3 - j2.$$

Then

$$A + B = 6 = E \quad \text{and} \quad j(A - B) = j(j4) = -4 = F.$$

(c) *If $D^2 + aD + b$ factorizes as $(D - h)^2$.* Consider

$$(D - h)^2 y = 0 \quad \text{and let} \quad (D - h)y = z, \tag{1}$$

then $$(D - h)(D - h)y = (D - h)z = 0.$$

$$\therefore \frac{dz}{dx} - hz = 0, \quad \text{giving} \quad z = Ae^{hx}. \tag{2}$$

From (1) and (2):

$$(D - h)y = Ae^{hx},$$

i.e. $$\frac{dy}{dx} - hy = Ae^{hx}.$$

This is linear and of first order.

The integrating factor is $e^{\int -h dx} = e^{-hx}$.

$$\therefore ye^{-hx} = \int Ae^{hx}e^{-hx}\,dx = \int A\,dx.$$

$$\therefore ye^{-hx} = (Ax + B),$$

and $$\underline{y = (Ax + B)e^{hx}.}$$

7·12. Linear equations of higher order

If the equation is linear with constant coefficients but of higher order than the second, the method is exactly the same. The only difficulty which may arise is in the factorization of the polynomial in D. Examples met with in this book will only be cases which easily factorize.

EXAMPLE

Solve $$(D^3 - 6D^2 + 11D - 6)y = 0.$$

On testing, it is easily seen that $D = 1$ makes $D^3 - 6D^2 + 11D - 6$ equal to zero, therefore $(D - 1)$ is a factor. On dividing out,

$$D^3 - 6D^2 + 11D - 6 = (D - 1)(D^2 - 5D + 6)$$
$$= (D - 1)(D - 2)(D - 3).$$

Thus $\qquad (D - 1)(D - 2)(D - 3)y = 0,$

and the general solution is

$$\underline{y = Ae^x + Be^{2x} + Ce^{3x}.}$$

Exercise 21

Verify that:

1. $(D^2 - 5D + 2) \sin x = (2 + D^2 - 5D) \sin x.$

2. $D^3\{D^2(e^{3x})\} = D^2\{D^3(e^{3x})\} = D^5(e^{3x}).$

3. $(D - 3)\{(D - 2) \sin 2x\} = (D - 2)\{(D - 3) \sin 2x\}$
$= (D^2 - 5D + 6) \sin 2x.$

4. Show that $\sin x D(x^2) \neq x^2 D(\sin x)$.

(This is an example which illustrates that the operator D does *not* obey the usual rules of algebra when combined with other than constants or integral powers of itself, see § 7·1.)

Solve the following equations:

5. $(D^2 - 10D + 16)y = 0.$ **6.** $(D^2 - 10D + 25)y = 0.$

7. $(D^2 + 3D - 10)y = 0$, given that $y = 2$ and $dy/dx = 1$ when $x = 0$.

8. $(D^2 + 6D + 10)y = 0.$

9. $(D^2 + 4D + 5)x = 0$, where $D \equiv d/dt$ and $x = 0$, $dx/dt = 3$, when $t = 0$.

10. Solve $(D^3 - 4D^2 + D + 6)y = 0$. ($D + 1$ is a factor.)

11. Solve $(D^3 + 8)y = 0$. (Factorize $D^3 + 8$ as the sum of two cubes.)

Answers

5. $y = Ae^{2x} + Be^{8x}.$ **6.** $y = (Ax + B)e^{5x}.$

7. $y = \frac{1}{7}(11e^{2x} + 3e^{-5x}).$

8. $y = e^{-3x}(A \cos x + B \sin x).$

9. $x = 3e^{-2t} \sin t.$

10. $y = Ae^{-x} + Be^{2x} + Ce^{3x}.$

11. $y = Ae^{-2x} + e^x(B \cos \sqrt{3}\, x + C \sin \sqrt{3}\, x).$

7·2. Complementary function and particular integral

So far only equations of the type $(D^2 + aD + b)y = 0$ have been considered. Now the more general type

$$(D^2 + aD + b)y = f(x)$$

will be dealt with, where $f(x)$ is some function of x.

As a simple example, consider

$$(D^2 - 3D + 2)y = 2x. \tag{1}$$

Now the solution of $(D^2 - 3D + 2)y = 0$ is given by

$$(D - 1)(D - 2)y = 0,$$

$$\therefore \quad \underline{y = Ae^x + Be^{2x}.} \tag{2}$$

This is called the *complementary function* (C.F.) of the original equation (1). It has the two arbitrary constants required by a general solution (G.S.), but if substituted into equation (1) it will make the right-hand side zero instead of $2x$.

Now substitute $y = x + \frac{3}{2}$ into the left-hand side of equation (1):

$$(D^2 - 3D + 2)(x + \tfrac{3}{2}) = -3 + 2x + 2 \,.\, \tfrac{3}{2} = 2x.$$

Hence $y = x + \frac{3}{2}$ is a solution of the complete equation (1).

There are no arbitrary constants in this solution. It is therefore called a *particular integral* (P.I.) *or particular solution.*

Finally, substitute $y = Ae^x + Be^{2x} + x + \frac{3}{2}$ into the left-hand side of equation (1). As substitution of $Ae^x + Be^{2x}$ gives zero, the complete substitution obviously gives $0 + 2x = 2x$. Thus $y = Ae^x + Be^{2x} + x + \frac{3}{2}$ satisfies the original equation and has the necessary two arbitrary constants; it is therefore the *general solution* of the equation.

Note that the general solution is the complementary function added to a particular integral, i.e.

$$\underline{\text{G.S.} = \text{C.F.} + \text{P.I.}} \quad \text{in symbolic form.}$$

The general solution is often called the *complete primitive* (C.P.).

General case. Consider the general equation

$$(D^2 + aD + b)y = f(x). \tag{1}$$

Let $y = u(x)$ be the complementary function; that is,

$$(D^2 + aD + b)u = 0. \tag{2}$$

Let $y = v(x)$ be a particular integral; that is

$$(D^2 + aD + b)v = f(x). \tag{3}$$

Adding (1) and (2):

$$(D^2 + aD + b)u + (D^2 + aD + b)v = f(x).$$

Simplifying the left-hand side

$$(D^2 + aD + b)(u + v) = f(x).$$

Thus $y = u(x) + v(x)$ satisfies the original equation (1); but $u(x)$ will contain two arbitrary constants, being the solution of $(D^2 + aD + b)y = 0$, therefore $y = u(x) + v(x)$ is the general solution of equation (1).

The method of finding the complementary function has been given in volume I and revised here in § 7·111. The sections which follow will be devoted to methods by which a particular integral can be found.

7·3. Calculation of particular integrals

There are many theorems concerning the operator D operating on various types of functions, all of which can be used to find particular integrals. Here the number of theorems will be reduced to a minimum, but the comparatively few methods used will be sufficient to solve all equations which the student is likely to meet at this stage.

In the following paragraphs $F(D)$ will be used to denote $(D^2 + aD + b)$ operating on the dependent variable, y, of the differential equation. Similar methods, however, will cope with any polynomial in D.

7·31. Theorem I. $F(D)[e^{kx}] = F(k)e^{kx}$, k a constant, real or complex

Now
$$D[e^{kx}] = ke^{kx},$$
$$D^2[e^{kx}] = k^2e^{kx}, \text{ and so on.}$$

Thus, if $F(D)$ is any polynomial in D,

$$F(D)[e^{kx}] = F(k)e^{kx},$$

e.g. $$(D^2 + aD + b)[e^{kx}] = (k^2 + ak + b)e^{kx},$$

i.e. k replaces D.

7·32. Theorem II. $F(D)[e^{kx}Y] = e^{kx}F(D+k)[Y]$

Here k is a constant, real or complex, and Y denotes any function of x. It should be noted that square brackets are often used to signify that the function within them is being operated on by the function of D which precedes them.

The theorem will be proved for $F(D) \equiv D^2 + aD + b$, but Leibniz's theorem can be used to prove it for any polynomial in D.

$$D^2[e^{kx}Y] = e^{kx}D^2Y + 2De^{kx}DY + Y \,.\, D^2e^{kx} \qquad \text{(from Leibniz)}$$

$$= e^{kx}(D^2Y + 2kDY + k^2Y)$$

$$= e^{kx}(D+k)^2Y.$$

Similarly, $\quad aD[e^{kx}Y] = a(e^{kx}DY + ke^{kx}Y)$

$$= ae^{kx}(D+k)Y,$$

and $\quad be^{kx}Y = be^{kx}Y.$

Adding these results

$$(D^2 + aD + b)[e^{kx}Y] = e^{kx}\{(D+k)^2 + a(D+k) + b\}[Y]$$

$$= e^{kx}F(D+k)[Y].$$

Note that the theorem indicates that e^{kx} can be taken out of the 'operating brackets' if D is replaced by $(D+k)$, e.g.

$$(D^2 + 4)[xe^{j2x}] = e^{j2x}\{(D+j2)^2 + 4\}[x]$$

$$= e^{j2x}(D^2 + j4D)[x]$$

$$= j4e^{j2x}.$$

Theorem I is used whenever applicable, and Theorem II only when Theorem I fails to give results.

In the following subsections the method of dealing with most of the standard functions will be given. These will refer to the equation

$$F(D)y \equiv (D^2 + aD + b)y = f(x),$$

but similar methods will deal with higher polynomials in D.

7·33. $f(x) = c$, a constant

This is a case which is easily memorized.

As all derivatives of a constant are zero, it is clear that $y = c/b$ satisfies the equation $(D^2 + aD + b)y = c$.

To verify:

$$(D^2 + aD + b)\left[\frac{c}{b}\right] = b\frac{c}{b} = c.$$

EXAMPLE

$$(3D^2 + 5D - 7)y = 9.$$

A particular integral is $y = \frac{9}{-7} = -\frac{9}{7}$.

7·341. $f(x) = le^{kx}$ (l, k constants), where $F(k) \neq 0$

Here, for a particular integral y is put equal to Ae^{kx} and A is calculated using Theorem I.

EXAMPLE 1

$$(D^2 - 3D + 2)y = 5e^{3x}.$$

Put $$y = Ae^{3x}.$$

From Theorem I:

$$(D^2 - 3D + 2)[Ae^{3x}] = A(3^2 - 3 \,.\, 3 + 2)e^{3x}$$
$$= 2Ae^{3x}.$$

Thus $$2Ae^{3x} \equiv 5e^{3x},$$

and $$A = \tfrac{5}{2}.$$

A particular integral is

$$\underline{y = \tfrac{5}{2}\, e^{3x}.}$$

The working can be reduced by noting a general case:

$$(D^2 + aD + b)y = le^{kx}.$$

Put $$y = Ae^{kx}.$$

Then $$(D^2 + aD + b)[Ae^{kx}] = A(k^2 + ak + b)e^{kx}.$$

If $$A(k^2 + ak + b)e^{kx} \equiv le^{kx},$$

$$A = \frac{l}{k^2 + ak + b}.$$

A particular integral is therefore $y = \dfrac{l}{(k^2 + ak + b)}\, e^{kx}$.

The particular integral in Example (1) can now be written on sight:

$$y = \frac{5}{3^2 - 3 \cdot 3 + 2} e^{3x} = \tfrac{5}{2} e^{3x}.$$

EXAMPLE 2

$$(3D^2 + 4D - 5)y = 7e^{2x} - 6e^{x}.$$

A particular integral is

$$y = \frac{7e^{2x}}{(3 \cdot 2^2 + 4 \cdot 2 - 5)} - \frac{6e^{x}}{(3 \cdot 1^2 + 4 \cdot 1 - 5)}$$

$$= \tfrac{7}{15} e^{2x} - 3e^{x}.$$

EXERCISE 22

Find particular integrals for the following equations:

1. $(D^2 - D + 1)y = 2e^{2x}$.
2. $(2D^2 + D - 3)y = 5e^{-x}$.
3. $(D^2 + 4D + 4)y = \frac{3}{2} e^{x}$.
4. $(D^2 - D + 3)y = 4$.
5. $(3D^2 - 2D + 5)y = 7$.
6. $(D^2 + 2D - 6)y = 4e^{x} - 2e^{-x}$.
7. $(10D^2 - 3D + 2)y = -5$.
8. $(5D^2 + D - 7)y = 3e^{\frac{3}{2}x} + 2e^{-\frac{1}{2}x}$.

ANSWERS

1. $\frac{2}{3}e^{2x}$. 2. $-\frac{5}{2}e^{-x}$. 3. $\frac{1}{6}e^{x}$. 4. $\frac{4}{3}$.

5. $\frac{7}{5}$. 6. $-\frac{4}{3}e^{x} + \frac{2}{7}e^{-x}$.

7. $-\frac{5}{2}$. 8. $\frac{12}{23}e^{\frac{3}{2}x} - \frac{8}{25}e^{-\frac{1}{2}x}$.

7·342. $f(x) = le^{kx}$, where $F(k) = 0$

In this case y is put equal to Axe^{kx} and Theorem II is used; if this still fails y is put equal to Ax^2e^{kx}, and so on.

EXAMPLE 1

$$(D^2 - 4)y = 3e^{2x}.$$

$$F(k) = 2^2 - 4 = 0.$$

Try $$y = Axe^{2x}.$$

$$(D^2 - 4)[Axe^{2x}] \equiv 3e^{2x}.$$

Using Theorem II,

$$Ae^{2x}\{(D + 2)^2 - 4\}[x] \equiv 3e^{2x},$$

i.e. $$Ae^{2x}(D^2 + 4D)[x] \equiv 3e^{2x},$$

giving $$Ae^{2x}4 \equiv 3e^{2x},$$

$$A = \tfrac{3}{4}.$$

A particular integral is $\underline{y = \tfrac{3}{4}xe^{2x}.}$

In general, $F(k)$ will be zero if $(D - k)$ is a factor of $F(D)$. Thus, in Example (1) above, $(D^2 - 4) = (D - 2)(D + 2)$. In this case $y = Axe^{kx}$ is tried.
If $(D - k)^2$ is a factor of $F(D)$, $y = Ax^2e^{kx}$ is tried, and so on.

EXAMPLE 2

$$(D^2 - 2D + 1)y = 5e^x,$$

i.e. $$(D - 1)^2y = 5e^x.$$

Try $$y = Ax^2e^x,$$

then $$(D - 1)^2 [Ax^2e^x] \equiv 5e^x,$$

$$Ae^x\{(D + 1) - 1\}^2 [x^2] \equiv 5e^x,$$

$$Ae^xD^2 [x^2] \equiv 5e^x,$$

i.e. $$A2 = 5,$$

$$A = \tfrac{5}{2}.$$

A particular integral is $\underline{y = \tfrac{5}{2}x^2e^x.}$

EXERCISE 23

Find particular integrals for:

1. $(D^2 - 1)y = 3e^x.$
2. $(D^2 + 1)y = 4e^{jx}.$
3. $(D + 1)^2y = 2e^{-x}.$
4. $(D^2 - 3D + 2)y = 5e^{2x}.$
5. $(D^2 + 6D + 9)y = 7e^{-3x}.$

ANSWERS

1. $\frac{3}{2}xe^{x}$. 2. $-j2xe^{jx}$. 3. x^2e^{-x}.

4. $5xe^{2x}$. 5. $\frac{7}{2}x^2e^{-3x}$.

7·35. $f(x) = l\cos nx + m\sin nx$, l, m, n constants

Usually either l or m is zero, but the method can be used when both are non-zero. $\cos nx$ and $\sin nx$ are treated respectively as the real and imaginary parts of e^{jnx}.

If $F(jn) \neq 0$, Theorem I is used.

EXAMPLE 1

$$(D^2 + 2D - 3)y = 4\cos 2x.$$

Treat $4\cos 2x$ as the r.p. of $4e^{j2x}$ and try $y =$ r.p. of Ae^{j2x}.

Then $$(D^2 + 2D - 3)[Ae^{j2x}] \equiv 4e^{j2x},$$

$$Ae^{j2x}\{(j2)^2 + 2(j2) - 3\} \equiv 4e^{j2x},$$

$$A(-4 + j4 - 3) = 4,$$

$$A = \frac{4}{(-7 + j4)} = \frac{-4(7 + j4)}{65}.$$

A particular integral is therefore

$$y = \text{r.p. of } -\frac{4(7 + j4)}{65}e^{j2x}$$

$$= \text{r.p. of } -\frac{4(7 + j4)}{65}(\cos 2x + j\sin 2x)$$

$$= \tfrac{4}{65}(4\sin 2x - 7\cos 2x).$$

If $F(jn) = 0$, Theorem II is used and y is tried equal to Axe^{jnx}, and if this still fails, Ax^2e^{jnx} and so on.

EXAMPLE 2

$$(D^2 + 4)y = 3\sin 2x.$$

Treat $3\sin 2x$ as the i.p. of $3e^{j2x}$.

As $(j2)^2 + 4 = 0$, try $y = Axe^{j2x}$.

Then
$$(D^2+4)[Axe^{j2x}] \equiv 3e^{j2x},$$
$$Ae^{j2x}\{(D+j2)^2+4\}[x] \equiv 3e^{j2x},$$
$$A(D^2+j4D)[x] = 3,$$
$$Aj4 = 3,$$
$$A = \frac{3}{j4} = -\tfrac{3}{4}j.$$

A particular integral is the i.p. of $-\frac{3}{4}jxe^{j2x}$

$$= \underline{-\tfrac{3}{4}x\cos 2x.}$$

EXAMPLE 3

$$(D^2+9)y = 2\cos x - 3\sin 3x.$$

Treat $2\cos x$ as the r.p. of $2e^{jx}$ and $3\sin 3x$ as the i.p. of $3e^{j3x}$.

Try $y =$ r.p. of Ae^{jx} + i.p. of Bxe^{j3x}.

(Note that D^2+9 is zero if $j3$ is written for D.)

Then

$$(D^2+9)[Ae^{jx}] + (D^2+9)[Bxe^{j3x}] \equiv 2e^{jx} - 3e^{j3x}.$$

Using Theorems I and II:

$$A8e^{jx} + Be^{j3x}\{(D+j3)^2+9\}[x] = 2e^{jx} - 3e^{j3x},$$
$$8Ae^{jx} + Be^{j3x}(D^2+j6D)[x] = 2e^{jx} - 3e^{j3x},$$
$$8Ae^{jx} + j6Be^{j3x} = 2e^{jx} - 3e^{j3x}.$$

$$\therefore\quad A = \tfrac{2}{8} = \tfrac{1}{4} \quad\text{and}\quad B = -\frac{3}{j6} = \frac{j}{2}.$$

A particular integral is

$$y = \text{r.p. of } \tfrac{1}{4}e^{jx} + \text{i.p. of } \frac{j}{2}xe^{j3x},$$

i.e.
$$\underline{y = \tfrac{1}{4}\cos x + \tfrac{1}{2}x\cos 3x.}$$

EXERCISE 24

Find particular integrals for:

1. $(D^2 + 1)y = 3 \sin 2x$.

2. $(D^2 + D - 1)y = 2 \sin x + 3 \cos x$.

3. $(D^2 + D - 2)y = 4 \sin 3x - \cos x$.

4. $(D^2 + 9)y = 4 \cos 3x$. 5. $(D^2 + 4)y = 2 \sin 2x$.

6. $(D^2 - 2D + 2)y = e^x + \sin 2x$.

ANSWERS

1. $-\sin 2x$. 2. $-\frac{1}{5}(\sin x + 8 \cos x)$.

3. $\frac{1}{10}(3 \cos x - \sin x) - \frac{1}{65}(6 \cos 3x + 22 \sin 3x)$.

4. $\frac{2}{3}x \sin 3x$. 5. $-\frac{1}{2}x \cos 2x$.

6. $e^x + \frac{1}{10}(2 \cos 2x - \sin 2x)$.

7·36. $f(x) = p(x)$, a polynomial in x

If the polynomial is of degree k, a polynomial of the same degree is tried. This will always give a result if $b \neq 0$ in

$$F(D) = D^2 + aD + b.$$

If $b = 0$, then a polynomial of degree x^{k+1} is tried, and so on.

EXAMPLE 1

$$(D^2 - 1)y = x^2.$$

Try $$y = px^2 + qx + r,$$

then $$(D^2 - 1)[px^2 + qx + r] \equiv x^2,$$

$$2p - px^2 - qx - r \equiv x^2,$$

$$-px^2 - qx + (2p - r) \equiv x^2.$$

Equating coefficients on each side:

$$-p = 1, \quad -q = 0, \quad 2p - r = 0.$$

These give $p = -1, \quad q = 0, \quad r = -2.$

A particular integral is

$$\underline{y = -x^2 - 2.}$$

In easy cases there is no objection to the student guessing a particular integral provided that he checks his solution.

EXAMPLE 2

$$(D^2 + D)y = x^2 - 2.$$

Try $$y = px^3 + qx^2 + rx + s.$$

Then $$(D^2 + D)[px^3 + qx^2 + rx + s] \equiv x^2 - 2,$$

$$6px + 2q + 3px^2 + 2qx + r \equiv x^2 - 2,$$

$$3px^2 + (6p + 2q)x + (2q + r) \equiv x^2 - 2.$$

Equating coefficients:

$$3p = 1, \quad 6p + 2q = 0, \quad 2q + r = -2,$$

giving $$p = \tfrac{1}{3}, \quad q = -1, \quad r = 0.$$

As s has not been found, it may be arbitrary; but as only a particular integral is required it is taken as zero.

A particular integral is

$$\underline{y = \tfrac{1}{3}x^3 - x^2.}$$

EXERCISE 25

Find particular integrals for:

1. $(D^2 - 5D + 6)y = 4x + 3$.
2. $(D^2 - 4)y = x^2$.
3. $(D^2 - 5D + 6)y = x^2$.
4. $(D^2 - 3D + 4)y = 2x - 1$.
5. $(D^2 + D)y = x^2$.
6. $(D^2 - 2D)y = 3x - 1$.

ANSWERS

1. $\tfrac{2}{3}x + \tfrac{19}{18}$.
2. $-\tfrac{1}{4}(x^2 + \tfrac{1}{2})$.
3. $\tfrac{1}{6}(x^2 + \tfrac{5}{3}x + \tfrac{19}{18})$.
4. $\tfrac{1}{4}(2x + \tfrac{1}{2})$.
5. $\tfrac{1}{3}x^3 - x^2 + 2x$.
6. $-\tfrac{1}{4}(3x^2 + x)$.

7·37. Some miscellaneous types

7·371. Hyperbolic functions

cosh nx and sinh nx may be treated as $\tfrac{1}{2}(e^{nx} + e^{-nx})$ and $\tfrac{1}{2}(e^{nx} - e^{-nx})$ respectively. The result may easily be expressed in terms of cosh nx and sinh nx using $e^{nx} = \cosh nx + \sinh nx$ and $e^{-nx} = \cosh nx - \sinh nx$.

EXAMPLE

$$(D^2 - 3D + 2)y = 5 \cosh 2x,$$

$$(D^2 - 3D + 2)y = \tfrac{5}{2}e^{2x} + \tfrac{5}{2}e^{-2x},$$

i.e. $$(D - 2)(D - 1)y = \tfrac{5}{2}e^{2x} + \tfrac{5}{2}e^{-2x}.$$

Put $$y = Axe^{2x} + Be^{-2x}.$$

Then $$(D^2 - 3D + 2)[Axe^{2x}] + (D^2 - 3D + 2)[Be^{-2x}] \equiv \tfrac{5}{2}e^{2x} + \tfrac{5}{2}e^{-2x},$$

$$Ae^{2x}\{(D + 2)^2 - 3(D + 2) + 2\}[x] + B\{(-2)^2 - 3(-2) + 2\}e^{-2x} \equiv \tfrac{5}{2}e^{2x} + \tfrac{5}{2}e^{-2x},$$

$$Ae^{2x}(D^2 + D)[x] + 12Be^{-2x} \equiv \tfrac{5}{2}e^{2x} + \tfrac{5}{2}e^{-2x},$$

$$Ae^{2x} + 12Be^{-2x} \equiv \tfrac{5}{2}e^{2x} + \tfrac{5}{2}e^{-2x}.$$

$$\therefore\ A = \tfrac{5}{2} \quad \text{and} \quad B = \tfrac{5}{24}.$$

A particular integral is

$$\underline{y = \tfrac{5}{2}xe^{2x} + \tfrac{5}{24}e^{-2x}},$$

or $$\underline{y = \tfrac{5}{2}x(\cosh 2x + \sinh 2x) + \tfrac{5}{24}(\cosh 2x - \sinh 2x).}$$

7·372. $f(x)$ the product of two functions

Products of the types $p(x) \sin nx$, $p(x) \cos nx$, $p(x)e^{nx}$, etc., where $p(x)$ is a polynomial in x, can be dealt with by using Theorem II.

Products such as $e^{ax} \sin nx$ can be dealt with, using $\sin nx$ as the i.p. of e^{jnx}.

EXAMPLE 1

$$(D^2 + 5D + 6)y = e^{-2x} \sin 2x.$$

Treat $e^{-2x} \sin 2x$ as i.p. of $e^{-2x}e^{j2x}$ = i.p. of $e^{(-2+j2)x}$.

Try y = i.p. of $Ae^{(-2+j2)x}$,

$$(D^2 + 5D + 6)[Ae^{(-2+j2)x}] \equiv e^{(-2+j2)x}.$$

Thus

$$\{(-2 + j2)^2 + 5(-2 + j2) + 6\}Ae^{(-2+j2)x} \equiv e^{(-2+j2)x}.$$

This gives $$(-4 + j2)A = 1.$$

$$\therefore\ A = \frac{1}{(-4+j2)} = \frac{-4-j2}{16+4} = -\tfrac{1}{10}(2+j).$$

A particular integral is

$$y = \text{i.p. of } -\tfrac{1}{10}(2+j)e^{(-2+j2)x}$$

$$= \text{i.p. of } -\tfrac{1}{10}e^{-2x}(2+j)(\cos 2x + j\sin 2x),$$

i.e. $$y = -\frac{e^{-2x}}{10}(2\sin 2x + \cos 2x).$$

EXAMPLE 2

$$(D^2+9)y = x^2\cos 3x.$$

Treat $x^2 \cos 3x$ as the r.p. of x^2e^{j3x}.
Try $y =$ r.p. of $(px^2+qx+r)e^{j3x}$,

$$(D^2+9)[(px^2+qx+r)e^{j3x}] \equiv x^2e^{j3x}.$$

Using Theorem II,

$$e^{j3x}\{(D+j3)^2+9\}[px^2+qx+r] \equiv x^2e^{j3x},$$

i.e. $$(D^2+j6D)[px^2+qx+r] \equiv x^2, \qquad (1)$$

giving $$2p + j6(2px+q) \equiv x^2.$$

The left-hand side has no term in x^2 and therefore does not give a result. Actually this could have been anticipated as $j3$ is a root of $D^2+9=0$.

Try $y =$ r.p. of $(px^3+qx^2+rx+s)e^{j3x}$, then

$$(D^2+j6D)[px^3+qx^2+rx+s] \equiv x^2,$$

going straight to stage (1) above.

This gives $$(6px+2q) + j6(3px^2+2qx+r) \equiv x^2,$$

i.e. $$j18px^2 + (6p+j12q)x + (2q+j6r) \equiv x^2.$$

Equating coefficients:

$$j18p = 1, \quad 6p + j12q = 0, \quad 2q + j6r = 0.$$

$$\therefore\ p = -\frac{j}{18}, \qquad q = \frac{j}{2}p, \qquad r = \frac{j}{3}q,$$

i.e. $$p = -\frac{j}{18}, \qquad q = \frac{1}{36}, \qquad r = \frac{j}{108}.$$

No definite value of s is given, so s is discarded as a particular integral only is required.

A particular integral is

$$y = \text{r.p. of}\left(-\frac{jx^3}{18}+\frac{x^2}{36}+\frac{jx}{108}\right)e^{j3x}$$

$$= \frac{x^3}{18}\sin 3x + \frac{x^2}{36}\cos 3x - \frac{x}{108}\sin 3x,$$

i.e. $$y = \frac{x^2}{36}\cos 3x + \frac{x}{108}(6x^2-1)\sin 3x.$$

Exercise 26

Find particular integrals for:

1. $(D^2+9)y = e^{3x}\sin x.$
2. $(D^2-2D-8)y = 6\cosh 2x.$
3. $(D^2+2)y = x^2\sin 2x.$
4. $(D^2+D+1)y = x^2-\cosh x.$
5. $(D^2-2D+2)y = x\cos 2x.$
6. $(D^2+2D-1)y = e^x\sin 2x.$

Answers

1. $\dfrac{e^{3x}}{325}(17\sin x - 6\cos x).$
2. $-\frac{3}{8}e^{2x}-\frac{1}{2}xe^{-2x}$ or $(\frac{1}{2}x-\frac{3}{8})\sinh 2x - (\frac{1}{2}x+\frac{3}{8})\cosh 2x.$
3. $\left(\dfrac{7-x^2}{2}\right)\sin 2x - 2x\cos 2x.$
4. $x^2-2x-\frac{1}{6}e^x-\frac{1}{2}e^{-x}$ or $x^2-2x+\frac{1}{3}(\sinh x - 2\cosh x).$
5. $-\frac{1}{50}(5x+11)\cos 2x - \frac{1}{25}(5x+1)\sin 2x.$
6. $-\frac{1}{34}e^x(4\cos 2x+\sin 2x).$

7·4. Examples of general solutions

Before reading through the following worked examples the student should read again § 7·2.

Example 1

Find the general solution of

$$(D-2)^2y = e^{2x}.$$

Auxiliary equation $(m - 2)^2 = 0.$

$$\therefore \; m = 2, \text{ a double root.}$$

Complementary function

$$y = (Ax + B)e^{2x}.$$

Particular integral. As $m = 2$ is a *double* root of the auxiliary equation, try $y = Cx^2e^{2x}$.

Then
$$(D - 2)^2 [Cx^2e^{2x}] \equiv e^{2x},$$

$$Ce^{2x}\{(D + 2) - 2\}^2 [x^2] \equiv e^{2x}.$$

$$\therefore \; CD^2[x^2] = 1,$$

$$C2 = 1,$$

$$C = \tfrac{1}{2}.$$

A particular integral is $y = \frac{1}{2}x^2e^{2x}$.

General solution.

$$\underline{y = (Ax + B + \tfrac{1}{2}x^2)e^{2x}.}$$

EXAMPLE 2

Solve $(D^2 + 2D + 2)y = x^2 + \sinh x.$ [L.U.]

A.E.
$$m^2 + 2m + 2 = 0.$$

$$\therefore \; m = -1 \pm j.$$

C.F.
$$y = e^{-x}(A \cos x + B \sin x).$$

P.I. Treat $\sinh x$ as $\frac{1}{2}(e^x - e^{-x})$ and try

$$y = px^2 + qx + r + se^x + te^{-x}.$$

The equation is

$$(D^2 + 2D + 2)y = x^2 + \tfrac{1}{2}(e^x - e^{-x}).$$

From § 7·341,

$$s = \frac{1}{2(1^2 + 2.1 + 2)} \quad \text{and} \quad t = \frac{-1}{2\{1^2 + 2(-1) + 2\}},$$

i.e.
$$s = \tfrac{1}{10} \quad \text{and} \quad t = -\tfrac{1}{2}. \tag{1}$$

Also $$(D^2 + 2D + 2)[px^2 + qx + r] \equiv x^2,$$

i.e. $$2p + 4px + 2q + 2px^2 + 2qx + 2r \equiv x^2.$$

Thus $$2p = 1, \quad 4p + 2q = 0, \quad 2p + 2q + 2r = 0.$$

These give $$p = \tfrac{1}{2}, \quad q = -1, \quad r = \tfrac{1}{2}.$$

A particular integral is

$$y = \tfrac{1}{2}x^2 - x + \tfrac{1}{2} + \tfrac{1}{10}e^x - \tfrac{1}{2}e^{-x}$$

or $$y = \tfrac{1}{2}(x-1)^2 + \tfrac{3}{5}\sinh x - \tfrac{2}{5}\cosh x.$$

G.S.

$$y = e^{-x}(A\cos x + B\sin x) + \tfrac{1}{2}(x-1)^2 + \tfrac{1}{5}(3\sinh x - 2\cosh x).$$

EXAMPLE 3

Solve the differential equation $(4D^2 - 4D + 5)y = e^x$, given that $y = 1$, $dy/dx = 1$ when $x = 0$. [L.U.]

A.E. $$4m^2 - 4m + 5 = 0.$$

$$\therefore m = \tfrac{1}{2} \pm j.$$

C.F. $$y = e^{\frac{1}{2}x}(A\cos x + B\sin x).$$

P.I. Putting $y = Ce^x$, from § 7·341,

$$C = \frac{1}{(4 \,.\, 1^2 - 4 \,.\, 1 + 5)} = \frac{1}{5}.$$

A particular integral is $y = \tfrac{1}{5}e^x$.

G.S. $$y = e^{\frac{1}{2}x}(A\cos x + B\sin x) + \tfrac{1}{5}e^x.$$

For the particular solution required: $y = 1$ when $x = 0$, therefore

$$1 = A + \tfrac{1}{5}, \quad \text{giving} \quad A = \tfrac{4}{5}.$$

Thus $$y = e^{\frac{1}{2}x}(\tfrac{4}{5}\cos x + B\sin x) + \tfrac{1}{5}e^x.$$

Also

$$dy/dx = \tfrac{1}{2}e^{\frac{1}{2}x}(\tfrac{4}{5}\cos x + B\sin x)$$
$$+ e^{\frac{1}{2}x}(B\cos x - \tfrac{4}{5}\sin x) + \tfrac{1}{5}e^x.$$

Now $dy/dx = 1$ when $x = 0$, therefore $1 = \frac{2}{5} + B + \frac{1}{5}$,

$$\text{i.e. } B = \tfrac{2}{5}.$$

The solution required is

$$y = \tfrac{2}{5}e^{\frac{1}{2}x}(2\cos x + \sin x) + \tfrac{1}{5}e^{x}.$$

Note. The student is warned against the fallacy of putting in the given conditions at the complementary function stage. Calculation of the special values of the arbitrary constants to satisfy given initial conditions must be left until the general solution has been found.

EXAMPLE 4

Solve the equation $(D^2 + 5D + 6)y = e^{-2x}\sin 3x$.

A.E. $\quad m^2 + 5m + 6 = (m + 2)(m + 3) = 0.$

$$\therefore m = -2 \text{ or } -3.$$

C.F. $\quad y = Ae^{-2x} + Be^{-3x}.$

P.I. Treat $e^{-2x}\sin 3x$ as the i.p. of $e^{(-2+j3)x}$.

Try $\quad y = \text{i.p. of } Ce^{(-2+j3)x}$.

Then $\quad (D^2 + 5D + 6)[Ce^{(-2+j3)x}] \equiv e^{(-2+j3)x}$,

i.e. $\quad \{(-2 + j3)^2 + 5(-2 + j3) + 6\}Ce^{(-2+j3)x} \equiv e^{(-2+j3)x}$.

This gives $\quad (-9 + j3)C = 1.$

$$\therefore \quad C = \frac{1}{(-9 + j3)} = \frac{1}{3(-3 + j)} = -\frac{(3 + j)}{3(3^2 + 1^2)}$$

$$= -\frac{(3 + j)}{30}.$$

A particular integral is

$$y = \text{i.p. of } -\tfrac{1}{30}(3 + j)e^{(-2+j3)x}$$

$$= \text{i.p. of } -\tfrac{1}{30}e^{-2x}(3 + j)(\cos 3x + j\sin 3x)$$

$$= -\tfrac{1}{30}e^{-2x}(\cos 3x + 3\sin 3x).$$

G.S. $\quad y = Ae^{-2x} + Be^{-3x} - \tfrac{1}{30}e^{-2x}(\cos 3x + 3\sin 3x).$

EXAMPLE 5

Solve the equation $\dfrac{d^2x}{dt^2} + 4\dfrac{dx}{dt} + 5x = 8\cos t$, given $x = 0$, $dx/dt = 3$ when $t = 0$. [L.U.]

A.E. $$m^2 + 4m + 5 = 0.$$

$$\therefore m = -2 \pm j.$$

C.F. $$x = e^{-2t}(A\cos t + B\sin t).$$

P.I. Treat $8\cos t$ as the r.p. of $8e^{jt}$ and try $x =$ r.p. of Ce^{jt}.

Then $$(D^2 + 4D + 5)[Ce^{jt}] \equiv 8e^{jt},$$

where $$D \equiv \frac{d}{dt},$$

i.e. $$C(-1 + j4 + 5)e^{jt} \equiv 8e^{jt},$$

giving $$4C(1 + j) = 8.$$

$$\therefore C = \frac{2}{1+j} = \frac{2(1-j)}{2} = (1-j).$$

A particular integral is

$$x = \text{r.p. of } (1-j)e^{jt}$$
$$= \text{r.p. of } (1-j)(\cos t + j\sin t)$$
$$= \cos t + \sin t.$$

G.S. $$x = e^{-2t}(A\cos t + B\sin t) + \cos t + \sin t.$$

When $$t = 0,\ x = 0.$$

$$\therefore 0 = A + 1,\ \underline{A = -1.}$$

Thus
$$x = e^{-2t}(B\sin t - \cos t) + \cos t + \sin t,$$
and
$$dx/dt = -2e^{-2t}(B\sin t - \cos t)$$
$$+ e^{-2t}(\sin t + B\cos t) - \sin t + \cos t.$$

When $$t = 0,\ dx/dt = 3.$$

$$\therefore\ 3 = 2 + B + 1,\ \underline{B = 0.}$$

The particular solution required is

$$x = -e^{-2t} \cos t + \cos t + \sin t,$$

or $$\underline{x = (1 - e^{-2t}) \cos t + \sin t.}$$

EXERCISE 27

Solve the following equations:

1. $(D^2 + D - 12)y = e^x$. [L.U.]
2. $(D^2 - 4)y = 10$. [L.U.]
3. $(D^2 - 3D + 2)y = 5 \sin 2x$.
4. $(D^2 - 5D + 6)y = x^2$.
5. $(D^2 + 5D + 6)y = 2x - 1$, given that $y = 0$, $dy/dx = \frac{4}{3}$ when $x = 0$.
6. $\dfrac{d^2i}{dt^2} + 10\dfrac{di}{dt} + 25i = 8e^{-3t}$.
7. $\dfrac{d^2x}{dt^2} + x = 3 \sin t$.
8. $(D^2 + 2aD + a^2)y = x^2e^{-ax}$. [L.U.]
9. $\dfrac{d^2i}{dt^2} + 6\dfrac{di}{dt} + 10i = 3 \sin 2t$.

   To what does i approximate when t is large?
10. $(D^2 + 4D + 5)y = \frac{1}{2} \sin x + 1$.
11. $(D^2 - 2aD + a^2)y = e^{ax}$. [L.U.]
12. $(D^2 - 5D + 4)y = e^{2x} + \sin x$. [L.U.]
13. $\dfrac{d^2y}{dx^2} + 6\dfrac{dy}{dx} + 8y = 1 - x^2$. [L.U.]
14. $3\dfrac{d^2x}{dt^2} - 2\dfrac{dx}{dt} - x = 2t - 1$, given that $x = 7$, $\dfrac{dx}{dt} = -2$ when $t = 0$. [L.U.]
15. $(D^2 - 4D + 5)y = e^{2x} \sin x$.
16. $(D^2 + D + 1)y = xe^x + e^x \sin x$. [L.U.]

Answers

1. $y = Ae^{3x} + Be^{-4x} - \frac{1}{10}e^{x}$.
2. $y = Ae^{2x} + Be^{-2x} - 2{\cdot}5$.
3. $y = Ae^{x} + Be^{2x} - \frac{1}{4}\sin 2x + \frac{3}{4}\cos 2x$.
4. $y = Ae^{3x} + Be^{2x} + \frac{1}{108}(18x^2 + 30x + 19)$.
5. $y = \frac{7}{3}e^{-2x} - \frac{17}{9}e^{-3x} + \frac{1}{3}x - \frac{4}{9}$.
6. $i = (A + Bt)e^{-5t} + 2e^{-3t}$.
7. $x = A\sin t + (B - \frac{3}{2}t)\cos t$.
8. $y = e^{-ax}(\frac{1}{12}x^4 + A + Bx)$.
9. $i = e^{-3t}(A\cos t + B\sin t) + \frac{1}{10}(\sin 2t - 2\cos 2t)$.
   i approaches $\frac{1}{10}(\sin 2t - 2\cos 2t)$, as $e^{-3t} \to 0$.
10. $y = e^{-2x}(A\cos x + B\sin x) + \frac{1}{16}(\sin x - \cos x) + \frac{1}{5}$.
11. $y = (\frac{1}{2}x^2 + Ax + B)e^{ax}$.
12. $y = Ae^{4x} + Be^{x} - \frac{1}{2}e^{2x} + \frac{1}{34}(3\sin x + 5\cos x)$.
13. $y = Ae^{-4x} + Be^{-2x} + \frac{1}{64}(1 + 12x - 8x^2)$.
14. $x = \frac{1}{2}e^{t} + \frac{3}{2}e^{-\frac{1}{3}t} + 5 - 2t$.
15. $y = e^{2x}(A\cos x + B\sin x - \frac{1}{2}x\cos x)$.
16. $y = e^{-\frac{1}{2}x}(A\cos\frac{1}{2}\sqrt{3}\,x + B\sin\frac{1}{2}\sqrt{3}\,x)$
    $+ \frac{1}{3}e^{x}(x - 1) + \frac{1}{13}e^{x}(2\sin x - 3\cos x)$.

7·5. The type $x^2\dfrac{d^2y}{dx^2} + ax\dfrac{dy}{dx} + by = f(x)$

This type of equation is called a homogeneous linear equation. It is seldom met with in engineering, but is included here as an example of transforming the independent variable, an important method in the solution of harder types of differential equations.

The equation transforms into a linear equation with constant coefficients by the substitution

$$x = e^{t}. \tag{1}$$

Then

$$\frac{dy}{dx} = \frac{dy}{dt}\frac{dt}{dx} = \frac{dy}{dt}\Big/\frac{dx}{dt} = \frac{dy}{dt}\Big/e^{t} = \frac{1}{x}\frac{dy}{dt}.$$

Thus

$$x\frac{dy}{dx} = \frac{dy}{dt}. \tag{2}$$

Differentiating (2) with respect to x:

$$x\frac{d^2y}{dx^2}+\frac{dy}{dx}=\frac{d}{dx}\left(\frac{dy}{dt}\right)=\frac{d}{dt}\left(\frac{dy}{dt}\right)\times\frac{dt}{dx},$$

i.e. $$x\frac{d^2y}{dx^2}+\frac{dy}{dx}=\frac{d^2y}{dt^2}\Big/\frac{dx}{dt}=\frac{d^2y}{dt^2}\Big/e^t=\frac{1}{x}\frac{d^2y}{dt^2}.$$

Thus $$x^2\frac{d^2y}{dx^2}+x\frac{dy}{dx}=\frac{d^2y}{dt^2}.$$

From (2), this gives

$$x^2\frac{d^2y}{dx^2}=\frac{d^2y}{dt^2}-\frac{dy}{dt}. \tag{3}$$

Equations (2) and (3) above are used in the transformation.

EXAMPLE

Solve $$x^3\frac{d^2y}{dx^2}+3x^2\frac{dy}{dx}-3xy=1.$$

Dividing through by x:

$$x^2\frac{d^2y}{dx^2}+3x\frac{dy}{dx}-3y=\frac{1}{x}.$$

Putting $x=e^t$ and using (2) and (3) above:

$$\left(\frac{d^2y}{dt^2}-\frac{dy}{dt}\right)+3\frac{dy}{dt}-3y=\frac{1}{e^t}=e^{-t},$$

i.e. $$\frac{d^2y}{dt^2}+2\frac{dy}{dt}-3y=e^{-t}.$$

A.E. $$m^2+2m-3=0,$$

$$(m+3)(m-1)=0.$$

C.F. $$y=Ae^t+Be^{-3t}.$$

P.I. Put $y=Ce^{-t}$, then

$$C=\frac{1}{(-1)^2+2(-1)-3}=-\frac{1}{4}.$$

G.S. $$y=Ae^t+Be^{-3t}-\tfrac{1}{4}e^{-t}.$$

But $x=e^t$.

$$\therefore\ e^{-3t}=\frac{1}{x^3}\quad\text{and}\quad e^{-t}=\frac{1}{x}.$$

The solution of the given equation is therefore

$$y = Ax + \frac{B}{x^3} - \frac{1}{4x}.$$

EXERCISE 28

Solve the following equations:

1. $$x^2 \frac{d^2y}{dx^2} + x\frac{dy}{dx} + y = x.$$

2. $$x \frac{d^2y}{dx^2} + 2 \frac{dy}{dx} = x^2.$$

3. $x \dfrac{d^2y}{dx^2} + 3 \dfrac{dy}{dx} + \dfrac{y}{x} = 8x^2$, given that $y = 0$, $dy/dx = 2$ when $x = 1$.

4. By means of the substitution $z = xy$, or otherwise, reduce the equation $x \dfrac{d^2y}{dx^2} + 2 \dfrac{dy}{dx} + a^2xy = 0$ to an equation with constant coefficients and hence solve it, given that when $x = \pi/a$, $y = 0$ and $dy/dx = -a$. [L.U.]

5. Change the independent variable in the differential equation $4x \dfrac{d^2y}{dx^2} + 2\dfrac{dy}{dx} + y = 0$ by means of the substitution $x = u^2$, and hence obtain its general solution.

ANSWERS

1. $y = A \cos (\log_e x) + B \sin (\log_e x) + \frac{1}{2}x.$

2. $y = A + \dfrac{B}{x} + \dfrac{x^3}{12}.$ **3.** $y = -\dfrac{1}{2x} + \dfrac{x^3}{2}.$

4. $y = \dfrac{\pi}{ax} \sin ax.$

The equation reduces to $\dfrac{d^2z}{dx^2} + a^2z = 0.$

5. $y = A \cos \sqrt{x} + B \sin \sqrt{x}.$

CHAPTER 8

SOME ELECTRICAL APPLICATIONS OF LINEAR DIFFERENTIAL EQUATIONS

8·1. Forced, damped, torsional oscillations

Because of its application to the suspension equipment of galvanometers, etc., the student of electrical engineering should make himself familiar with forced, damped mechanical oscillations. A typical example of a coil whose angular deflexion is controlled by a spring will be considered. The equation of motion will be deduced and that equation solved using the methods of the preceding chapter.

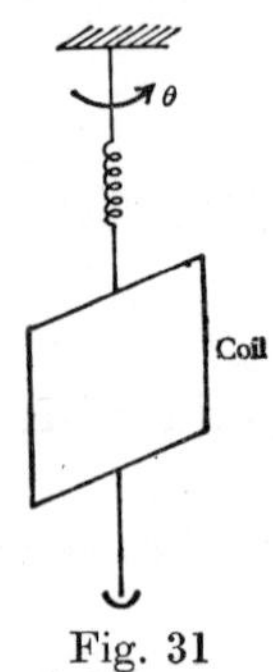

Fig. 31

Fig. 31 shows a simple diagram of such an arrangement. θ is taken as the measure of the angular deflexion. In the motion of the coil several damping factors may occur, for example, those due to friction, air resistance and electromagnetic damping due to the motion of the coil which itself is in a magnetic field. Such damping factors are often proportional to the angular velocity of the coil and this will be assumed in this case. There is also an accelerating torque, in this case due to a current flowing through the coil.

Let T be the torque acting on the coil.

Let I be the moment of inertia of the coil about the axis of rotation.

Let k be the resisting torque per unit angular velocity, due to damping.

Let c be the (resisting) torque per unit angular displacement, due to the elasticity of the suspension.

The equation of motion is given by

Net accelerating torque

$$= \text{Moment of inertia} \times \text{angular acceleration},$$

i.e.
$$T - k\frac{d\theta}{dt} - c\theta = I\frac{d^2\theta}{dt^2},$$

giving $$I\frac{d^2\theta}{dt^2}+k\frac{d\theta}{dt}+c\theta=T. \tag{1}$$

Writing $$2a=\frac{k}{I},\quad n^2=\frac{c}{I}, \tag{2}$$

this reduces to

$$\frac{d^2\theta}{dt^2}+2a\frac{d\theta}{dt}+n^2\theta=\frac{T}{I}. \tag{3}$$

Solving this equation:

A.E. $$m^2+2am+n^2=0,$$

$$m=-a\pm\sqrt{(a^2-n^2)}.$$

The complementary function can be of three types:

(i) $a>n$. *Heavily damped.* Both values of m will be negative, say $-\lambda_1$ and $-\lambda_2$, and the C.F. will be

$$\theta=Ae^{-\lambda_1 t}+Be^{-\lambda_2 t}. \tag{4}$$

(ii) $a=n$. *Critically damped.* Both values of m are $-a$, and the C.F. will be

$$\theta=(A+Bt)e^{-at}. \tag{5}$$

(iii) $a<n$. *Lightly damped.* Here

$$m=-a\pm j\sqrt{(n^2-a^2)}$$

$$=-a\pm jp \text{ say.}$$

The C.F. will be

$$\theta=e^{-at}(A\cos pt+B\sin pt). \tag{6}$$

In all three cases it should be noted that the negative exponential factor ensures that 'free' motion dies away with time. If the damping factor, a, is large, the free motion will die out extremely rapidly.

For this reason the complementary function part of the general solution is called the *transient* (cf. forced, damped electrical oscillations).

The particular integral gives the *steady-state* nature of the motion, that is, the motion after a suitable time-interval.

The particular integral, of course, depends on the torque T.

Two cases of interest are:

(a) T = *constant.* In equation (3) let T/I = constant = $n^2\theta_0$ say.

Then the particular integral is given by $n^2\theta_0/n^2 = \theta_0$.

The possible general solutions, corresponding to C.F.'s in (4), (5) and (6) are:

$\theta = Ae^{-\lambda_1 t} + Be^{-\lambda_2 t} + \theta_0$, heavily damped.

$\theta = (A + Bt)e^{-at} + \theta_0$, critically damped.

$\theta = e^{-at}(A \cos pt + B \sin pt) + \theta_0$, lightly damped.

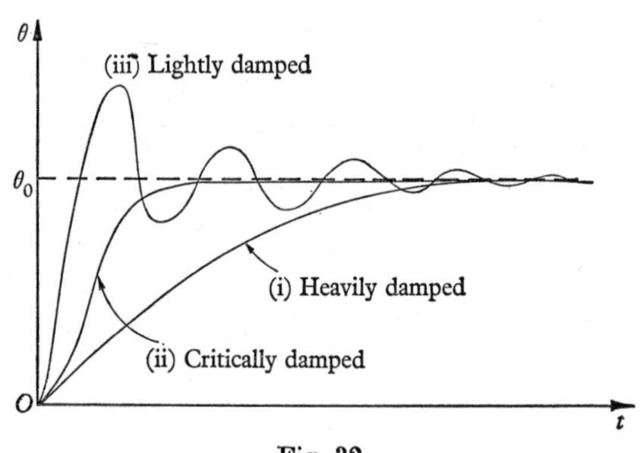

Fig. 32

Fig. 32 shows the deflexion-time curves for the three cases. It is seen that the steady deflexion, θ_0, is reached in each case.

A nearly critical damping is the ideal for a normal galvanometer suspension system.

(b) T *periodic.* In the case of an alternating current flowing through the coil, the torque T would be periodic.

In equation (3) above let $T/I = n^2\theta_0 \cos \omega t$.

Treating $n^2\theta_0 \cos \omega t$ as the r.p. of $n^2\theta_0 e^{j\omega t}$, try θ = r.p. of $Ce^{j\omega t}$ as a particular integral.

Then $$(D^2 + 2aD + n^2)\, Ce^{j\omega t} \equiv n^2\theta_0 e^{j\omega t},$$

$$C(-\omega^2 + j2a\omega + n^2) = n^2\theta_0.$$

$$\therefore C = \frac{n^2\theta_0}{(n^2 - \omega^2) + j2a\omega}$$

$$= \frac{n^2\theta_0\{(n^2 - \omega^2) - j2a\omega\}}{(n^2 - \omega^2)^2 + 4a^2\omega^2}.$$

A particular integral is therefore

$$\theta = \text{r.p. of } \frac{n^2\theta_0\{(n^2 - \omega^2) - j2a\omega\}}{(n^2 - \omega^2)^2 + 4a^2\omega^2} e^{j\omega t}$$

$$= \frac{n^2\theta_0}{\{(n^2 - \omega^2)^2 + 4a^2\omega^2\}} \{(n^2 - \omega^2) \cos \omega t + 2a\omega \sin \omega t\}.$$

Using the form $R \cos (\omega t - \alpha)$, this reduces to

$$\theta = \frac{n^2\theta_0}{\{(n^2 - \omega^2)^2 + 4a^2\omega^2\}^{\frac{1}{2}}} \cos (\omega t - \alpha), \qquad (7)$$

where $$\alpha = \tan^{-1} \left(\frac{2a\omega}{n^2 - \omega^2}\right). \qquad (8)$$

This represents a forced oscillation and, as long as the damping factor, a, is present, the deflexion θ will take this form after a short time lapse.

The amplitude, $\dfrac{n^2\theta_0}{\{(n^2 - \omega^2)^2 + 4a^2\omega^2\}^{\frac{1}{2}}}$, obviously varies with the frequency, $\omega/2\pi$, of the applied torque. The full mathematical treatment of the variation is long, but fig. 33 shows the salient features.

(The student is again reminded of the close analogy with forced electrical oscillations, see § 8·32.)

The figure shows the variation of the amplitude A with ω. The different curves are for different values of the damping factor a. The frequency of the 'free' oscillations, $n/2\pi$, has been taken with $n = 5$. These free oscillations are the oscillations which would occur with no damping and no applied torque. Equation (3) would then be

$$\frac{d^2\theta}{dt^2} + n^2\theta = 0 \quad \text{(s.h.m. of frequency } n/2\pi).$$

It is seen that when the damping factor is small a very sharp maximum of amplitude occurs when $\omega \doteqdot n$. This phenomenon is called *resonance* and occurs when the frequencies of the free and applied oscillations are approximately the same. This forms the basis of the use of the vibration galvanometer as a detector instrument in a.c. circuits.

On the other hand, resonance is often a thing to be avoided. For example, sensitive instruments may need to be protected from vibrations of the building or room in which they are to

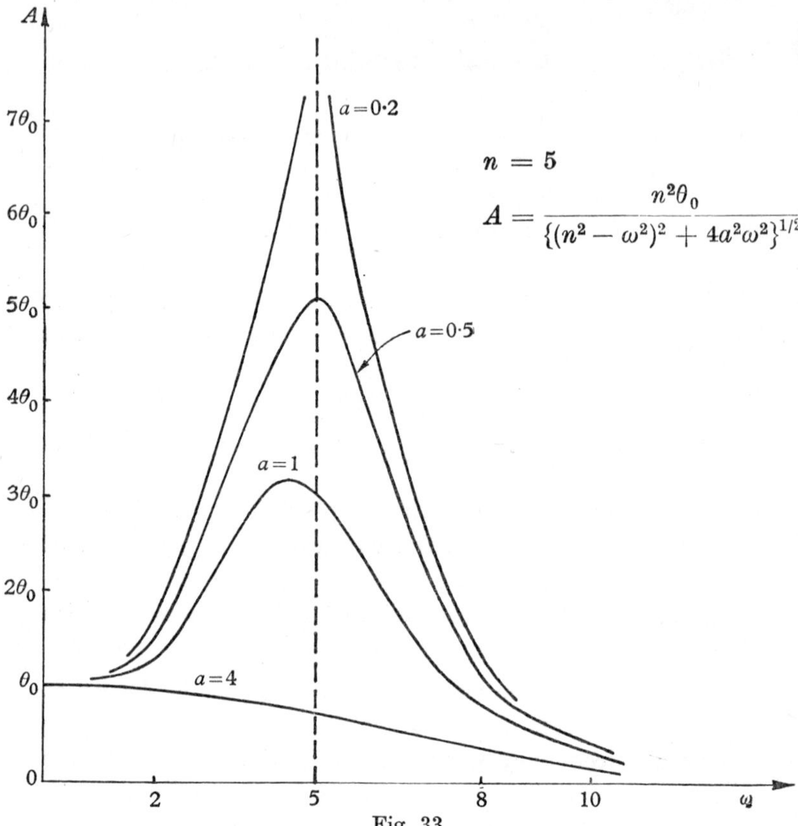

Fig. 33

be used. In this case they may be placed on a heavy platform suspended from the ceiling by springs which have a low natural frequency compared with the vibrations of the building. The instruments are then isolated from the vibrations of the room. This isolation principle is also used in reverse. If a heavy electric motor or generator is likely to cause unwanted vibrations in the building in which it is housed, it is mounted on a spring system which has a low natural frequency compared with the operating speed of the motor or generator.

8·2. Some typical galvanometers

A brief account of the mathematical theory of three main types of galvanometers will now be given.

8·21. D'Arsonval type

This is a common deflecting type of instrument for steady currents. It is mainly used in the many methods of measuring resistances, in potentiometer work, etc. Basically it consists of a coil suspended between the poles of a permanent magnet. Often there is an iron core inside the coil and the coil wires are situated in the air gaps between this iron core and the magnet poles. The pole faces are curved so as to give a radial field. The suspension arrangement (usually a fine strip of phosphor-bronze) carries a small mirror on to which a beam of light is focused. This beam of light is reflected on to a scale which measures the deflexion of the coil. Damping is provided by winding the coil on a light metal former. When this former rotates in the magnetic field, currents are induced in it which in turn produce their own magnetic field and the interaction of the two magnetic fields produces a torque opposing the motion of the coil. Damping can also be obtained by connecting a resistance in parallel across the galvanometer terminals. The damping effect then depends upon the size of this resistance, which may be adjusted to make the damping critical if so desired.

Referring to fig. 34, let

h = height of coil,

w = width of coil,

i = current flowing in the coil,

B = magnetic flux density (constant),

N = number of turns in the coil.

The force on each edge of the coil is $(Nh)\,Bi$.

The torque on the coil is therefore $(NhBi)\,w$.

In equation (1) of §8·1, $T = (NhBi)\,w$.

Let $A = hw$ = area of coil, then

$$T = NABi. \qquad (1)$$

Let
$$G = NBA, \tag{2}$$
then
$$T = Gi. \tag{3}$$

G is actually the flux-turns of the coil in respect of the magnetic field in which it lies.

Assuming the coil to turn about a central axis, the speed of

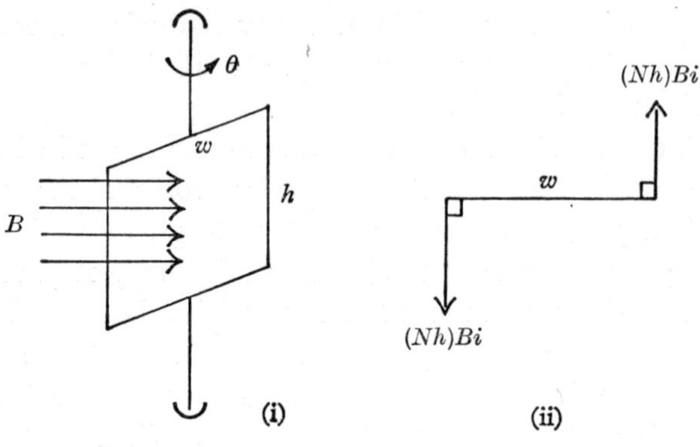

Fig. 34

an edge is $\frac{w}{2}\frac{d\theta}{dt}$. Thus the back e.m.f. generated in one edge, due to the rotation of the coil, is $NhB\,\frac{w}{2}\frac{d\theta}{dt}$.

$$\left(\text{The total lines cut by one edge per unit time is } Bh\frac{w}{2}\frac{d\theta}{dt}.\right)$$

Due to the two sides of the coil the total back e.m.f. generated is

$$2NhB\frac{w}{2}\frac{d\theta}{dt} = NAB\,\frac{d\theta}{dt} = G\frac{d\theta}{dt}. \tag{4}$$

Assuming that the inductance of the coil is negligible, that the applied voltage is e, and that R is the resistance of the coil the current is given by

$$i = \frac{1}{R}\left(e - G\frac{d\theta}{dt}\right). \tag{5}$$

Using (3) and (5) above, equation (1) of §8·1 becomes

$$I\frac{d^2\theta}{dt^2}+k\frac{d\theta}{dt}+c\theta-Gi=\frac{G}{R}\left(e-G\frac{d\theta}{dt}\right). \tag{6}$$

The term $k\dfrac{d\theta}{dt}$ here represents the damping effect of the eddy currents in the metal former on which the coil may be mounted plus any other damping effects such as friction, air resistance, etc.

These latter effects are usually quite small, and if the coil is wound on a non-conductive former the term $k\dfrac{d\theta}{dt}$ can be omitted.

Rearranging equation (6):

$$I\frac{d^2\theta}{dt^2}+\left(k+\frac{G^2}{R}\right)\frac{d\theta}{dt}+c\theta=\frac{Ge}{R}. \tag{7}$$

The solution to this equation has been discussed in §8·1 (*a*) above.

The damping is critical when

$$\left(k+\frac{G^2}{R}\right)^2=4Ic. \tag{8}$$

The damping is increased when R is decreased.

Sensitivity. The sensitivity to a steady current is given by putting $d\theta/dt=0=d^2\theta/dt^2$, which is the position when the deflexion is steady.

Then from equation (7)

$$c\theta=\frac{Ge}{R}=Gi_0,$$

where $i_0=e/R$ is the value of the steady current when $d\theta/dt=0$ (see equation (5) above).

Thus, the deflexion per unit (steady) current is given by

$$\frac{\theta}{i_0} = \frac{G}{c}. \tag{9}$$

The sensitivity of the instrument to a steady current is therefore given by G/c.

Undamped period of oscillation. Putting $k + \frac{G^2}{R} = 0$, which is the condition for no damping, equation (7) becomes

$$I\frac{d^2\theta}{dt^2} + c\theta = \frac{Ge}{R}.$$

This gives the undamped period of oscillation as $2\pi/n$, where

$$n = \sqrt{\frac{c}{I}}. \tag{10a}$$

The undamped frequency is then given by

$$f = \frac{n}{2\pi} = \frac{1}{2\pi}\sqrt{\frac{c}{I}}. \tag{10b}$$

Critical damping when damping is wholly electromagnetic. Assuming that the coil is wound on a non-conductive former and that there is no friction or air resistance, the damping term is $\frac{G^2}{R}\frac{d\theta}{dt}$.

From (8) above, the condition for critical damping is then given by

$$\left(\frac{G^2}{R}\right)^2 = 4Ic,$$

i.e.
$$R = \frac{G^2}{2\sqrt{(Ic)}}. \tag{11}$$

Note. If there is a shunt R_s, and R_c is the coil resistance, R is replaced by $R_c + R_s$.

PRACTICAL EXAMPLES

(1) The composite rectangular coil in a moving-coil meter has an active length of coil of 12 mm and a width of 9 mm. It moves in a uniform radial flux density of 0·3 tesla against spiral springs producing a torque control of 12×10^{-6} Nm per radian of coil deflexion. The composite coil comprises two identical windings side-by-side, namely, AB of $25\frac{1}{2}$ turns and CD of $25\frac{1}{2}$ turns, each having a resistance of 30 Ω. Ends A and C are connected together and to one terminal of a 2-volt supply. Connexions to the other terminal of the supply are from end B via an 80 Ω resistor, and from end D via a nickel wire coil having a resistance at 0 °C of 80 Ω and a temperature coefficient of resistance at 0 °C of 0·006 per degC, the two windings being connected to produce opposing fluxes. Determine the angle of deflexion of the meter coil when the temperature of the nickel wire changes from 0 to 125 °C, all resistances other than that of the nickel being assumed constant. [L.U.]

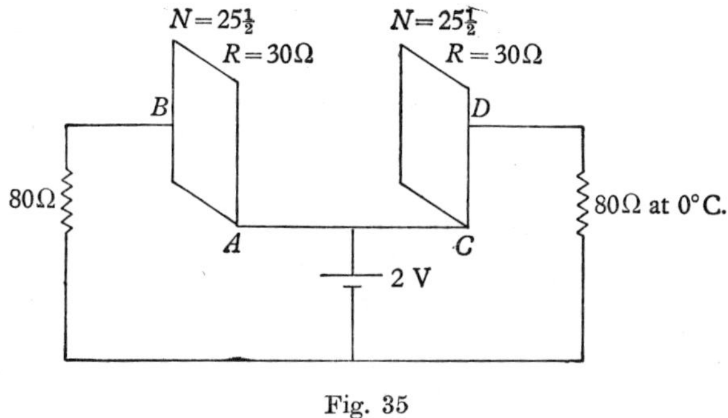

Fig. 35

Fig. 35 shows a sketch of the circuit of the meter.

With the usual notation, and using SI units,

$$N = 25\tfrac{1}{2},$$

$$A = 1{\cdot}08 \times 10^{-4}\,\text{m}^2,$$

$$B = 0{\cdot}3\,\text{tesla}.$$

In this case,

$$G = NAB = \tfrac{51}{2} \times 1{\cdot}08 \times 10^{-4} \times 0{\cdot}3.$$

At 0 °C the deflexion is obviously zero as the winding is such as to produce equal but opposite fluxes.

Left-hand coil. The steady current is

$$i_0 = \frac{e}{R} = \frac{2}{110}\ \text{A}.$$

Also $$c = 12 \times 10^{-6}\ \text{Nm}.$$

Ignoring the effect of the right-hand coil, the deflexion produced would be given by

$$\frac{G}{c} i_0 \quad (\text{see } \S 8{\cdot}21(9)) = \frac{51 \times 1{\cdot}08 \times 10^{-4} \times 0{\cdot}3}{2 \times 12 \times 10^{-6}} \times \frac{2}{110}\ \text{rad}$$

$$\simeq 1{\cdot}2517\ \text{rad}.$$

Right-hand coil. Ignoring the effect of the left-hand coil:

resistance of coil $= 30\ \Omega$,

external resistance at 125 °C is $80(1 + 0{\cdot}006 \times 125)\ \Omega$

$$= 80 \times 1{\cdot}75\ \Omega$$

$$= 140\ \Omega.$$

The steady current is therefore

$$\frac{2}{30 + 140}\ \text{A} = \frac{2}{170}\ \text{A}.$$

G and c are the same, giving the steady deflexion produced as

$$\frac{51 \times 1{\cdot}08 \times 10^{-4} \times 0{\cdot}3}{2 \times 12 \times 10^{-6}} \times \frac{2}{170}\ \text{rad}$$

$$= 0{\cdot}81\ \text{rad}.$$

The coils are in opposition, thus the actual deflexion shown on the meter would be

$$(1{\cdot}2517 - 0{\cdot}81)\ \text{rad} = 0{\cdot}4417\ \text{rad}$$

$$\simeq 25^\circ\ 19'.$$

(2) The moving coil of a galvanometer is wound with 300 turns on a non-metallic former 0·02 m wide and 0·02 m long. It is suspended in a uniform radial magnetic field of flux density 0·12 tesla. The coil has a moment of inertia of $1 \cdot 5 \times 10^{-7}$ kg m² and the control constant is 3×10^{-9} newton metre per degree. Calculate the conditions for critical damping and find the value of the shunt necessary to give critical damping with this galvanometer, assuming that the moving coil has a resistance of 10 ohms.

Working in SI units:

$$A = 4 \times 10^{-4}\ \text{m}^2,$$

$$N = 300,$$

$$B = 0 \cdot 12 \text{ tesla}.$$

$$\therefore G = NAB = 4 \times 10^{-4} \times 300 \times 0 \cdot 12$$

$$= 144 \times 10^{-4}.$$

$$c = 3 \times 10^{-9} \text{ newton metre per degree}$$

$$\simeq 3 \times 57 \cdot 3 \times 10^{-9} \text{ newton metre per radian}$$

$$(57 \cdot 3^\circ \simeq 1 \text{ radian}),$$

$$I = 1 \cdot 5 \times 10^{-7}\ \text{kg m.}^2$$

Now for critical damping, $R = \dfrac{G^2}{2\sqrt{(Ic)}}$ (see §8·21 (11) above), where R is the (total) resistance of the galvanometer circuit.

Thus $$R = \frac{144^2 \times 10^{-8}}{2\sqrt{(1 \cdot 5 \times 10^{-7} \times 3 \times 57 \cdot 3 \times 10^{-9})}}\ \Omega$$

$$= \frac{144^2}{2\sqrt{(4 \cdot 5 \times 57 \cdot 3)}}\ \Omega$$

$$\simeq 646\ \Omega.$$

Let S be the shunting resistance required, then

$$10 + S = 646.$$

$$\therefore S = 636\ \Omega.$$

8·22. Vibration type

This type is used as a detecting instrument in a.c. circuits. It is susceptible to mechanical vibrations whose frequency is close to that to which it has been tuned. For this reason some form of support is often given which isolates the instrument from external vibrations (see remarks at end of §8·1).

Some vibration-type galvanometers have a moving magnet but only those employing a fixed magnet and moving coil will be considered here.

The magnet is a permanent magnet with a very strong field, and the moving coil is designed to be very light with a small moment of inertia. In consequence the coil has a very short natural period of oscillation. The damping is made very small in order to give a sharp resonance curve (see fig. 33). Thus the deflexion produced by a current only a little different from the frequency to which the coil has been tuned will cause a comparatively small deflexion; whilst a current possessing the tuned frequency will cause a large deflexion. The galvanometer is tuned by adjusting the length and tension of the suspended system, which often consists only of a single turn of wire.

With these adjustments then, the instrument gives a large steady vibration for a current of the frequency to which it has been tuned whilst remaining insensitive to any harmonics that may be present.

The equation of motion is

$$I\frac{d^2\theta}{dt^2} + \alpha\frac{d\theta}{dt} + c\theta = Gi, \tag{1}$$

where $\alpha(d\theta/dt)$ is the damping term.

If the current i is alternating, and is given by $i = i_m \cos\omega t$, equation (1) becomes

$$I\frac{d^2\theta}{dt^2} + \alpha\frac{d\theta}{dt} + c\theta = Gi_m \cos\omega t. \tag{2}$$

This type of equation has been discussed fully in §8·1 (*b*).

The full solution consists of a transient term which soon dies away and the steady state is given by the particular integral, which in this case is a steady oscillation given by

$$\theta = \frac{Gi_m}{\{(c - I\omega^2)^2 + \alpha^2\omega^2\}^{\frac{1}{2}}} \cos(\omega t - \phi), \tag{3}$$

where
$$\phi = \tan^{-1}\frac{\alpha\omega}{(c - I\omega^2)}. \tag{4}$$

(These are easily obtained by writing α/I for $2a$, c/I for n^2 and Gi_m/I for $n^2\theta_0$ in equations (7) and (8) of §8·1. Alternatively the particular integral can be worked out from first principles.)

For resonance $c = I\omega^2$ and the steady oscillations are then given by

$$\theta = \frac{Gi_m}{\alpha\omega}\sin\omega t, \tag{5}$$

for at resonance
$$\phi = \tan^{-1}\infty = \frac{\pi}{2}.$$

As the damping is small, α is small and therefore (except near the resonant frequency) $c - I\omega^2$ is much greater than $\alpha\omega$. Thus, from equation (3) above, only small deflexions are obtained when the frequency is different from the resonant frequency.

Practical Example

Derive an expression for the sensitivity of the vibration galvanometer in terms of its moment of inertia, damping constant, etc., and calculate the ratio of its sensitivities to the fundamental and to the third harmonic of the supply when it is tuned to 50 Hz.

The inertia constant may be taken as 10^{-12} kg m^2 and the damping constant as 25×10^{-13} Nm per rad/s. [L.U.]

Let c = control constant in newton metres per radian,

G = displacement constant,

I = moment of inertia of moving system,

α = damping constant,

$i = i_m \cos \omega t$ be the current supplied.

Then, from equations (3) and (4) above, the steady oscillations are given by

$$\theta = \frac{Gi_m}{\{(c - I\omega^2)^2 + \alpha^2\,\omega^2\}^{\frac{1}{2}}} \cos(\omega t - \phi),$$

where
$$\phi = \tan^{-1} \frac{\alpha\omega}{(c - I\omega^2)}.$$

The sensitivity is thus given by

$$\frac{\theta}{i_m} = \frac{G}{\{(c - I\omega^2)^2 + \alpha^2\omega^2\}^{\frac{1}{2}}}. \qquad (1)$$

Fundamental.

$$\omega = 2\pi f = 100\pi.$$

The instrument is tuned for resonance at this frequency, therefore

$$c = I\omega^2 = 10^{-12} \times (100\pi)^2$$
$$= \pi^2 \times 10^{-8}. \qquad (2)$$

Now $\alpha = 25 \times 10^{-13}$, thus the sensitivity to the fundamental is given by

$$\frac{G}{\{0^2 + (25 \times 10^{-13}.100\pi)^2\}^{\frac{1}{2}}} = \frac{G}{25\pi.10^{-11}} \text{ rad/ampere}, \quad (3).$$

Third harmonic. Here $\omega = 300\pi$.

Using (1) and (2), the sensitivity is given by

$$\frac{G}{\{(\pi^2 \times 10^{-8} - 10^{-12} \times 300^2\pi^2)^2 + (25 \times 10^{-13} \times 300\pi)^2\}^{\frac{1}{2}}}$$

$$= \frac{G}{\left\{\left(\frac{\pi^2}{10^8} - \frac{9\pi^2}{10^8}\right)^2 + \frac{9{\cdot}625\pi^2}{10^{22}}\right\}^{\frac{1}{2}}}$$

$$= \frac{G}{\pi\left\{\frac{64\pi^2}{10^{16}} + \frac{5{\cdot}625}{10^{19}}\right\}^{\frac{1}{2}}}$$

$$\simeq \underline{\frac{G \times 10^8}{\pi \times 25{\cdot}14}} \text{ rad/ampere.} \qquad (4)$$

From (3) and (4), the ratio of the sensitivities is

$$\frac{G.10^{11}}{25\pi} \Big/ \frac{G.10^8}{25{\cdot}14\pi} \simeq \underline{1006.}$$

This example is ideal in showing how very selective the instrument is to the resonant frequency.

8·23. Ballistic type

This type of galvanometer measures small *quantities* of electricity, that is, charges. The time of discharge through the coil has to be small compared with the natural period of the instrument. The 'throw' is then proportional to the charge. Thus the moment of inertia of the moving system of the galvanometer is made large whilst the controlling torque of the suspension is made small. In this way the natural period of oscillation is often made as long as 10–15 sec. Damping is reduced to a minimum so that the first throw is large and the instrument is made almost 'dead-beat'. In this way the reading of the initial throw can be easily taken. The damping present must be electromagnetic so that it can be accurately calculated from the known constants of the instrument; other forms of damping, such as air damping, are not capable of accurate calculation.

Taking the discharge as almost instantaneous, the torque acting, before the instrument has time to deflect, is solely due to the current and is Gi (see § 8·21 (3)).

With the symbols previously used, the initial equation of motion is

$$I\frac{d^2\theta}{dt^2} = Gi.$$

Integrating this expression:

$$I\frac{d\theta}{dt} = \int Gi\,dt = Gq, \tag{1}$$

taking $d\theta/dt = 0$ when $q = 0$.

After the discharge (when $i = 0$), the *subsequent* motion is given by

$$I\frac{d^2\theta}{dt^2} + \alpha\frac{d\theta}{dt} + c\theta = 0,$$

where $\alpha(d\theta/dt)$ is the damping term.

Solving in the usual way, this gives

$$\theta = e^{-(\alpha/2I)t}(A\cos\omega t + B\sin\omega t), \tag{2}$$

where $$\omega = \sqrt{\left(\frac{c}{I} - \frac{\alpha^2}{4I^2}\right)}. \tag{3}$$

Taking $\theta = 0$ at $t = 0$, from (2) above $A = 0$.

Thus $$\theta = e^{-(\alpha/2I)t}\,.\,B\sin\omega t$$

and $$d\theta/dt = Be^{-(\alpha/2I)t}\left(\omega\cos\omega t - \frac{\alpha}{2I}\sin\omega t\right).$$

At $t = 0$, therefore, $d\theta/dt = B\omega$. But from (1) above $\dfrac{d\theta}{dt} = \dfrac{Gq}{I}$. Thus, at $t = 0$, $\dfrac{Gq}{I} = B\omega$. This gives

$$\theta = \frac{Gq}{I\omega}e^{-(\alpha/2I)t}\sin\omega t. \tag{4}$$

If the damping is very small and the moment of inertia of the moving parts is large, then $\alpha/2I \simeq 0$ and

$$\theta = \frac{Gq}{I\omega}\sin\omega t. \tag{5}$$

Also $$\omega \simeq \sqrt{\frac{c}{I}} \text{ as } \frac{\alpha^2}{4I} \simeq 0 \text{ in (3) above.} \tag{6}$$

The maximum readings in the positive direction are therefore reached when $\omega t = \frac{1}{2}\pi, \frac{5}{2}\pi, \frac{9}{2}\pi, \ldots$, and in the negative direction when $\omega t = \frac{3}{2}\pi, \frac{7}{2}\pi, \ldots$. The maximum deflexion, either way, is given by $\theta_m = \frac{Gq}{\omega I} = \frac{Gq}{\sqrt{(Ic)}}$, from (5) and (6).

The sensitivity is given by

$$\frac{\theta_m}{q} = \frac{G}{\sqrt{(Ic)}}. \tag{7}$$

The undamped (natural) period is of course $2\pi\sqrt{(I/c)}$ (see § 8·21, (10a)).

From the full equation, equation (4) above, successive maxima, one positive, one negative, occur when the values of ωt differ by π (e.g. $\omega t = \frac{1}{2}\pi$ and $\omega t = \frac{3}{2}\pi$). If θ_1 and θ_2 are two such successive values, the numerical value of their ratio is given by

$$\frac{\theta_1}{\theta_2} = \frac{Gq}{I\omega} e^{-\frac{\alpha}{2I}\frac{\pi}{2\omega}} \Big/ \frac{Gq}{I\omega} e^{-\frac{\alpha}{2I}\frac{3\pi}{2\omega}}$$

$$= e^{\pi\alpha/2\omega I} = e^{[\pi\alpha/2\sqrt{(Ic)}]}.$$

Thus $$\log_e \frac{\theta_1}{\theta_2} = \frac{\pi}{2}\frac{\alpha}{\sqrt{(Ic)}}.$$

This is called the logarithmic decrement and is often denoted by λ. (The student should also refer to vol. I, § 14·22.)

Practical Example

Find the sensitivity of a ballistic galvanometer given the following data:

Coil: 400 turns; 30 mm long by 20 mm wide; moment of inertia about vertical axis 5×10^{-7} kg m^2.
Suspension: torsional stiffness, $1{\cdot}5 \times 10^{-7}$ Nm per radian.
Radial magnetic field: 8×10^{-2} tesla.
Damping: negligible. [L.U.]

$$B = 8 \times 10^{-2}, \quad A = 6 \times 10^{-4}, \quad N = 400.$$

$$\therefore G = 400 \times 48 \times 10^{-6} = 192 \times 10^{-4},$$

$$c = 1{\cdot}5 \times 10^{-7},$$

$$I = 5 \times 10^{-7}.$$

From §8·23 (7) the sensitivity is

$$\frac{G}{\sqrt{(Ic)}} = \frac{192 \times 10^{-4}}{\sqrt{(5 \times 1{\cdot}5 \times 10^{-14})}}$$

$$= 0{\cdot}0701 \times 10^{6} \text{ rad per coulomb.}$$

$$\simeq 0{\cdot}07 \text{ rad per microcoulomb.}$$

Exercise 29

1. A moving-coil galvanometer (ballistic) has a sensitivity of 50 mm per microampere with a scale distance of 1 metre. Its open-circuit periodic time is 3 sec. The movement, which has a resistance of 200 Ω, is critically damped when a 3000 Ω resistor is connected across its terminals. Calculate its moment of inertia. [L.U.]

2. The length of the coil of such an instrument (a moving-coil instrument) is 35 mm and its width is 25 mm, the number of turns is 75, the field strength is 8×10^{-2} tesla, and the controlling toque exerted by the springs is $0{\cdot}5 \times 10^{-7}$ Nm per degree. What current will be required to give a deflexion of 45°? [L.U.]

3. If the first swing of a ballistic galvanometer is 0·35 m and the second swing in the same direction is 0·28 m, what would the first swing have been in the absence of damping? Prove any formula used. [L.U.]

4. A moving-coil ammeter has the following constants:

Air gap flux density $= 5 \times 10^{-2}$ tesla.

Turns on moving coil = 100.

Moment of inertia of moving coil $= 2 \times 10^{-7}$ kg m^2.

Control constant of moving spring $= 3{\cdot}19 \times 10^{-5}$ Nm/radian.

The rectangular moving coil is 20 mm by 50 mm. It is wound on a non-conducting former and the long sides may be assumed to be entirely within the air-gap. The resistance of the winding is 7 ohms, and it is connected to a shunt, the resistance of which may be neglected. Frictional forces may also be neglected. Sketch a curve of angular displacement of the moving coil to a time base, for the first two or three oscillations, when a current of 5 mA. is suddenly established in the moving coil. The maxima and minima, and the points at which the displacement instantaneously possesses its ultimate steady value, should be calculated. [L.U.]

5. Derive expressions connecting the moment of inertia, the control constant and the damping coefficient of a moving-coil meter which correspond to the cases of:

(*a*) over-damped movement,

(*b*) critically damped movement and

(*c*) under-damped movement.

Draw deflexion/time graphs which illustrate these three conditions.

If the moment of inertia of a milliammeter is 5×10^{-9} kg m^2 and the control constant is 2×10^{-6} Nm/radian, find the value of the damping coefficient which will ensure critical damping. State the units in which it is expressed. [L.U.]

6. Give the equation of motion of the coil of a moving-coil meter, and derive the relationship that must exist if the damping of the movement is to be critical.

A moving-coil instrument has a rectangular coil wound on an aluminium former of resistivity $2{\cdot}7 \times 10^{-2}$ $\mu\Omega$ m. The sides of the former, each of effective length 30 mm, move in a radial field of uniform flux density 0·15 tesla against a control torque of 18×10^{-6} newton metre per radian. The width of the former is 20 mm, the ends being effectively outside the magnet field, and the moment of inertia of the moving system is 8×10^{-8} kg m^2.

Determine the cross-sectional area of the aluminium former if it is to provide critical damping, other sources of damping being neglected. [L.U.]

Answers

1. $2{\cdot}78 \times 10^{-7}$ kg m^2. **2.** 0·43 milliamp. **3.** 0·37 m.

5. 2×10^{-7} Nm per rad/s.

6. 0·8 mm.^{2} [*Note.* Torque due to eddy-current damping $= \dfrac{B^2l^2d^2}{R}\dfrac{d\theta}{dt}$ newton metres, where SI units are used.

$$R = \frac{2{\cdot}7 \times 10^{-8} \times (2l + 2d)}{a} \text{ ohms,}$$

where l, d are now in metres and a is cross-sectional area in square metres.]

8·3. Some common a.c. circuits

In vol. I electrical circuits were considered which had only zero or constant applied e.m.f.'s (driving voltages). In the cases now to be considered it will be seen that the application of any given driving voltage is equivalent, mathematically, to the appearance of a particular integral in the solution of the differential equation for the circuit.

This may be a suitable occasion to remind the student of the sign conventions adopted for the direction of voltages and currents in electrical circuits. Reference will be made to fig. 36 below. For the purposes of analysis one of the terminals, say A, is selected as positive. The voltage is then taken positive whenever A is at a higher potential than B. The current i is taken as positive when flowing away from the positive terminal A. If, on calculation, the value obtained its negative, this simply means that the current is flowing towards A when A is at a higher potential than B. The positive plate of the capacitor C is taken as the plate to which i is flowing. In this case $i = dq/dt$. When i is positive, q is increasing. If dq/dt, when calculated, is negative, then i is flowing out of the positive side of the capacitor. It is important that correct signs are attached to initial values of i and q.

8·31. The L-C series circuit with applied alternating e.m.f.

This is a typical example of undamped, forced electrical oscillations.

Fig. 36 shows a capacitance C in series with an inductance L. The applied alternating e.m.f. is taken as $E \cos\omega t$. Let q be the charge on the capacitor, and i the current flowing at any time t. Then

$$i = \frac{dq}{dt}. \qquad (1)$$

Equating the passive voltages (voltage drops) to the driving voltage:

$$L\frac{di}{dt} + q\frac{1}{C} = E\cos\omega t.$$

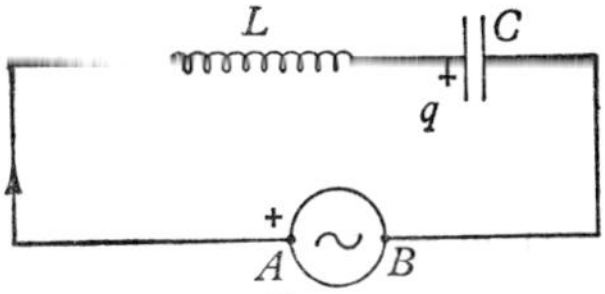

Using (1):

$$L\frac{d^2q}{dt^2} + \frac{q}{C} = E\cos\omega t,$$

Fig. 36

giving $$\frac{d^2q}{dt^2} + \frac{q}{LC} = \frac{E}{L}\cos\omega t.$$

Putting $$n^2 = \frac{1}{LC}, \tag{2}$$

$$\frac{d^2q}{dt^2} + n^2q = \frac{E}{L}\cos\omega t. \tag{3}$$

Solving this equation:

A.E. $m^2 + n^2 = 0$, giving $m = \pm jn$.

C.F. $$q = B\cos nt + F\sin nt. \tag{4}$$

With no applied e.m.f. the free, or natural, oscillations have period

$$\frac{2\pi}{n} = 2\pi\sqrt{(LC)}.$$

The particular integral is calculated for the two cases $\omega \neq n$ and $\omega = n$; that is, when the applied frequency is not equal to, or is equal to, the natural frequency.

(i) $\omega \neq n$. Write $\frac{E}{L}\cos\omega t$ as the real part of $\frac{E}{L}e^{j\omega t}$ and put $q =$ r.p. of $Ae^{j\omega t}$.

Then $$(-\omega^2 + n^2)Ae^{j\omega t} \equiv \frac{E}{L}e^{j\omega t},$$

giving $$A = \frac{E}{(n^2 - \omega^2)L}.$$

A particular integral is the real part of $\dfrac{E}{(n^2 - \omega^2)L} e^{j\omega t}$, i.e.

$$q = \frac{E}{(n^2 - \omega^2)L} \cos \omega t. \tag{5}$$

(ii) $\omega = n$. The equation becomes $(D^2 + n^2)q = \dfrac{E}{L} \cos nt$.

Put $q =$ r.p. of Ate^{jnt}, then

$$Ae^{jnt}\{(D + jn)^2 + n^2\}[t] \equiv \frac{E}{L} e^{jnt},$$

$$A(D^2 + 2jnD)[t] = \frac{E}{L},$$

$$A.2jn = \frac{E}{L}, \quad \text{giving} \quad A = -\frac{jE}{2nL}.$$

Thus $\qquad q = \text{r.p. of } -\dfrac{jEt}{2nL} e^{jnt} = \dfrac{Et}{2nL} \sin nt.$

A particular integral is

$$q = \frac{Et}{2nL} \sin nt. \tag{6}$$

The general solutions in the two cases are:

(i) $\omega \neq n$.

$$q = B \cos nt + F \sin nt + \frac{E}{(n^2 - \omega^2)L} \cos \omega t. \tag{7}$$

(ii) $\omega = n$.

$$q = B \cos nt + F \sin nt + \frac{Et}{2nL} \sin nt. \tag{8}$$

If the initial conditions are $q = 0$ and $i = dq/dt = 0$ when $t = 0$, then equation (7) gives $B = -\dfrac{E}{(n^2 - \omega^2)L}$, $F = 0$ and for equation (8), $B = 0 = F$.

Thus:

(i) $\omega \neq n$.

$$q = \frac{E}{L(n^2 - \omega^2)} (\cos \omega t - \cos nt). \tag{9}$$

(ii) $\omega = n$.

$$q = \frac{Et}{2nL}\sin nt. \tag{10}$$

By differentiation the currents in the two cases are:

(i) $\omega \neq n$.

$$i = \frac{E}{L(n^2-\omega^2)}(n\sin nt - \omega\sin\omega t). \tag{11}$$

(ii) $\omega = n$.

$$i = \frac{E}{2nL}(\sin nt + nt\cos nt). \tag{12}$$

It is seen that when the frequency of the applied e.m.f. is equal to that of the natural oscillations (i.e. $\omega = n$), both the current and charge become very large as the time increases due to the occurrence of t as a factor in equations (10) and (12). Such a condition is called one of *resonance*.

A useful exercise for the student is to obtain equations (10) and (12) as limiting cases of equations (9) and (11) as ω approaches n, treating ω as the variable.

Fig. 37 shows sketches of the cases when $\omega \neq n$ and when $\omega = n$, for the charge q.

Fig. 37 (i) is also interesting from the point of view of beat phenomena, when charges or currents of slightly different frequencies are combined; as, for example, in the beat-frequency stage of radio reception sets.

8·32. The *L-C-R* series circuit with applied alternating e.m.f.

This is an example of forced, damped electrical oscillations and should be compared with § 8·1 which dealt with forced damped mechanical oscillations. It will also be of advantage if the student reads again the account of the closed *L-C-R* circuit given in vol. I, § 14·42.

Fig. 38 shows a *L-C-R* series circuit with an applied alternating e.m.f. of amount $E\sin\omega t$. If i is the current flowing and q is the charge on the capacitor at time t, the equating of the passive and driving voltages gives

$$L\frac{di}{dt} + Ri + \frac{q}{C} = E\sin\omega t. \tag{1}$$

Using $i = dq/dt$ and $di/dt = d^2q/dt^2$ this becomes

$$L\frac{d^2q}{dt^2} + R\frac{dq}{dt} + \frac{q}{C} = E \sin \omega t. \tag{2}$$

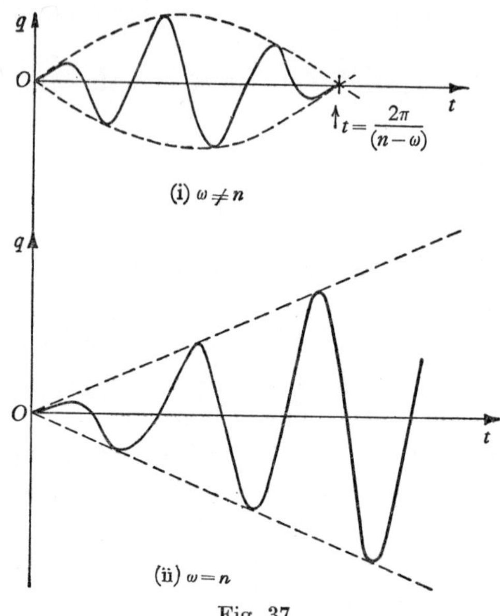

Fig. 37

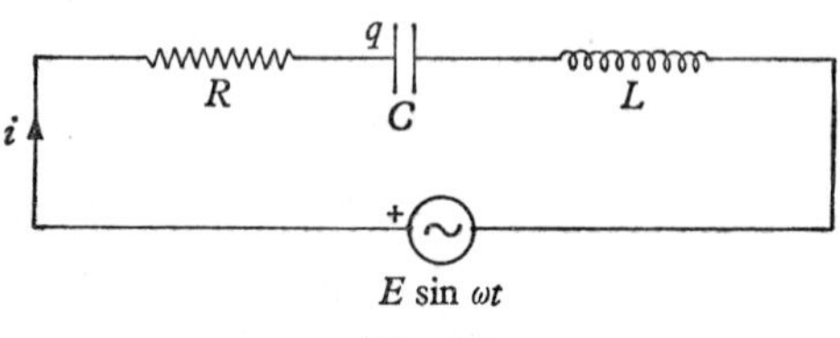

Fig. 38

If solved, equation (2) will give a formula for the charge q. From this the current can be deduced as $i = dq/dt$.

An alternative method, which gives i directly, is to differentiate each side of equation (1) with respect to t. This gives

$$L\frac{d^2i}{dt^2} + R\frac{di}{dt} + \frac{i}{C} = \omega E \cos \omega t. \tag{3}$$

Here, equation (3) will be solved.

Dividing through by L:

$$\left(D^2 + \frac{RD}{L} + \frac{1}{CL}\right) i = \frac{\omega E}{L} \cos \omega t, \tag{4}$$

where $$D \equiv \frac{d}{dt}.$$

Putting $$k = \frac{R}{2L} \quad \text{and} \quad n^2 = \frac{1}{LC}, \tag{5}$$

equation (4) becomes

$$(D^2 + 2kD + n^2)\, i = \frac{\omega E}{L} \cos \omega t. \tag{6}$$

A.E. $$m^2 + 2km + n^2 = 0.$$

$$\therefore\ m = -k \pm \sqrt{(k^2 - n^2)}.$$

This has been thoroughly studied in vol I, § 14·42. There it was shown that e^{-kt} ($= e^{-Rt/2L}$) is a factor in all three possible cases of the roots for m. Thus the complementary function part of the general solution dies away with time. It is called the *transient* for this reason. The part of the solution given by the particular integral does not die away with time and gives the *steady-state* value of the current.

Note the difference between this type of circuit and that of § 8·31 above. There, no resistance was present and therefore no damping factor. In consequence the part of the general solution contributed by the complementary function was always present. It is the presence of a resistance (R) which causes any natural oscillations to die away.

To solve equation (6) for a particular integral in order to find the steady-state current treat $\dfrac{\omega E}{L} \cos \omega t$ as the real part of $\dfrac{\omega E}{L} e^{j\omega t}$ and put $i =$ r.p. of $Ae^{j\omega t}$; then

$$-A\omega^2 e^{j\omega t} + j2k\omega A e^{j\omega t} + n^2 A e^{j\omega t} \equiv \frac{\omega E}{L} e^{j\omega t},$$

giving $$A\{(n^2 - \omega^2) + j2k\omega\} = \frac{\omega E}{L},$$

$$A = \frac{\omega E}{L\{(n^2 - \omega^2) + j2k\omega\}}$$

$$= \frac{\omega E}{L}\frac{\{(n^2 - \omega^2) - j2k\omega\}}{\{(n^2 - \omega^2)^2 + 4k^2\omega^2\}}.$$

Thus a particular integral is given by

$$i = \text{r.p. of } \frac{\omega E}{L}\frac{\{(n^2 - \omega^2) - j2k\omega\}}{\{(n^2 - \omega^2)^2 + 4k^2\omega^2\}}e^{j\omega t}$$

$$= \frac{\omega E}{L}\frac{\{2k\omega \sin \omega t + (n^2 - \omega^2)\cos \omega t\}}{\{(n^2 - \omega^2)^2 + 4k^2\omega^2\}}.$$

This can be put in the form $I \sin(\omega t + \phi)$, giving as the steady-state current

$$i = \frac{\omega E}{L\{(n^2 - \omega^2)^2 + 4k^2\omega^2\}^{\frac{1}{2}}}\sin(\omega t + \phi), \tag{7}$$

where $$\tan \phi = \frac{(n^2 - \omega^2)}{2k\omega}.$$

The amplitude, I, of this steady-state current is given by

$$I = \frac{\omega E}{L\{(n^2 - \omega^2)^2 + 4k^2\omega^2\}^{\frac{1}{2}}}, \tag{8}$$

and the phase lead over the impressed e.m.f. is given by

$$\phi = \tan^{-1}\frac{(n^2 - \omega^2)}{2k\omega}. \tag{9}$$

The frequency, $\omega/2\pi$, is the same as that of the impressed e.m.f.

If k and n, and therefore L, R and C, are kept constant and ω is varied, the amplitude I will vary. Its greatest value will occur when $\dfrac{\omega}{\{(n^2 - \omega^2)^2 + 4k^2\omega^2\}^{\frac{1}{2}}}$ is a maximum, that is when $\dfrac{1}{\left\{\left(\dfrac{n^2}{\omega}-\omega\right)^2+4k^2\right\}^{\frac{1}{2}}}$ is a maximum (dividing numerator and denominator by ω). This will occur when the denominator is least, that is, when $\dfrac{n^2}{\omega}-\omega=0$, or $n=\omega$.

The maximum value of the steady-state current is thus $E/2kL$ and occurs when $\omega=n=1/\sqrt{(LC)}$.

As $k=R/2L$, the maximum amplitude is E/R and occurs when $\omega=1/\sqrt{(LC)}$.

Resonance is then said to occur. It will be noted that the circuit behaves as a pure resistance and the current is in phase with the applied voltage $\left(\phi=\tan^{-1}\dfrac{0}{2k\omega}=0\right)$. The current is in fact $\dfrac{E}{R}\sin\omega t$ and the frequency $\dfrac{\omega}{2\pi}=\dfrac{1}{2\pi\sqrt{(LC)}}$. At resonance, therefore, the effects of the inductance and capacitance are absent, having cancelled each other.

The student will have dealt with this before, but from a different aspect. The vector method, with use of the operator j (and calculation of impedances), will already be familiar to him in his electrical technology lectures. It may be pointed out that this method does give effectively the steady-state condition which has been reached here from the purely mathematical point of view as a particular integral of a differential equation. However, the vector method does not give the form of the transient, which may, in some special cases, be important.

Equation (8) may be written

$$I=\frac{E}{L\left\{\left(\dfrac{n^2}{\omega}-\omega\right)^2+4k^2\right\}^{\frac{1}{2}}}.$$

Putting in the values $k=R/2L$, $n^2=1/LC$, this becomes

$$\frac{E}{L\left\{\left(\dfrac{1}{\omega CL}-\omega\right)^2+\dfrac{R^2}{L^2}\right\}^{\frac{1}{2}}}=\frac{E}{\left\{\left(\dfrac{1}{\omega C}-\omega L\right)^2+R^2\right\}^{\frac{1}{2}}}. \tag{10}$$

Also equation (9) gives the phase lead as

$$\phi = \tan^{-1}\left(\frac{1}{\omega C} - \omega L\right)\Big/ R. \tag{11}$$

Equation (11) serves to show that if R is small the phase angle may approach $\frac{1}{2}\pi$.

Fig. 39 shows a sketch of the amplitude of the steady-state current against the value of ω for an applied e.m.f. of amount

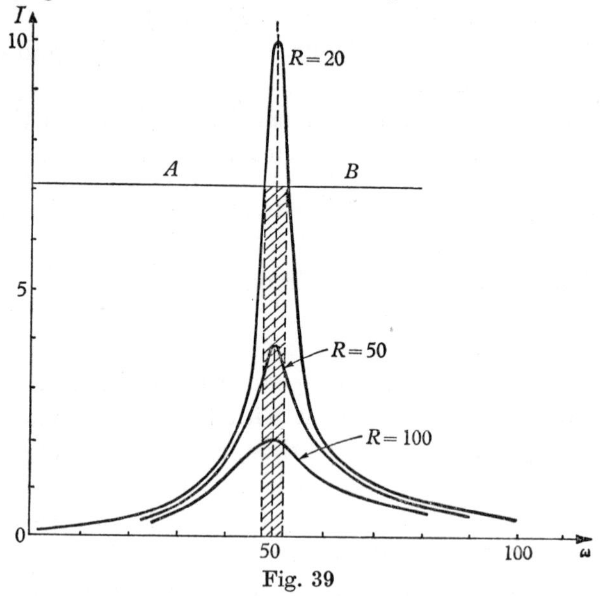

Fig. 39

200 sin ωt, when $C = 0{\cdot}00008$ and $L = 5$. At resonance $\omega = 50$. The curves are for three values of the resistance R.

It is seen that with a smallish resistance the effect of resonance is very pronounced; the current has a sharp peak and falls off in value very steeply for values of ω not near resonance. As the value of the resistance is increased the current maximum falls and the curve flattens out until no sharp peak is present. A series L-C-R circuit in which R is small is thus very sensitive to resonance. This property has important practical uses.

Curves such as those in fig. 39 are called *resonance curves*. If a horizontal line such as AB is drawn to cut any particular curve, the range of periodicity cut off is called the *pass band*.

Normally it is defined for a height of the horizontal AB of $E/\sqrt{2R}$. This has been done in fig. 39.

For the values of ω, ω_1 and ω_2 say, corresponding to the pass band width:

$$\frac{E}{\sqrt{2R}} = \frac{E}{\left\{\left(\frac{1}{\omega C} - \omega L\right)^2 + R^2\right\}^{\frac{1}{2}}},$$

giving
$$2R^2 = R^2 + \left(\frac{1}{\omega C} - \omega L\right)^2,$$

$$R^2 = \left(\frac{1}{\omega C} - \omega L\right)^2.$$

$$\therefore \quad \frac{1}{\omega C} - \omega L = \pm R.$$

Thus
$$\omega^2 LC \pm \omega CR - 1 = 0.$$

This gives
$$\omega = \pm \frac{CR \pm \sqrt{(C^2R^2 + 4LC)}}{2LC}.$$

For a selective circuit R is small. Neglecting the term in R^2 gives

$$\omega = \pm \frac{CR \pm 2\sqrt{(LC)}}{2LC} = \pm \frac{R}{2L} \pm \frac{1}{\sqrt{(LC)}}.$$

Now $1/\sqrt{(LC)} = n$ at resonance.

$$\left.\begin{aligned} \therefore \quad \omega_2 &= +\frac{R}{2L} + n \\ \text{and} \quad \omega_1 &= -\frac{R}{2L} + n \end{aligned}\right\} \quad \text{taking} \quad +\frac{1}{\sqrt{(LC)}}.$$

The band width is $\underline{\omega_2 - \omega_1 = R/L.}$

The smaller the value of R, the sharper the peak of the resonance curve. The selectivity of the circuit is therefore inversely proportional to R. It is arbitrarily defined as nL/R, where n is the resonant period. For *any* period ω the expression $\omega L/R$ is called the Q of the circuit.

8·33. An example of a parallel circuit

The case considered is that of an inductance L and resistance R in one branch and a capacitance C in the other. An alternating e.m.f. of amount $E \sin \omega t$ is applied. Fig. 40 shows a diagram of the circuit.

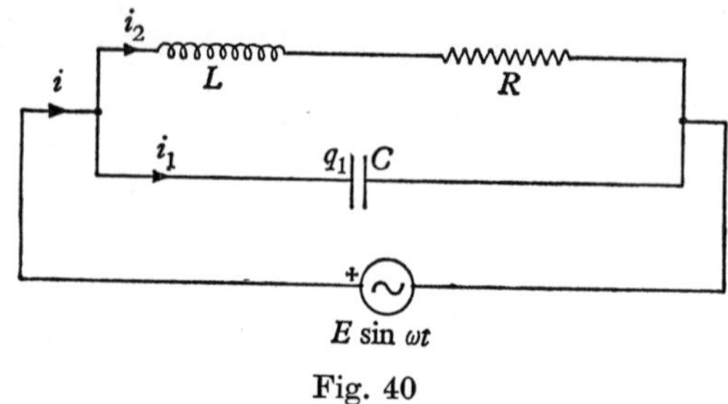

Fig. 40

Let i_1, i_2 be the currents flowing in each branch at time t and q_1 be the charge on the capacitor at this time: then $q_1 = C.E \sin \omega t$ (from $q = CV$).

Now,
$$i_1 = \frac{dq_1}{dt}.$$

$$\therefore i_1 = \omega CE \cos \omega t. \tag{1}$$

The voltage across the top branch is also $E \sin \omega t$, thus

$$L\frac{di_2}{dt} + Ri_2 = E \sin \omega t,$$

giving
$$\left(D + \frac{R}{L}\right)i_2 = \frac{E}{L} \sin \omega t, \tag{2}$$

where
$$D \equiv \frac{d}{dt}.$$

Solving equation (2)

A.E.
$$m + \frac{R}{L} = 0, \quad m = -\frac{R}{L}.$$

C.F.
$$i_2 = B.e^{-Rt/L}. \tag{3}$$

This dies away with time and is a transient. The particular integral therefore gives the steady-state value of i_2.

P.I. Treat $(E/L) \sin \omega t$ as the imaginary part of $(E/L)\, e^{j\omega t}$ and put $i_2 =$ i.p. of $Ae^{j\omega t}$ in (2).

Then
$$\left(j\omega + \frac{R}{L}\right)Ae^{j\omega t} \equiv \frac{E}{L}\,e^{j\omega t},$$

giving
$$A = \frac{E}{L(j\omega + R/L)} = \frac{E(R - j\omega L)}{(R^2 + \omega^2 L^2)}.$$

A particular integral is given by

$$i_2 = \text{i.p. of } \frac{E(R - j\omega L)}{(R^2 + \omega^2 L^2)}\,e^{j\omega t}$$

$$= \frac{E}{(R^2 + \omega^2 L^2)}\,(R \sin \omega t - \omega L \cos \omega t), \qquad (4)$$

i.e.
$$i_2 = \frac{E}{\sqrt{(R^2 + \omega^2 L^2)}} \sin(\omega t - \phi), \qquad (5)$$

where
$$\phi = \tan^{-1} \frac{\omega L}{R}. \qquad (6)$$

But the total current $i = i_1 + i_2$, therefore the steady-state value of i is given by

$$i = \omega CE \cos \omega t + \frac{E}{\sqrt{(R^2 + \omega^2 L^2)}} \sin(\omega t - \phi). \qquad (7)$$

If $R = 0$, that is, if only L and C are in parallel, then from (6), $\phi = \tan^{-1} \infty = \frac{1}{2}\pi$ and equation (7) gives

$$i = \omega CE \cos \omega t + \frac{E}{\omega L} \sin(\omega t - \tfrac{1}{2}\pi)$$

$$= E\left(\omega C - \frac{1}{\omega L}\right) \cos \omega t.$$

Thus, if $\omega C = 1/\omega L$, that is, $\omega = 1/\sqrt{(LC)}$, the total current flowing is zero although the currents in the individual branches are not zero. This condition is called current resonance.

It is obvious, then, that a parallel circuit can be tuned to *reject* one certain frequency, whilst a series circuit can be tuned to *accept* only one certain frequency.

It may be noted that the differential equation (2) is of first-order linear type which can be solved by the usual method (see vol. I, § 13·24). Here it has been solved by the general method for linear equations with constant coefficients.

8·34. Some numerical worked examples

(1) Find a complete expression for the value of the current flowing in a circuit consisting of a resistance R in series with a capacitance C due to the application at time $t = 0$ of a voltage $V \sin \omega t$.

If the values of R, C and V are 500 ohms, 40 microfarads and 200 volts in that order, and the frequency is 50 Hz, find the value of the current when $t = 0{\cdot}03$ sec. [L.U.]

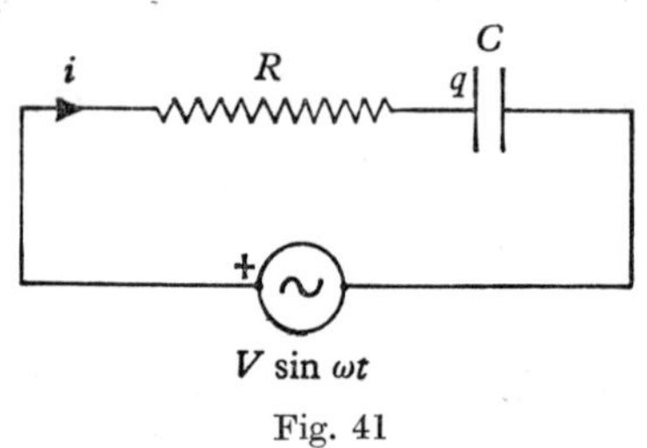

Fig. 41

As a 'complete expression' is required for the current, this means that the vector method of finding the impedance, etc., is not sufficient. The transient must be found.

Let i be the current flowing and q the charge on the capacitor, at time t (see fig. 41).

Then $$Ri + \frac{q}{C} = V \sin \omega t. \tag{1}$$

Differentiating equation (1) and using $i = dq/dt$:

$$R\frac{di}{dt} + \frac{i}{C} = \omega V \cos \omega t.$$

Solving this equation:

A.E. $$Rm + \frac{1}{C} = 0, \quad m = -\frac{1}{RC}.$$

C.F. $$\underline{i = B.e^{-t/RC}.} \tag{2}$$

P.I. Treat $\omega V \cos \omega t$ as the real part of $\omega V e^{j\omega t}$ and put $i =$ r.p. of $Ae^{j\omega t}$.

Then $$\left(Rj\omega + \frac{1}{C}\right)Ae^{j\omega t} \equiv \omega V e^{j\omega t}.$$

$$\therefore\ A = \frac{\omega VC}{1 + j\omega CR} = \frac{\omega VC(1 - j\omega CR)}{(1 + \omega^2 C^2 R^2)}.$$

Thus a particular integral is

$$i = \text{r.p. of } \frac{\omega VC(1 - j\omega CR)}{(1 + \omega^2C^2R^2)} e^{j\omega t}$$

$$= \frac{\omega VC(\cos \omega t + \omega CR \sin \omega t)}{(1 + \omega^2C^2R^2)} . \qquad (3)$$

A complete expression for the current is

$$i = Be^{-t/CR} + \frac{\omega VC(\cos \omega t + \omega CR \sin \omega t)}{(1 + \omega^2C^2R^2)} .$$

Now at $t = 0$, $q = 0$, therefore from (1), $i = 0$.

Thus $$0 = B + \frac{\omega VC}{(1 + \omega^2C^2R^2)} .$$

$$\therefore\ i = \frac{\omega VC}{(1 + \omega^2C^2R^2)} (\cos \omega t + \omega CR \sin \omega t - e^{-t/CR}). \quad (4)$$

If $R = 500, C = 40 \times 10^{-6}, V = 200$ and $\omega = 2\pi f = 100\pi$, then

$$\omega VC = 100\pi \times 200 \times 40 \times 10^{-6} = \tfrac{4}{5}\pi,$$

$$\omega CR = 100\pi \times 40 \times 10^{-6} \times 500 = 2\pi,$$

and $$\frac{1}{CR} = \frac{10^6}{40 \times 500} = \frac{100}{2} = 50 .$$

Equation (4) then gives

$$i = \frac{4\pi}{5(1 + 4\pi^2)} (\cos 100\pi t + 2\pi \sin 100\pi t - e^{-50t}).$$

Substituting $t = 0{\cdot}03$ gives

$$i = \frac{4\pi}{5(1 + 4\pi^2)} (\cos 3\pi + 2\pi \sin 3\pi - e^{-1\cdot5}),$$

i.e. $$i \simeq 0{\cdot}0621\ (-1 + 0 - 0{\cdot}2231)$$

$$\simeq -\ 0{\cdot}076.$$

The current after 0·03 sec. is approx. 76 milliamp.

(2) An electric circuit consists of a generator voltage $200 \sin 100t$, an open switch, a resistance of 300 ohms, an inductance of 1 henry and an uncharged capacitor of 50 microfarads capacitance, all in series. At time $t = 0$ the switch is closed. If q is the charge on a plate of the capacitor at time t, show that

$$\frac{d^2q}{dt^2} + 300\frac{dq}{dt} + 20{,}000q = 200 \sin 100t.$$

Hence obtain a complete expression for the current flowing at time t. [L.U.]

The voltage drops across the inductance, resistance and capacitance are respectively $1\frac{di}{dt}$, $300i$ and $\frac{q}{50 \times 10^{-6}}$.

The back e.m.f. equation is therefore

$$\frac{di}{dt} + 300i + \frac{q}{50 \times 10^{-6}} = 200 \sin 100t.$$

Now $$i = \frac{dq}{dt}.$$

$$\therefore \frac{d^2q}{dt^2} + 300\frac{dq}{dt} + 20{,}000q = 200 \sin 100t. \quad (1)$$

A.E. $$m^2 + 300m + 20{,}000 = 0,$$

$$m = -150 \pm \{150^2 - 20{,}000\}^{\frac{1}{2}}$$
$$= -150 \pm 50$$
$$= -100 \quad \text{or} \quad -200.$$

C.F. $$q = Be^{-100t} + Ce^{-200t}. \quad (2)$$

P.I. Treat $200 \sin 100t$ as the imaginary part of $200e^{j100t}$ and put $q =$ i.p. of Ae^{j100t} in (1).

Then $$(-10{,}000 + j30{,}000 + 20{,}000)Ae^{j100t} = 200e^{j100t},$$

giving $$A = \frac{2}{(100 + j300)}.$$

Thus a particular integral is given by

$$q = \text{i.p. of } \frac{1}{(50 + j150)} e^{j100t}$$

$$= \text{i.p. of } \frac{1}{50} \frac{(1 - j3)}{10} (\cos 100t + j \sin 100t),$$

i.e.
$$q = \frac{1}{500} (\sin 100t - 3 \cos 100t). \qquad (3)$$

G.S. $q = Be^{-100t} + Ce^{-200t} + \frac{1}{500} (\sin 100t - 3 \cos 100t).$ (4)

$$\therefore\ i = \frac{dq}{dt} = -100Be^{-100t} - 200Ce^{-200t} + \tfrac{1}{5} (3 \sin 100t + \cos 100t). \qquad (5)$$

Now when t $= 0$, $q = 0 = i$.

Therefore from (4) and (5)

$$0 = B + C - \tfrac{3}{500},$$

and
$$0 = -100B - 200C + \tfrac{1}{5}.$$

These give $B = \frac{1}{100}, \quad C = -\frac{1}{250}.$ (6)

The current flowing at time t is given (from (5) and (6)) by

$$\underline{i = \tfrac{4}{5}e^{-200t} - e^{-100t} + \tfrac{1}{5}(3 \sin 100t + \cos 100t).}$$

Exercise 30

1. An alternating e.m.f. of amount $E \sin \omega t$ is applied to a circuit consisting of a capacitance C and a resistance R in series. Show that in the steady state the current i is given by

$$i = \frac{E}{\sqrt{\left(R^2 + \dfrac{1}{\omega^2 C^2}\right)}} \sin (\omega t + \phi), \quad \text{where} \quad \tan \phi = \frac{1}{\omega CR}.$$

2. A circuit contains inductance L and resistance R in series. At $t = 0$, when the instantaneous current is i_0, a constant voltage V is applied for a time τ, after which the voltage remains zero. Form the differential equation of the circuit, determine the current for $0 < t < \tau$ and show that for $t > \tau$

$$i = \left[i_0 + \frac{V}{R}(e^{R\tau/L} - 1)\right] e^{-Rt/L}.$$

Determine the limiting form of the solution for $t > \tau$ when τ and V are changed in such a way that $\tau \to 0$ while $V\tau$ remains constant and equal to J. [I.E.E.]

3. Derive the differential equation showing the time variation of current in a circuit containing resistance and inductance, to which a sinusoidal voltage is suddenly applied.

Show that there are initial switching conditions under which there is no switching transient and draw a graph of the current waveform under these conditions. [L.U.]

4. Derive an expression from which the instantaneous current flowing in a circuit of inductance L and resistance R may be calculated at any time t after applying a voltage $V \cos(\omega t + \phi)$.

From the expression find the ratio of the maximum value to which the current rises to the steady-state maximum value, when the voltage is applied at the instant when it is zero, taking $R = 20$ ohms, $L = 0{\cdot}1$ henry, and $\omega = 1000\pi$. [L.U.]

5. An alternating e.m.f. $E \sin \omega t$ is supplied to a circuit containing inducance L, resistance R and capacitance C in series. Obtain the differential equation satisfied by the current i. Find the resistance if it is just large enough to prevent natural oscillations. For this value of R and $\omega = (LC)^{-\frac{1}{2}}$ prove that $i = (E/2k)(\sin \omega t - \omega t e^{-\omega t})$, where $k^2 = L/C$, when the current and charge on the capacitor are both zero at time $t = 0$. [L.U.]

6. An uncharged capacitor of capacitance C is charged by applying an e.m.f. of $E \sin\left(\frac{1}{\sqrt{(LC)}} t\right)$ through leads of self-inductance L and negligible resistance. Prove that at time t the charge on one of the plates is

$$\frac{1}{2} EC\left[\sin \frac{t}{\sqrt{(LC)}} - \frac{t}{\sqrt{(LC)}} \cos \frac{t}{\sqrt{(LC)}}\right].$$

If, in addition, there is a small resistance, in what respect is the mathematical form of the above result altered? [L.U.]

7. An alternating e.m.f. of amplitude E and frequency $\omega/2\pi$ is supplied to a coil of inductance L and resistance R.

Write down the differential equation for the current in the coil and solve it, indicating the transient term.

If the coil is shunted by a capacitor of capacitance C and resistance S, show that the circuit can be replaced, as far as the permanent current is concerned, by a non-inductive resistance, provided that $CR^2 - L = \omega^2 CL(CS^2 - L)$.

[L.U.]

8. The terminals of a generator, producing a variable voltage $V(t)$, are connected through a capacitor of capacitance C, a wire of resistance R and a coil of self-inductance L in series. Show that the charge q on the capacitor satisfies the equation

$$LC\frac{d^2q}{dt^2} + RC\frac{dq}{dt} + q = C\,.\,V(t).$$

Given that $2L = CR^2$, $V = 0$ for $t < 0$, $V = E$ for $0 < t < \pi/n$, $V = 0$ for $t > \pi/n$, where E is constant and $RCn = 1$, show that the current I flowing round the circuit is given by $I = ([2E/R]\,e^{-nt})\sin nt$ for $0 < t < \pi/n$. Find q for $t > \pi/n$. [L.U.]

9. If a 50 Hz generator has a source impedance equivalent to an inductance of 31·8 mH in series with a resistance of 0·1 Ω, and its generated e.m.f. is sinusoidal with an r.m.s. value of 70·7 V, find approximately the maximum peak value of the current when a short-circuit is suddenly applied.

Answers

2. $i = (i_0 - [V/R])\,e^{-Rt/L} + V/R;\ 0 < t < \tau;$

$i \to (i_0 + [J/L])\,e^{-Rt/L}.$

Note. This is an example of an *impulsive* voltage.

3. If L, R and $E \sin \omega t$ are the inductance, resistance and applied voltage, the differential equation is

$$\frac{di}{dt} + \frac{R}{L} i = \frac{E}{L} \sin \omega t.$$

The full solution is

$$i = \frac{E}{\sqrt{(R^2 + \omega^2 L^2)}} \sin (\omega t - \alpha) + Ae^{-Rt/L},$$

where $\tan \alpha = \omega L/R$.

If the switch is closed when the voltage has the value $E \sin \alpha$, then the value of the constant A is zero and there is no switching transient.

4. $i = \dfrac{V}{\sqrt{(R^2 + \omega^2 L^2)}} \sin (\omega t + \phi + \alpha) + Ae^{-Rt/L}$, where $\tan \alpha = R/\omega L$ and the value of A depends on the initial conditions.

Ratio is 1·8 (approx.). [Max. value of i occurs when $\omega t \simeq \pi$.]

5. $R = 2\sqrt{(L/C)}$ or $2k$.

6. Neglecting terms in R^2 as R is small, the new value of q is

$$E \,.\, e^{-Rt/2L} \left(\frac{\sqrt{(LC)}}{R} \cos \frac{t}{\sqrt{(LC)}} + \frac{C}{2} \sin \frac{1}{\sqrt{(LC)}} t \right) - \frac{E\sqrt{(LC)}}{R} \cos \frac{t}{\sqrt{(LC)}}.$$

There is now a transient term and no term with t as a factor; thus the charge does not increase steadily with time.

7. $i = \dfrac{E}{(R^2 + \omega^2 L^2)} (R \sin \omega t - \omega L \cos \omega t) + Ae^{-Rt/L}$.

$Ae^{-\frac{Rt}{L}}$ is the transient term.

8. $q = (E/nR)(e^{\pi} + 1)e^{-nt}(\cos nt + \sin nt)$.

9. About 19·7 amperes.

8·4. The stability of oscillatory circuits

8·41. Introduction

The student should acquire a working knowledge of determinants from Chapter 13 before proceeding with this section.

In the practical cases reviewed so far most of the resulting linear differential equations have given auxiliary equations having real negative roots or complex roots with negative real parts. These have given rise to exponential terms with negative indices in the solution (cf. $e^{-t/CR}$ in equation (4) of § 8·34). These 'transient' terms have died away exponentially or have represented damped oscillations which are very soon negligible. If, on the other hand, any of the roots of the auxiliary equation are positive or have positive real parts, the solution will contain a term (or terms) which increases its amplitude exponentially. Any circuit or system which gives rise to such terms is called *unstable*; one whose auxiliary equation has *all* roots with negative real parts is called *stable*.

This question of stability is important. For example, systems such as servo-mechanisms and amplifier circuits are characterized by the fact that a small 'input' controls the supply of energy to a large 'output'. It is important for such systems to be stable, otherwise oscillations of large amplitude may be set up by any accidental disturbance. On the other hand some systems are required to be unstable. For example, in a triode-valve oscillator circuit large oscillations *are* required. From the (approximate) linear equation for the system the circuit parameters are chosen to give instability, so that large oscillations are produced.

8·42. Tests for stability

As explained above, the criterion for stability is that *all* the roots of the auxiliary equation (often called the subsidiary equation when referring to servo-mechanisms) have *negative real parts*. Tests for this are simple if the auxiliary equation is a quadratic. Thus the roots of $a_0m^2 + a_1m + a_2 = 0$, arranged so that a_0 is positive, has roots

$$\frac{-a_1 \pm \sqrt{(a_1{}^2 - 4a_0a_2)}}{2a_0}$$

and are therefore real and negative if $a_1 > 0$, $a_2 > 0$ and $a_1^2 > 4a_0a_2$, and have negative real parts if $a_1 > 0$, $a_2 > 0$ and $a_1^2 < 4a_0a_2$. In either case stability exists if

$$a_1 > 0, \quad a_2 > 0. \tag{1}$$

Obviously with auxiliary equations of higher orders it is impractical to attempt to find the values of all the roots.

Tests for the general nth order case have been developed. They are often called the *Hurwitz-Routh stability conditions*. The proof of these conditions is beyond the scope of this book; they will merely be stated.

Before giving these conditions a simple necessary condition for stability will be obtained. The nth order auxiliary equation can be written

$$a_0m^n + a_1m^{n-1} + a_2m^{n-2} + \dots + a_{n-1}m + a_n = 0, \tag{2}$$

where it will be supposed that the equation has been arranged so that $a_0 > 0$. Now suppose that the left-hand side has been resolved into its linear and quadratic factors (this is theoretically possible):

$$a_0(m+b_1)(m+b_2)\dots(m^2+c_1m+d_1)(m^2+c_2m+d_2)\dots = 0. \tag{3}$$

For stability $b_1, b_2, \dots$ must be positive and, from (1) above $c_1, d_1, c_2, d_2, \dots$ must all be positive. Multiplying out the left-hand side of (3), it follows that $a_1, a_2, a_3, \dots, a_n$ must all be positive. This is a very simple condition to apply, but unfortunately it is only a necessary condition; it is not both necessary and sufficient. It merely gives the information that if $a_1, a_2, \dots, a_n$ are *not* all positive, the system cannot be stable. An equation with one or more coefficients negative has at least one root with a positive real part, and the corresponding system is unstable. If all the coefficients are positive, the system may be stable or unstable. The complete necessary and sufficient conditions (the Hurwitz-Routh conditions) are,

$$\text{given } a_0 > 0, \tag{4}$$

then *all of the following determinants must be positive:*

$$|a_1| \quad \begin{vmatrix} a_1 & a_0 \\ a_3 & a_2 \end{vmatrix}, \quad \begin{vmatrix} a_1 & a_0 & 0 \\ a_3 & a_2 & a_1 \\ a_5 & a_4 & a_3 \end{vmatrix}, \quad \begin{vmatrix} a_1 & a_0 & 0 & 0 \\ a_3 & a_2 & a_1 & a_0 \\ a_5 & a_4 & a_3 & a_2 \\ a_7 & a_6 & a_5 & a_4 \end{vmatrix}, \tag{5}$$

$$\begin{vmatrix} a_1 & a_0 & 0 & 0 & \cdots & \cdots & 0 \\ a_3 & a_2 & a_1 & a_0 & 0 & \cdots & 0 \\ a_5 & a_4 & a_3 & a_2 & a_1 & \cdots & 0 \\ \vdots & \vdots & \vdots & \vdots & \vdots & \vdots & \vdots \\ a_{2n-1} & a_{2n-2} & a_{2n-3} & a_{2n-4} & a_{2n-5} & \cdots & a_n \end{vmatrix}. \tag{6}$$

Note that in writing down these determinants for any particular case, the element a_r is replaced by zero if $r > n$.

It can be shown that these conditions include the condition that $a_1, a_2, \ldots, a_n$ are all positive.

As an example, the conditions for a fourth-order equation ($n = 4$) would be, with $a_0 > 0$,

$$\begin{vmatrix} a_1 & a_0 & 0 & 0 \\ a_3 & a_2 & a_1 & a_0 \\ 0 & a_4 & a_3 & a_2 \\ 0 & 0 & 0 & a_4 \end{vmatrix} > 0, \quad \begin{vmatrix} a_1 & a_0 & 0 \\ a_3 & a_2 & a_1 \\ 0 & a_4 & a_3 \end{vmatrix} > 0, \quad \begin{vmatrix} a_1 & a_0 \\ a_3 & a_2 \end{vmatrix} > 0, \quad a_1 > 0.$$

8·43. Worked example of a simple servo-mechanism

Fig. 42 shows a block diagram of a simple servo-mechanism. A heavy rotor R is to be turned about its axis so that it always follows the motion of a pointer P. θ_i is the angular displacement of the pointer and is called the input displacement. It is a given function of the time. For this example θ_i will be taken as equal to ωt, ω a constant. The corresponding displacement of the rotor is θ_0 and is called the output displacement. The value of $\theta_0 - \theta_i$ is for obvious reasons called the error. $\theta_i - \theta_0$ is called the correction. (In some books $\theta_i - \theta_0$ is defined to be the error, but this does not follow the normal practice as applied to standard measuring instruments.)

The torque T, which drives the rotor, is provided by a motor. The correction, $\theta_i - \theta_0$, is fed into a controlling device which controls the torque provided by the motor. T is thus a function of $\theta_i - \theta_0$. An amplification stage may also be

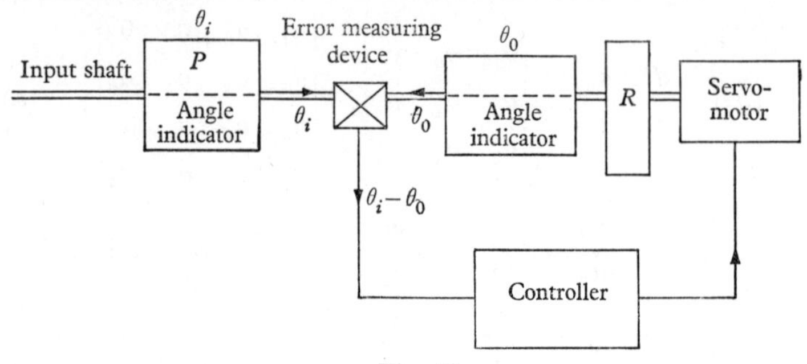

Fig. 42

included in this part of the loop. The torque T may be quite complicated, but in this case will be taken to be

$$T = \lambda(\theta_i - \theta_0) + \lambda_1 \int_0^t (\theta_i - \theta_0)\, dt, \tag{1}$$

where λ, λ_1 are positive constants. Thus T has a term proportional to the correction and a term proportional to the time integral of the correction.

Let I be the moment of inertia of the rotor, and let there be a frictional resistance torque of amount $k\dot{\theta}_0$.

The equation of motion of the rotor is then

$$T - k\dot{\theta}_0 = I\ddot{\theta}_0.$$

Using (1), this becomes

$$\lambda(\theta_i - \theta_0) + \lambda_1 \int_0^t (\theta_i - \theta_0)\, dt - k\dot{\theta}_0 = I\ddot{\theta}_0.$$

Differentiating with respect to time and collecting up terms in θ_0 and θ_i on opposite sides of the equation:

$$I\dddot{\theta}_0 + k\ddot{\theta}_0 + \lambda\dot{\theta}_0 + \lambda_1\theta_0 = \lambda\dot{\theta}_i + \lambda_1\theta_i,$$

i.e. $$(ID^3 + kD^2 + \lambda D + \lambda_1)\theta_0 = (\lambda D + \lambda_1)\theta_i. \tag{2}$$

Taking $\theta_i = \omega t$, the particular integral of (2) is $\theta_0 = \omega t$, showing that if the system is stable the rotor follows the pointer faithfully, with no constant lag or lead.

The auxiliary equation is

$$Im^3 + km^2 + \lambda m + \lambda_1 = 0.$$

The conditions for stability, from § 8·42, (5) and (6), are therefore

$$k > 0, \qquad \begin{vmatrix} k & I \\ \lambda_1 & \lambda \end{vmatrix} > 0 \qquad \text{and} \qquad \begin{vmatrix} k & I & 0 \\ \lambda_1 & \lambda & k \\ 0 & 0 & \lambda_1 \end{vmatrix} > 0.$$

Now k, λ_1, λ and I are all positive, and the conditions reduce to $\underline{k\lambda > I\lambda_1}$.

This sets a limit to the values of λ and λ_1.

Note. (1) The above example is an extremely simplified version of a servo-mechanism. In practice the controller is likely to be a complicated amplifier circuit and the 'control' equation is not likely to be as simple as equation (1) above.

(2) The tests for stability are of little use in deciding the *degree* of stability of the system. The degree of stability is, however, very important. Obviously a sluggish, overdamped response is just as bad as a highly oscillatory, too lightly damped, response. The methods by which the degree of stability is established are beyond the scope of this book. They can be found in specialist text-books on servo-mechanisms.

Exercise 31

1. If the subsidiary equation for a certain servo-mechanism is

$$\frac{T^2}{\omega^2} D^4 + \frac{2T}{\omega^2} D^3 + \frac{1}{\omega^2} D^2 + \frac{2}{\omega} D + 1 = 0,$$

find the condition for stability.

2. In the worked example of § 8·43 above, find the condition for stability if the torque were given by $T = -\lambda(\theta_0 - \theta_i)$. Also show that in this case, for large values of the time, θ_0 lags behind θ_i by $k\omega/\lambda$.

3. A simple remote-position-control servo-mechanism has the following subsidiary equation:

$$\frac{T}{\omega_0^2} D^3 + \frac{1}{\alpha\omega_0^2} D^2 + TD + 1 = 0, \tag{1}$$

where T is a time constant.

Find the restriction on the value of α for stability.

If the term $\frac{T}{\omega_0^2} D^3$ of equation (1) above is negligible and $\alpha = \frac{1}{25}$, $\omega_0 = 60\,\text{rad/s}$, find the value of T which would make the transient response critically damped. (Take this as being the value of T for which equation (1) has equal roots.)

4. Fig. 43 shows a simple triode oscillator.

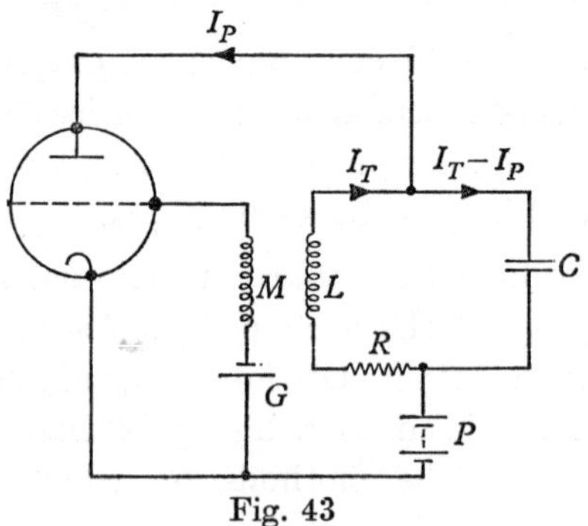

Fig. 43

G is the grid-voltage supply and P the plate-voltage supply. I_T is the current in the inductance and I_P the total plate current. Assuming that the current to the grid is negligible and that the characteristic of the valve is linear, it can be shown that the differential equation for I_T is

$$[LCD^2 + \{RC + g(L + \mu M)\}D + (1 + Rg)]I_T = g(P + \mu G) + K,$$

where K is a constant and μ, g are the usual valve characteristics.

The condition for the valve to act as a generator of oscillations is that the oscillations represented by the complementary function of this equation are unstable. Show that this is so if $M < -\frac{L}{\mu} - \frac{RC}{g\mu}$.

(*Note.* (i) This condition can be satisfied, as M, the mutual inductance, can be negative.

(ii) the P.I. of the differential equation represents the direct current to the plate.)

ANSWERS

1. $T\omega < 1$ and $T\omega < 0{\cdot}5$; both are satisfied if $T\omega < 0{\cdot}5$.

2. $k > 0$, $\lambda > 0$, which are in this case inherently satisfied. The particular integral, with $\theta_i = \omega t$, is $\theta_0 = \omega t - k\omega/\lambda$.

3. (i) $T/\alpha\omega_0^2 > T/\omega_0^2$, giving $\alpha < 1$.

(ii) $T = \frac{1}{6}$ sec.

4. For instability at least one of the roots of the auxiliary equation must have positive real part. Thus if the auxiliary equation is $a_0m^2 + a_1m + a_2 = 0$, $a_0 > 0$, the condition is that $a_1 < 0$.

CHAPTER 9

SIMULTANEOUS LINEAR DIFFERENTIAL EQUATIONS AND SOME APPLICATIONS

9·1. Method of solution

If x and y are both functions of a third variable t and $D \equiv d/dt$, and if

$$(a_1 D + b_1)x + (c_1 D + d_1)y = f_1(t),$$

$$(a_2 D + b_2)x + (c_2 D + d_2)y = f_2(t)$$

have to be satisfied by the same general solutions for x and y, they are said to be a pair of simultaneous differential equations. If each equation is linear with constant coefficients, y or x can be eliminated as in ordinary algebra to give a linear equation with only one dependent variable.

EXAMPLE 1

Solve the simultaneous equations

$$\frac{dx}{dt} + 6\frac{dy}{dt} + x = 0, \quad -3\frac{dx}{dt} + 2\frac{dy}{dt} + 2y = 0,$$

being given that $x = 0$ and $y = a$ when $t = 0$. [L.U.]

Writing $D \equiv d/dt$,

$$(D+1)x + 6Dy = 0, \tag{1}$$

$$-3Dx + 2(D+1)y = 0. \tag{2}$$

Eliminate y by operating on (1) by $(D+1)$ and on (2) by $3D$ and then subtracting:

$$(D+1)^2 x + 6D(D+1)y = 0,$$

$$\underline{-9D^2 x + 6D(D+1)y} = \underline{0},$$

$$(D+1)^2 x + 9D^2 x \qquad = 0,$$

giving

$$(10D^2 + 2D + 1)x = 0. \tag{3}$$

Solving equation (3):

A.E. $$10m^2 + 2m + 1 = 0.$$

$$\therefore m = \frac{-1 \pm j3}{10} = -0{\cdot}1 \pm j0{\cdot}3.$$

G.S. $$\underline{x = e^{-0\cdot 1t}(A \cos 0{\cdot}3t + B \sin 0{\cdot}3t).} \qquad (4)$$

Returning to equations (1) and (2), x could now be eliminated in a similar manner and the resulting second-order equation for y solved to give y. However, this would involve a further two arbitrary constants, say C and E, and a relationship between A, B, C and E would have to be found. This can be avoided by using equations (1) and (2) to obtain y in terms of x and its derivatives; then equation (4) can be used to find y, keeping the original arbitrary constants A and B only. Thus:

Multiply (2) by 3 and subtract from (1), thereby eliminating Dy:

$$(D+1)x + 6Dy = 0,$$
$$\underline{-9Dx + 6(D+1)y} = \underline{0},$$
$$(D+1)x + 9Dx - 6y = 0.$$

This gives $$6y = (10D + 1)x$$

or $$y = \tfrac{1}{6}(10D + 1)x.$$

Using (4):

$$Dx = e^{-0\cdot 1t}(0{\cdot}3B \cos 0{\cdot}3t - 0{\cdot}3A \sin 0{\cdot}3t) - 0{\cdot}1e^{-0\cdot 1t}(A \cos 0{\cdot}3t + B \sin 0{\cdot}3t)$$
$$= e^{-0\cdot 1t}\{(0{\cdot}3B - 0{\cdot}1A) \cos 0{\cdot}3t - (0{\cdot}3A + 0{\cdot}1B) \sin 0{\cdot}3t\}.$$

Thus $$y = \tfrac{1}{6}[e^{-0\cdot 1t}\{(3B - A) \cos 0{\cdot}3t - (3A + B) \sin 0{\cdot}3t\} + e^{-0\cdot 1t}(A \cos 0{\cdot}3t + B \sin 0{\cdot}3t)],$$

giving $$y = \frac{e^{-0\cdot 1t}}{6}(3B \cos 0{\cdot}3t - 3A \sin 0{\cdot}3t),$$

i.e. $$\underline{y = \frac{e^{-0\cdot 1t}}{2}(B \cos 0{\cdot}3t - A \sin 0{\cdot}3t).} \qquad (5)$$

Now when $t = 0$, $x = 0$ and $y = a$, therefore from (4) and (5):

$$0 = A \quad \text{and} \quad a = \tfrac{1}{2}B.$$

Thus
$$\left.\begin{aligned} &\underline{x = 2ae^{-0\cdot 1t}\sin 0\cdot 3t} \\ &\underline{y = ae^{-0\cdot 1t}\cos 0\cdot 3t} \end{aligned}\right\}.$$
and

EXAMPLE 2

Solve
$$2\frac{d^2x}{dt^2} + x - y = 3\cos 2t,$$

$$3\frac{d^2y}{dt^2} - x + 2y = 0. \qquad \text{[L.U.]}$$

Writing $D \equiv d/dt$:

$$(2D^2 + 1)x - y = 3\cos 2t, \tag{1}$$

$$-x + (3D^2 + 2)y = 0. \tag{2}$$

Operate on (2) by $(2D^2 + 1)$ and add to (1):

$$\begin{aligned} (2D^2+1)x \qquad\qquad\qquad - y &= 3\cos 2t \\ -(2D^2+1)x + (2D^2+1)(3D^2+2)y &= \underline{\quad 0 \quad}, \\ \hline (2D^2+1)(3D^2+2)y - y \qquad &= 3\cos 2t \end{aligned}$$

giving
$$(6D^4 + 7D^2 + 1)y = 3\cos 2t. \tag{3}$$

Solving equation (3):

A.E.
$$6m^4 + 7m^2 + 1 = 0,$$

$$\therefore (6m^2 + 1)(m^2 + 1) = 0,$$

$$\therefore m^2 = -1 \quad \text{or} \quad -\tfrac{1}{6}.$$

$$\therefore m = \pm j \quad \text{or} \quad \pm j/\sqrt{6}.$$

C.F. $y = A\cos t + B\sin t + C\cos\dfrac{1}{\sqrt{6}}t + E\sin\dfrac{1}{\sqrt{6}}t.$ (4)

P.I. Write $3\cos 2t$ as the real part of $3e^{j2t}$ and put $y =$ r.p. of ae^{j2t} in (3):

$$(6.16 - 7.4 + 1)ae^{j2t} \equiv 3e^{j2t},$$

giving $$a = \tfrac{3}{69} = \tfrac{1}{23}.$$

A particular integral is $y =$ r.p. of $\tfrac{1}{23}e^{j2t}$, i.e.

$$y = \tfrac{1}{23}\cos 2t. \tag{5}$$

G.S. $$\underline{y = A\cos t + B\sin t + C\cos\frac{1}{\sqrt{6}}t}$$
$$\underline{+ E\sin\frac{1}{\sqrt{6}}t + \tfrac{1}{23}\cos 2t.} \tag{6}$$

Without further elimination, equation (2) gives x in terms of y, viz. $$x = (3D^2 + 2)y.$$

Using (6), this gives

$$x = -3A\cos t - 3B\sin t - \frac{3C}{6}\cos\frac{1}{\sqrt{6}}t$$
$$-\frac{3E}{6}\sin\frac{1}{\sqrt{6}}t - \tfrac{12}{23}\cos 2t + 2A\cos t + 2B\sin t$$
$$+ 2C\cos\frac{1}{\sqrt{6}}t + 2E\sin\frac{1}{\sqrt{6}}t + \tfrac{2}{23}\cos 2t,$$

which gives

$$\underline{x = -A\cos t - B\sin t + \tfrac{3}{2}C\cos\frac{1}{\sqrt{6}}t}$$
$$\underline{+ \tfrac{3}{2}E\sin\frac{1}{\sqrt{6}}t - \tfrac{10}{23}\cos 2t.} \tag{7}$$

Points to note. (i) Any of the rules already given for the use of the operator D may be used here.

(ii) The number of arbitrary constants in the general solution is the sum of the orders of the two equations. Thus in Example 1 there were two arbitrary constants and in Example 2 there were four.

(iii) If sufficient initial conditions are given, the constants can be evaluated to fit these conditions.

(iv) When operating on any equation, do not forget to operate on the right-hand side as well as the left-hand side.

Exercise 32

1. Solve the differential equations:

$$\frac{dx}{dt} + 2x + y = 0; \quad \frac{dy}{dt} + x + 2y = 0,$$

subject to the conditions that $x = 1$ and $y = 0$ at $t = 0$. [L.U.]

2. Solve the simultaneous differential equations:

$$\frac{d^2x}{dt^2} + x = \frac{dy}{dt}, \quad 4\frac{dx}{dt} + 2x = \frac{dy}{dt} + 2y.$$

If x, y are the Cartesian coordinates of a point in a plane, and if, when $t = 0$, $x = 0$, $y = 1$, $dx/dt = 2$, prove that the point lies on the parabola $(5x - 2y)^2 = 4(y - 2x)$. [L.U.]

3. Solve the simultaneous differential equations:

$$\frac{dx}{dt} + 3x - 2y = 1, \quad \frac{dy}{dt} - 2x + 3y = e^t,$$

given that when $t = 0$, $x = y = 0$. [L.U.]

4. Solve the equations:

$$\frac{dx}{dt} + y = \sin t, \quad \frac{dy}{dt} + x = \cos t,$$

subject to the conditions $x = 2$, $y = 0$ when $t = 0$. [L.U.]

5. Solve:

$$\frac{dx}{dt} + 6\frac{dy}{dt} + x = 0, \quad -2\frac{dx}{dt} + 2\frac{dy}{dt} - 3y = 0,$$

given that $x = 0$, and $y = a$ when $t = 0$.

6. Determine the solution of the simultaneous equations

$$\frac{dx}{dt} + ax = \omega y, \quad \frac{dy}{dt} + ay = \omega(k - x),$$

which satisfies the conditions $x = y = 0$ at $t = 0$; ω, a and k being constants. [I.E.E.]

7. Determine x and y as functions of t, given that

$$\frac{dx}{dt} = x + y, \quad \frac{dx}{dt} + \frac{dy}{dt} = \cos t - x,$$

and $x = 0$ when $t = 0$, $y = 0$ when $t = \frac{1}{2}\pi$. [L.U.]

8. x and y are functions of t satisfying the simultaneous differential equations:

$$3x + y + \frac{dx}{dt} = 0, \quad 4x + 3y + \frac{dy}{dt} = 0.$$

If at $t = 0$, $x = 0$ and $y = -1$, show that $x = \frac{1}{4}(e^{-t} - e^{-5t})$. Show that the maximum value of x is $5^{-5/4}$, and find the value of y at the time when x has this value. [I.E.E.]

9. Solve the equations:

$$m\frac{du}{dt} = eE - evH; \quad m\frac{dv}{dt} = euH,$$

where m, e, E, H are constants. If $u = dx/dt$ and $v = dy/dt$, find x and y as functions of t. Show that if $x = y = u = v = 0$ when $t = 0$, then

$$x = \frac{E}{\omega H}(1 - \cos \omega t), \quad y = \frac{E}{\omega H}(\omega t - \sin \omega t),$$

where $\omega = eH/m$.

(*Note.* These equations occur in an experiment which determines m/e, the ratio of the mass of an electron to its charge.)

10. Find the solution of the simultaneous equations:

$$3\frac{dx}{dt} + 5x - y = 0, \quad 2x - 3\frac{dy}{dt} - 4y = 0$$

for which $x = 0$ and $y = 3$ at $t = 0$.

Show that both x and y are positive for all positive values of t and find the maximum value of x. [I.E.E.]

Answers

1. $x = \frac{1}{2}(e^{-t} + e^{-3t}), \quad y = \frac{1}{2}(e^{-3t} - e^{-t})$.

2. $x = Ae^{2t} + Be^{t} + Ce^{-t}; \quad y = \frac{5}{2}Ae^{2t} + 2Be^{t} - 2Ce^{-t}$.

With the given conditions:

$$x = 2e^{2t} - 2e^{t}; \quad y = 5e^{2t} - 4e^{t}.$$

It is easy to verify by substitution that these values always satisfy the equation $(5x - 2y)^2 = 4(y - 2x)$.

3. $x = \frac{3}{5} + \frac{1}{6}e^t - \frac{3}{4}e^{-t} - \frac{1}{60}e^{-5t};$

$$y = \frac{2}{5} + \frac{1}{3}e^t - \frac{3}{4}e^{-t} + \frac{1}{60}e^{-5t}.$$

4. $x = e^t + e^{-t}$ or $2 \cosh t, \quad y = \sin t + e^{-t} - e^t.$

5. $x = \frac{18}{13}\, a(e^{-\frac{3}{7}t} - e^{\frac{1}{2}t}); \quad y = \dfrac{a}{13}\,(4e^{-\frac{3}{7}t} + 9e^{\frac{1}{2}t}).$

6. $x = \dfrac{\omega k}{(a^2 + \omega^2)}\,\{e^{-at}(\omega \cos \omega t + a \sin \omega t) + \omega\},$

$$y = \frac{\omega k}{(a^2 + \omega^2)}\,\{- e^{-at}\,(a \cos \omega t + \omega \sin \omega t) + a\}.$$

7. It will shorten the working a little by putting $z = x + y$, though this is not essential.

$$x = \tfrac{1}{2}(1 - \tfrac{1}{2}\pi + t) \sin t,$$

$$y = \tfrac{1}{2}(1 - \tfrac{1}{2}\pi + t) \cos t + \tfrac{1}{2}(\tfrac{1}{2}\pi - t) \sin t.$$

8. $- 3{\cdot}5^{-5/4}$

9. $u = - A \sin (\omega t - \alpha)$, A, α arbitrary,

$$v = A \cos (\omega t - \alpha) + E/H,$$

where $\omega = eH/m$.

$$x = \frac{A}{\omega} \cos (\omega t - \alpha) + B,$$

$$y = \frac{A}{\omega} \sin (\omega t - \alpha) + \frac{Et}{H} + C.$$

10. $x = e^{-t} - e^{-2t}, \quad y = 2e^{-t} + e^{-2t}; \quad$ Max. $x = \frac{1}{4}.$

9·2. Electrical applications

Coupled electrical circuits give rise to simultaneous differential equations. The coupling may be by direct resistance, capacitance or inductance, or by a combination of these three.

Three examples of electrical applications are given below.

9·21. Inductive coupling, transformer type

Fig. 44 shows two L-R circuits coupled by mutual inductance M. An alternating e.m.f. of amount $E \sin \omega t$ is

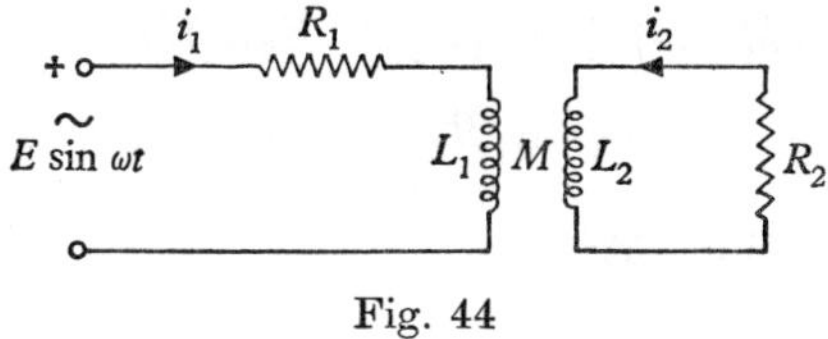

Fig. 44

applied to the first (primary) circuit and the second circuit (the secondary) is closed.

Let i_1, i_2 be the primary and secondary currents at any time t.

Due to the mutual inductance the back e.m.f. in the primary due to the current i_2 in the secondary is of amount $M(di_2/dt)$; the induced e.m.f. in the secondary due to the current i_1 in the primary is $M(di_1/dt)$.

The equations for the two circuits are thus:

Primary.

$$L_1 \frac{di_1}{dt} + R_1 i_1 + M \frac{di_2}{dt} = E \sin \omega t. \tag{1}$$

Secondary.

$$M \frac{di_1}{dt} + L_2 \frac{di_2}{dt} + R_2 i_2 = 0. \tag{2}$$

These two simultaneous equations can be solved by the method of § 9·1.

Writing $$D \equiv \frac{d}{dt},$$

$$(L_1 D + R_1) i_1 + M D i_2 = E \sin \omega t, \tag{3}$$

$$M D i_1 + (L_2 D + R_2) i_2 = 0. \tag{4}$$

In practice it is often the secondary current i_2 which is required. Eliminate i_1 by operating on (3) by MD, on (4) by $(L_1D + R_1)$ and subtracting:

$$\begin{aligned} MD(L_1D + R_1)i_1 + M^2D^2i_2 &= M\omega E \cos \omega t \\ MD(L_1D + R_1)i_1 + (L_1D + R_1)(L_2D + R_2)i_2 &= 0 \\ \hline M^2D^2i_2 - (L_1D + R_1)(L_2D + R_2)i_2 &= M\omega E \cos \omega t. \end{aligned}$$

This reduces to

$$\{(L_1L_2 - M^2)D^2 + (R_1L_2 + R_2L_1)D + R_1R_2\}i_2 = -M\omega E \cos \omega t. \quad (5)$$

Solving equation (5):

A.E.

$$(L_1L_2 - M^2)m^2 + (R_1L_2 + R_2L_1)m + R_1R_2 = 0.$$

In the majority of cases $L_1L_2 > M^2$.

Put
$$\left.\begin{aligned} a &= L_1L_2 - M^2 \ (a \text{ positive}), \\ 2b &= R_1L_2 + R_2L_1 \ (b \text{ positive}), \\ c &= R_1R_2 \ (c \text{ positive}). \end{aligned}\right\} \quad (6)$$

Then
$$am^2 + 2bm + c = 0,$$

giving
$$m = \frac{-b \pm \sqrt{(b^2 - ac)}}{a}.$$

Now as a, c are both positive, $b^2 - ac < b^2$, thus both $-b + \sqrt{(b^2 - ac)}$ and $-b - \sqrt{(b^2 - ac)}$ are negative.

The complementary function will always have, therefore, a predominant factor $e^{-(b/a)t}$ which will cause this part of the general solution to be a transient, dying away with time.

The steady state is given by the particular integral.

P.I. In equation (5) treat $-M\omega E \cos \omega t$ as the real part of $-M\omega Ee^{j\omega t}$ and put i_2 = r.p. of $Ae^{j\omega t}$, then

$$\{-(L_1L_2 - M^2)\omega^2 + (R_1L_2 + R_2L_1)j\omega + R_1R_2\}Ae^{j\omega t} \equiv -M\omega Ee^{j\omega t}.$$

This gives, using (6),

$$\{(c - a\omega^2) + j2b\omega\}A = -M\omega E.$$

$$\therefore\ A = -\frac{\{(c - a\omega^2) - j2b\omega\}M\omega E}{\{(c - a\omega^2)^2 + 4b^2\omega^2\}}.$$

The particular integral is thus

$$i_2 = \text{r.p. of } \frac{-M\omega E}{\{(c - a\omega^2)^2 + 4b^2\omega^2\}}\{(c - a\omega^2) - j2b\omega\}e^{j\omega t},$$

i.e.

$$i_2 = \frac{-M\omega E}{\{(c - a\omega^2)^2 + 4b^2\omega^2\}}\{2b\omega \sin \omega t - (a\omega^2 - c)\cos \omega t\}. \tag{7a}$$

This may be written

$$i_2 = \frac{-M\omega E}{\{(a\omega^2 - c)^2 + 4b^2\omega^2\}^{1/2}} \sin(\omega t - \phi), \tag{7}$$

where

$$\tan \phi = \frac{(a\omega^2 - c)}{2b\omega}. \tag{8}$$

This means that for the secondary,

$$i_2 = -\frac{E}{z}\sin(\omega t - \phi),$$

where

$$z = \frac{\{(a\omega^2 - c)^2 + 4b^2\omega^2\}^{1/2}}{M\omega} = \frac{1}{M}\left\{a^2\omega^2 - 2ac + 4b^2 + \frac{c^2}{\omega^2}\right\}^{1/2},$$

and the phase lag is given by $\phi = \tan^{-1}\left(\frac{a\omega^2 - c}{2b\omega}\right)$.

Putting in the values of a, b and c:

$$z = \frac{1}{M}\left\{(L_1L_2 - M^2)^2\omega^2 - 2R_1R_2(L_1L_2 - M^2) + (R_1L_2 + R_2L_1)^2 + \frac{R_1^2R_2^2}{\omega^2}\right\}^{\frac{1}{2}}$$

and

$$\phi = \tan^{-1}\left\{\frac{(L_1L_2 - M^2)\omega^2 - R_1R_2}{\omega(R_1L_2 + R_2L_1)}\right\}.$$

16

If it is known from the start that only the steady-state values are required the following is an alternative method. It derives the P.I. for i_2 from the original equations (1) and (2). In these equations treat $E \sin \omega t$ as the imaginary part of $Ee^{j\omega t}$ and put

$$i_1 = \text{i.p. of } i_1' e^{j\omega t}, \quad i_2 = \text{i.p. of } i_2' e^{j\omega t},$$

then
$$\{(R_1 + j\omega L_1)i_1' + j\omega M i_2'\}e^{j\omega t} = Ee^{j\omega t},$$

$$\{j\omega M i_1' + (R_2 + j\omega L_2)i_2'\}e^{j\omega t} = 0.$$

Thus

$$(R_1 + j\omega L_1)i_1' + j\omega M i_2' = E,$$

$$j\omega M i_1' + (R_2 + j\omega L_2)i_2' = 0.$$

Eliminating i_1' gives

$$\{-\omega^2 M^2 - (R_1 + j\omega L_1)(R_2 + j\omega L_2)\}i_2' = j\omega M E.$$

$$\therefore \; i_2' = \frac{j\omega M E}{\{\omega^2(L_1 L_2 - M^2) - R_1 R_2 - j\omega(R_1 L_2 + R_2 L_1)\}} \tag{9}$$

$$= \frac{j\omega M E}{\{(a\omega^2 - c) - j2b\omega\}}, \quad \text{using (6).}$$

Thus
$$i_2 = \text{i.p. of } \frac{j\omega M E}{\{(a\omega^2 - c) - j2b\omega\}} e^{j\omega t}$$

$$= \text{i.p. of } \frac{\omega M E\{j(a\omega^2 - c) - 2\,b\omega\}e^{j\omega t}}{(a\omega^2 - c)^2 + 4b^2\omega^2},$$

i.e.
$$i_2 = \frac{\omega M E\{(a\omega^2 - c)\cos \omega t - 2b\omega \sin \omega t\}}{(a\omega^2 - c)^2 + 4b^2\omega^2},$$

$$i_2 = -\frac{\omega M E\{2b\omega \sin \omega t - (a\omega^2 - c)\cos \omega t\}}{(a\omega^2 - c)^2 + 4b^2\omega^2}, \tag{10}$$

which is the same result as equation (7*a*).

The above method can be compared with the symbolic (operator j) method of dealing with a.c. circuits in general, which the student learns in his electrical technology lectures.

Thus, with the usual notation, using dots to denote complex voltages or currents:

$$\left.\begin{aligned} &\textit{Primary.} & \dot{E} &= \dot{I}_1(R_1 + j\omega L_1) + j\omega M\dot{I}_2, \\ &\textit{Secondary.} & 0 &= \dot{I}_2(R_2 + j\omega L_2) + j\omega M\dot{I}_1, \end{aligned}\right\} \text{ where } \dot{E} = Ee^{j\omega t}.$$

If solved for $\dot{I}_2$, these equations give

$$\dot{I}_2 = \frac{j\omega M\dot{E}}{\{\omega^2(L_1L_2 - M^2) - R_1R_2 - j\omega(R_1L_2 + R_2L_1)\}}. \tag{11}$$

As the actual applied e.m.f. is $E \sin \omega t$, i_2 is calculated from (11) taking the imaginary part of the right-hand side.

This will obviously give the result as previously obtained. In fact it is hoped that the student has already noticed in the previous work that the method used in this book for finding particular integrals for differential equations in which trigonometric functions occur is effectively the same as the symbolic method of calculation used in a.c. circuit theory.

9·22. A simple single-inductance-coupled circuit

In this example the full solution, including transients, will be given.

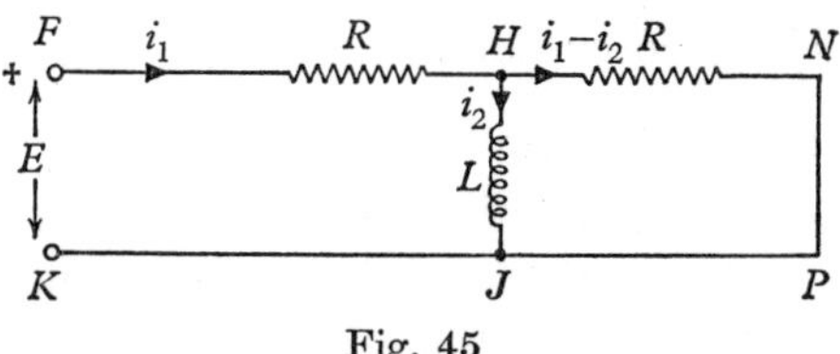

Fig. 45

Fig. 45 is self-explanatory. The applied e.m.f., E, is constant. For the circuit $FHJK$

$$Ri_1 + L\frac{di_2}{dt} = E.$$

For the closed circuit $HNPJ$

$$R(i_1 - i_2) - L\frac{di_2}{dt} = 0.$$

Writing $D \equiv d/dt$, these give

$$Ri_1 + LDi_2 = E, \tag{1}$$

$$Ri_1 - (LD + R)i_2 = 0. \tag{2}$$

Subtracting (2) from (1) to eliminate i_1,

$$(2LD + R)i_2 = E.$$

This is easily solved, giving

$$\underline{i_2 = \frac{E}{R} + Ae^{(-R/2L)t}}, \quad A \text{ arbitrary.} \tag{3}$$

From (1) and (3):

$$Ri_1 + L\left(-\frac{R}{2L}Ae^{-(R/2L)t}\right) = E,$$

$$Ri_1 = E + A\frac{R}{2}e^{-(R/2L)t},$$

$$\underline{i_1 = \frac{E}{R} + \frac{A}{2}e^{-(R/2L)t}}. \tag{4}$$

The transient terms, involving $e^{-(R/2L)t}$, die out with time and leave $i_1 = i_2 = E/R$ and $i_1 - i_2 = 0$.

This steady state is obvious from fig. 45. Although this example is extremely simple, the same method can be used for more complicated circuits.

9·23. Line transmission

Transmission of current along a cable having resistance and leakage only will be considered here. The inclusion of capacitance and inductance introduce time effects which have the result of producing partial differential equations. These are dealt with in Chapter 11.

Referring to fig. 46 let:

R = resistance per unit length of cable,

G = leakage conductance per unit length of cable,

v and i be the voltage and current at distance x from the sending end,

and $v + \delta v$, $i + \delta i$ their values at distance $x + \delta x$.

The resistance of the element δx is $R\delta x$; the voltage drop across it is given by $v - (v + \delta v) = -\delta v$.

$$\therefore \quad -\delta v = (R\delta x)i. \tag{1}$$

The leakage conductance of the element is $G\delta x$; the leakage current through the dielectric of the element is therefore $(G\delta x)v$.

Thus at the point A:

$$i = (i + \delta i) + (G\delta x)v,$$

giving
$$-\delta i = (G\delta x)v. \tag{2}$$

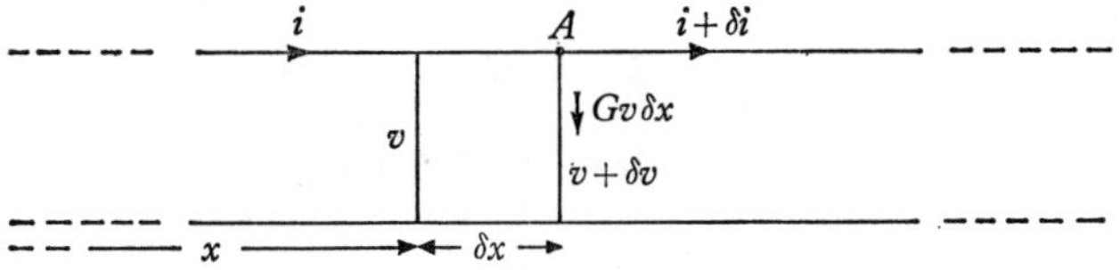

Fig. 46

Dividing both sides of equations (1) and (2) by δx, and letting $\delta x \to 0$:

$$-\frac{dv}{dx} = Ri, \tag{3}$$

and
$$-\frac{di}{dx} = Gv. \tag{4}$$

These are a simple pair of simultaneous equations. They give

$$\frac{d^2v}{dx^2} = -R\frac{di}{dx} = RGv,$$

i.e.
$$\frac{d^2v}{dx^2} - n^2v = 0, \tag{5}$$

where
$$n^2 = RG. \tag{6}$$

Solving equation (5) gives

$$v = be^{nx} + ce^{-nx}, \quad b, c \text{ arbitrary}$$

or
$$\underline{v = B\cosh nx + C\sinh nx}, \quad B, C \text{ arbitrary.} \tag{7}$$

Equations (3) and (7) then give

$$i = -\frac{1}{R}\frac{dv}{dx} = -\frac{n}{R}(B \sinh nx + C \cosh nx). \qquad (8)$$

A pair of initial conditions suffices to calculate values for B and C.

For example, let the voltage at the sending end ($x = 0$) be V_s, and at the receiving end ($x = l$) let it be zero. Then, from (7),

$$V_s = B$$

and $$0 = B \cosh nl + C \sinh nl.$$

These give $$C = -V_s \frac{\cosh nl}{\sinh nl}.$$

Putting these values into (7) gives

$$v = V_s\left(\cosh nx - \frac{\cosh nl}{\sinh nl}\sinh nx\right)$$

$$= \frac{V_s}{\sinh nl}(\cosh nx \sinh nl - \cosh nl \sinh nx),$$

i.e. $$v = \frac{V_s \sinh n(l - x)}{\sinh nl}. \qquad (9)$$

Similarly

$$i = \frac{nV_s}{R}\frac{\cosh n(l - x)}{\sinh nl}. \qquad (10)$$

Exercise 33

1. The currents x and y in two coupled circuits are given by

$$L\frac{dx}{dt} + Rx + R(x - y) = E,$$

$$L\frac{dy}{dt} + Ry - R(x - y) = 0,$$

where L, R and E are constants. Find x and y in terms of t, given that $x = 0 = y$ at $t = 0$. [L.U.]

2. Sketch a circuit which satisfies the differential equations of question 1.

3. Using fig. 44 of § 9·21, if the applied e.m.f. is constant and of value E, and if $R_1 = R_2 = R$, $L_1 = L_2 = L$, show that the currents in the primary and secondary are respectively

$$i_1 = \frac{E}{2R}\{2 - e^{-Rt/(L+M)} - e^{-Rt/(L-M)}\},$$

$$i_2 = \frac{E}{2R}\{e^{-Rt/(L-M)} - e^{-Rt/(L+M)}\}.$$

Take $i_1 = 0 = i_2$ when $t = 0$.

4. An electric cable has resistance R per unit length, while the resistance of unit length of insulation is S. If v is the voltage and i the current at a distance x from one end, show that

$$\text{(i)}\ \frac{dv}{dx} + Ri = 0, \quad \text{(ii)}\ \frac{d^2v}{dx^2} = \frac{Rv}{S}.$$

If the cable, of length L, is supplied at one end at a voltage V and is insulated at the other, find the voltage drop on the cable and show that the current entering is

$$\frac{V}{\sqrt{(RS)}} \tanh L\sqrt{\frac{R}{S}}. \qquad \text{[L.U.]}$$

5. Fig. 47 shows a capacitance-coupled circuit with an applied alternating voltage $E \sin \omega t$.

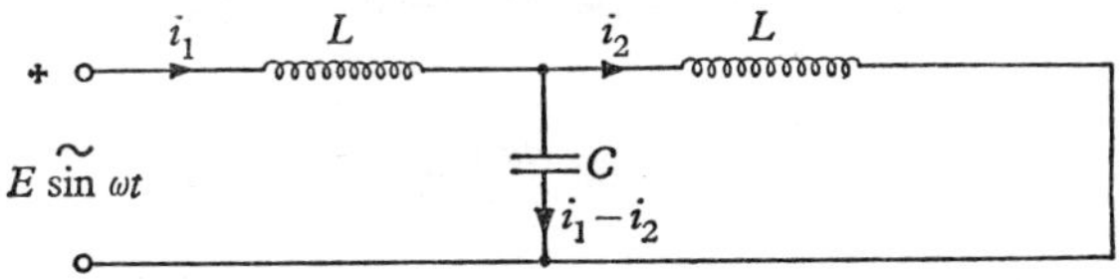

Fig. 47

Show that the currents i_1, i_2 satisfy the equations

$$L\frac{d^2i_1}{dt^2} + \frac{(i_1 - i_2)}{C} = \omega E \cos \omega t,$$

$$L\frac{d^2i_2}{dt^2} - \frac{(i_1 - i_2)}{C} = 0.$$

Find a complete expression for the current i_2.

6. The currents i_1, i_2 in the primary and secondary windings of a transformer are given by

$$L\frac{di_1}{dt} + M\frac{di_2}{dt} - Ri_1 = Ee^{jpt},$$

$$M\frac{di_1}{dt} + N\frac{di_2}{dt} - Si_2 = 0,$$

where L, M, N, R, S, E and p are given constants and t is the time. If for sufficiently large values of t it can be assumed that $i_1 = Ae^{jpt}$, $i_2 = Be^{jpt}$, where A and B are complex constants, show that

$$\left|\frac{B}{A}\right| = \frac{Mp}{\sqrt{(S^2 + N^2p^2)}}.$$

Find A and B when $R = 0$ and $2LN = M^2$. [L.U.]

7. The primary winding of a transformer has an inductance of 2 H and negligible resistance. The secondary winding has an inductance of 3 H and is in series with a capacitance of 0·04 F and a resistance of 6 Ω. The mutual inductance between the windings is 2 H. A constant e.m.f. of 50 V is applied across the primary at $t = 0$; find the value of the current in the secondary circuit for $t > 0$.

Answers

1. $$x = \frac{E}{6R}\left\{4 - 3e^{-Rt/L} - e^{-3Rt/L}\right\},$$

$$y = \frac{E}{6R}\left\{2 - 3e^{-Rt/L} + e^{-3Rt/L}\right\}.$$

The steady-state currents would be $x = 2E/3R$, $y = E/3R$.

2.

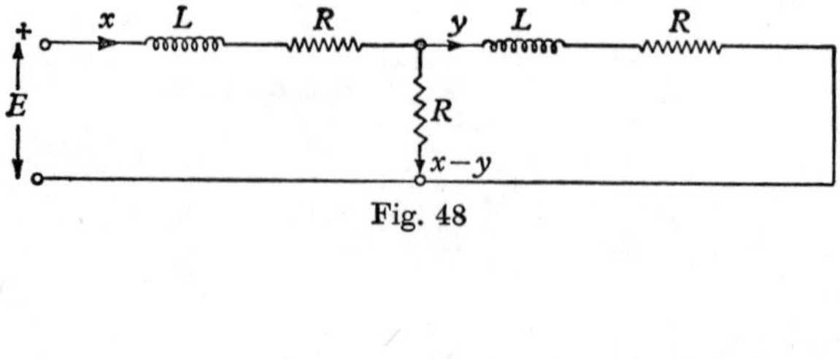

Fig. 48

4. $$V(1 - \operatorname{sech} L \sqrt{(R/S)}).$$

Note that at the insulated end the current must be zero.

5. $$i_2 = A + Bt + H \cos \sqrt{2}nt + F \sin \sqrt{2}nt + \frac{E \cos \omega t}{\omega L(LC\omega^2 - 2)},$$

where $n^2 = 1/LC$ and A, B, H, F are arbitrary constants.

6. $$A = \frac{E}{Lp(S^2 + N^2p^2)} \{-2NpS + j(N^2p^2 - S^2)\},$$

$$B = \frac{EM}{L} \frac{(S - jpN)}{(S^2 + p^2N^2)}.$$

[*Note*. For first part the ratio B/A can be obtained from the second given equation; thus $|B/A|$ can easily be found.]

7. $\frac{25}{2} e^{-3t} \sin 4t$ amperes.

CHAPTER 10

SOLUTION IN SERIES. THE LAPLACE TRANSFORMATION

10·1. Solution in series

For linear differential equations whose coefficients are polynomials in x, but which are not homogeneous, none of the methods of the previous chapters apply. To solve such equations a solution is sought in the form of infinite series. As a start, a solution is assumed of the form

$$y = x^c(a_0 + a_1x + a_2x^2 + a_3x^3 + \dots + a_nx^n + \dots),$$

where c, a_0, a_1, a_2, ... are constants and a_0 can be taken *not* zero as the series must start somewhere. This series is substituted into the differential equation and, by equating the coefficients of successive powers of x on either side of the equation, values for the constants are obtained. Various difficulties may occur in the general use of this method; matters such as the existence of solutions of this type and the justification of the use of the method are amongst such difficulties. The student can find these dealt with in an advanced text-book on differential equations, but here only the practical process of finding a solution will be given. This is demonstrated in the following two examples.

EXAMPLE 1

Solve the equation

$$2x\frac{d^2y}{dx^2} + 3\frac{dy}{dx} + y = 0.$$

Assume that

$$y = x^c(a_0 + a_1x + a_2x^2 + \dots + a_nx^n + \dots), \qquad (1)$$

where $a_0 \neq 0$.

Putting in the values of the derivatives:

$$y = a_0x^c + a_1x^{c+1} + \dots + a_rx^{c+r} + \dots,$$

$$3\frac{dy}{dx} = 3a_0cx^{c-1} + 3a_1(c+1)x^c + 3a_2(c+2)x^{c+1} + \dots + 3a_{r+1}(c+r+1)x^{c+r} + \dots,$$

$$2x\frac{d^2y}{dx^2} = 2a_0c(c-1)x^{c-1} + 2a_1(c+1)cx^c + 2a_2(c+2)(c+1)x^{c+1} + \dots + 2a_{r+1}(c+r+1)(c+r)x^{c+r} + \dots.$$

The successive coefficients of powers of x in

$2x\frac{d^2y}{dx^2} + 3\frac{dy}{dx} + y$ are therefore:

x^{c-1}: $\quad 3a_0c + 2a_0c(c-1) = a_0c(2c+1).$

x^c: $a_0 + 3a_1(c+1) + 2a_1(c+1)c = a_0 + a_1(c+1)(2c+3).$

x^{c+1}: $\quad a_1 + 3a_2(c+2) + 2a_2(c+2)(c+1) = a_1 + a_2(c+2)(2c+5)$

... ...

x^{c+r}: $\quad a_r + 3a_{r+1}(c+r+1) + 2a_{r+1}(c+r+1)(c+r) = a_r + a_{r+1}(c+r+1)(2c+2r+3).$

As the right-hand side of the differential equation is zero, all these coefficients of powers of x must be zero.

Thus

$$a_0c(2c+1) = 0, \tag{2}$$

$$a_0 + a_1(c+1)(2c+3) = 0, \tag{3}$$

$$a_1 + a_2(c+2)(2c+5) = 0, \tag{4}$$

... ...

$$a_r + a_{r+1}(c+r+1)(2c+2r+3) = 0. \tag{5}$$

Equation (5) is for the general term, where r is any positive integer.

Equation (2), obtained from the coefficient of the lowest power of x, is called the *indicial equation* and gives values for the index c.

In this case $\quad \underline{c = 0 \text{ or } c = -\tfrac{1}{2}},\quad$ as $a_0 \neq 0.$ (6)

These values for c give two possible solutions of the differential equation and, as a_0 is arbitrary and can thus be made different in the two solutions, a complete solution is found by adding the two solutions.

Taking the two possible values of c separately:

(i) $c = 0$. From (3), $a_0 + a_1.1.3 = 0$, $a_1 = -\dfrac{a_0}{1.3}$. (7)

From (4), $a_1 + a_2.2.5 = 0$, $a_2 = -\dfrac{a_1}{2.5} = \dfrac{a_0}{(1.3)(2.5)}$. (8)

In general, from (5),

$$a_r + a_{r+1}(r+1)(2r+3) = 0, \quad a_{r+1} = \frac{-a_r}{(r+1)(2r+3)}. \tag{9}$$

By taking $r = 0, 1, 2, \ldots$, equation (9) gives all the values of the a's.

One solution is therefore of the form

$$y = a_0 x^0 \left[1 - \frac{x}{1.3} + \frac{x^2}{(1.3)(2.5)} - \frac{x^3}{(1.3)(2.5)(3.7)} + \ldots\right],$$

i.e. $$\underline{y = a_0 \left[1 - \frac{x}{1.3} + \frac{x^2}{(1.3)(2.5)} - \frac{x^3}{(1.3)(2.5)(3.7)} + \ldots\right].} \tag{10}$$

(ii) $c = -\frac{1}{2}$. From (3),

$$a_0 + a_1\tfrac{1}{2}.2 = 0, \quad a_1 = -a_0. \tag{11}$$

From (4), $a_1 + a_2\frac{3}{2}.4 = 0$, $a_2 = \dfrac{-a_1}{2.3} = \dfrac{a_0}{2.3}$. (12)

In general, from (5),

$$a_r + a_{r+1}(-\tfrac{1}{2} + r + 1)(-1 + 2r + 3) = 0,$$

i.e. $$a_r + a_{r+1}\left(\frac{2r+1}{2}\right)2(r+1) = 0,$$

giving $$a_{r+1} = \frac{-a_r}{(r+1)(2r+1)} \quad (r = 0, 1, 2, \ldots), \tag{13}$$

e.g. $$a_3 = \frac{-a_2}{3.5}, \quad \text{putting } r = 2.$$

$$= \frac{-a_0}{(2.3)(3.5)}, \quad \text{from (12).}$$

This gives a second solution (writing b_0 for a_0 as a_0 is arbitrary):

$$y = b_0 x^{-\frac{1}{2}} \left[1 - x + \frac{x^2}{2.3} - \frac{x^3}{(2.3)(3.5)} + \frac{x^4}{(2.3)(3.5)(4.7)} - \dots\right]. \quad (14)$$

A general solution, with two arbitrary constants, is therefore

$$y = A\left[1 - \frac{x}{1.3} + \frac{x^2}{(1.3)(2.5)} - \frac{x^3}{(1.3)(2.5)(3.7)} + \dots\right] + Bx^{-\frac{1}{2}}\left[1 - x + \frac{x^2}{2.3} - \frac{x^3}{(2.3)(3.5)} + \frac{x^4}{(2.3)(3.5)(4.7)} - \dots\right].$$

EXAMPLE 2

Find a solution in series for the equation

$$\frac{d^2y}{dx^2} + \frac{1}{x}\frac{dy}{dx} + \left(1 - \frac{n^2}{x^2}\right)y = 0, \quad \text{where } n \text{ is a constant.}$$

This is Bessel's equation of order n (see § 3·33).

Multiplying through by x^2:

$$x^2\frac{d^2y}{dx^2} + x\frac{dy}{dx} + (x^2 - n^2)y = 0. \quad (1)$$

It is conventional to use n only if the constant is an integer and to use ν if the constant is a fraction. The importance of this equation in engineering has already been mentioned.

Assume a solution of the form

$$y = x^c(a_0 + a_1x + a_2x^2 + \dots + a_rx^r + \dots) \quad (a_0 \neq 0). \tag{2}$$

Then

$$x^2\frac{d^2y}{dx^2} = a_0c(c-1)x^c + a_1(c+1)cx^{c+1} + \dots$$
$$+ a_r(c+r)(c+r-1)x^{c+r} + \dots,$$
$$x\frac{dy}{dx} = a_0cx^c + a_1(c+1)x^{c+1} + \dots$$
$$+ a_r(c+r)x^{c+r} + \dots,$$
$$x^2y = a_0x^{c+2} + \dots + a_{r-2}x^{c+r} + \dots,$$
$$-n^2y = -a_0n^2x^c - a_1n^2x^{c+1} - \dots - a_rn^2x^{c+r} - \dots .$$

Equating to zero successive coefficients of x gives

$$x^c: \quad a_0\{c(c-1) + c - n^2\} = 0, \quad \therefore\ a_0(c^2 - n^2) = 0. \tag{3}$$

$$x^{c+1}: \quad a_1\{(c+1)c + (c+1) - n^2\} = 0,$$
$$\therefore\ a_1\{(c+1)^2 - n^2\} = 0. \tag{4}$$

And the general coefficient:

$$x^{c+r}: \quad a_r\{(c+r)(c+r-1) + (c+r) - n^2\} + a_{r-2} = 0,$$

i.e.
$$a_r\{(c+r)^2 - n^2\} + a_{r-2} = 0. \tag{5}$$

This equation holds true for $r \geqslant 2$.

Equation (3) is the indicial equation and gives

$$\underline{c = \pm n.} \tag{6}$$

(i) *Assume n is a fraction, say ν.* Equation (6) then becomes

$$c = \pm \nu.$$

Equation (4) gives

$$a_1\{(\nu+1)^2 - \nu^2\} = 0 \quad \text{or} \quad a_1\{(-\nu+1)^2 - \nu^2\} = 0.$$

In either case $a_1 = 0$. (For special case $\nu = \frac{1}{2}$, see Ex. 34, note in Question 3).

Equation (5) then shows that $a_1 = a_3 = a_5 = \dots = 0$, i.e. *all the coefficients of odd powers of x are zero.*

For $c = +\nu$. Equation (5) gives, on putting $r = 2, 4, \ldots$,

$$a_2 = \frac{-a_0}{(\nu+2)^2 - \nu^2} = \frac{a_0}{\nu^2 - (\nu+2)^2}, \tag{7}$$

$$a_4 = \frac{-a_2}{(\nu+4)^2 - \nu^2} = \frac{a_0}{\{\nu^2 - (\nu+2)^2\}\{\nu^2 - (\nu+4)^2\}},$$

and so on.

Using the difference of two squares, these become

$$a_2 = \frac{-a_0}{2^2.(\nu+1)}, \quad a_4 = \frac{a_0}{2^4.2(\nu+1)(\nu+2)},$$

$$a_6 = \frac{-a_0}{2^6.2.3(\nu+1)(\nu+2)(\nu+3)},$$

and in general

$$a_{2r} = \frac{(-1)^r a_0}{2^{2r}(\nu+1)(\nu+2)\ldots(\nu+r)r!}.$$

The series for $c = +\nu$ is therefore

$$y = a_0 x^\nu\left[1 - \frac{x^2}{2^2(\nu+1)} + \frac{x^4}{2^4(\nu+1)(\nu+2)2!} - \ldots\right],$$

or $$y = a_0 x^\nu\left\{1 + \sum_{r=1}^{\infty} \frac{(-1)^r\left(\frac{x}{2}\right)^{2r}}{(\nu+1)(\nu+2)\ldots(\nu+r)r!}\right\}. \tag{8}$$

For $c = -\nu$. This gives the same result with $-\nu$ replacing ν. Thus another solution is

$$y = a_0 x^{-\nu}\left\{1 + \sum_{r=1}^{\infty} \frac{(-1)^r\left(\frac{x}{2}\right)^{2r}}{(-\nu+1)(-\nu+2)\ldots(-\nu+r)r!}\right\}.$$

i.e.

$$y = a_0 x^{-\nu}\left\{1 + \sum_{r=1}^{\infty} \frac{\left(\frac{x}{2}\right)^{2r}}{(\nu-1)(\nu-2)\ldots(\nu-r)r!}\right\}, \tag{9}$$

noting that $(-1)^r \div (-1)^r = +1$.

a_0 being arbitrary, the general solution for $n = \nu$, a fraction, is

$$y = Ax^{\nu}\left\{1 + \sum_{r=1}^{\infty} \frac{(-1)^r \left(\frac{x}{2}\right)^{2r}}{(\nu+1)(\nu+2)\ldots(\nu+r)r!}\right\}$$

$$+ Bx^{-\nu}\left\{1 + \sum_{r=1}^{\infty} \frac{\left(\frac{x}{2}\right)^{2r}}{(\nu-1)(\nu-2)\ldots(\nu-r)r!}\right\}. \qquad (10)$$

The cases in which n is zero or a positive integer are those which arise most in practice. If n is a positive integer or zero the above process leads to only one solution which cannot be the general solution as it will have only one arbitrary constant.

(ii) *Assume* $n = 0$. From (6) above $c = 0$.

Equation (4) still gives $a_1 = 0$, and from (5) this leads again to

$$a_1 = a_3 = a_5 = \ldots = 0.$$

For the even coefficients, equation (5) gives

$$a_2 = \frac{-a_0}{2^2}, \quad a_4 = \frac{-a_2}{4^2} = \frac{a_0}{2^2 . 4^2} = \frac{a_0}{2^4 . (2!)^2},$$

$$a_6 = \frac{-a_4}{6^2} = \frac{-a_0}{2^4 . (2!)^2 . 6^2} = \frac{-a_0}{2^6 . (3!)^2}, \text{ etc.}$$

A solution is therefore given by

$$y = a_0\left\{1 - \frac{x^2}{2^2} + \frac{x^4}{2^4 . (2!)^2} - \frac{x^6}{2^6 . (3!)^2} + \ldots\right\}$$

$$= a_0\left\{1 + \sum_{r=1}^{\infty} \frac{(-1)^r (\frac{1}{2}x)^{2r}}{(r!)^2}\right\}. \qquad (11)$$

(iii) *Assume* $n = m$, *a positive integer*. From equation (6),

$$c = \pm m.$$

Taking $c = +m$ merely gives the solution of equation (8) with m replacing ν, i.e.

$$y = a_0 x^m \left\{ 1 + \sum_{r=1}^{\infty} \frac{(-1)^r(\frac{1}{2}x)^{2r}}{(m+1)(m+2)\dots(m+r)r!} \right\}. \quad (12)$$

Taking $c = -m$, equations (4) and (5) still ensure $a_1 = a_3 = a_5 = \dots = 0$, and equation (5) gives

$$a_{2r}\{(-m+2r)^2 - m^2\} + a_{2r-2} = 0.$$

Now when $r = m$ (possible, as m is an integer),

$$a_{2m}.0 + a_{2m-2} = 0, \quad \text{therefore} \quad a_{2m-2} = 0.$$

Working backwards, using equation (5) with $r = 2m-2$, $2m-4$, ..., this means that $a_{2m-2} = a_{2m-4} = a_{2m-6} = \dots = a_2 = 0$.

Thus, only a_{2m} and higher even coefficients are not zero.

From (5), $a_{2m+2}\{(-m+2m+2)^2 - m^2\} + a_{2m} = 0.$

This gives $$a_{2m+2} = \frac{-a_{2m}}{(m+2)^2 - m^2}.$$

This is identical with equation (7), as a_0 is arbitrary. Thus the case of $c = -m$ will merely lead to a constant multiple of the solution given in (12), the case for $c = +m$.

Equation (12) is therefore the only distinct solution given by this method.

(iv) $n = -m$, *m a positive integer*. This case, when n is a negative integer, again gives rise to only one distinct solution which is of the form of equation (9) with m replacing ν.

Bessel functions. The solution in (12), for n a positive integer, can be written

$$y = a_0 x^m \sum_{r=0}^{\infty} \frac{(-1)^r(\frac{1}{2}x)^{2r}}{(m+1)(m+2)\dots(m+r)r!},$$

as $x^0 = 1$ and $0! = 1$.

Taking the value of the arbitrary constant a_0 as $\dfrac{1}{2^m.m!}$

(i.e. giving a_0 this special value), the series becomes

$$y = \sum_{r=0}^{\infty} \frac{(-1)^r(\frac{1}{2}x)^{2r}\, x^m}{(m+1)(m+2)\ldots(m+r)r\,!2^m m!}.$$

Now $\quad m!\, m+1)(m+2)\ldots(m+r) = (m+r)!$

and $\quad \left(\frac{x}{2}\right)^{2r}\left(\frac{x^m}{2^m}\right) = \left(\frac{x}{2}\right)^{m+2r}.$

Thus the series becomes

$$y = \sum_{r=0}^{\infty} \frac{(-1)^r(\frac{1}{2}x)^{m+2r}}{(m+r)!\, r!}. \tag{13}$$

This is a special solution of the differential equation

$$\frac{d^2y}{dx^2} + \frac{1}{x}\frac{dy}{dx} + \left(1 - \frac{m^2}{x^2}\right)y = 0,$$

m a positive integer.

It is called the Bessel function of the first kind of order m and is denoted by $J_m(x)$ (see § 3·33). Tables of values for various values of m and x have been drawn up and are available in sets of advanced mathematical tables.

Note. For a more rigorous treatment including a full discussion of the case when only one series is found, as in Example 2 (ii) above, the student may refer to any standard textbook on Differential Equations.

Exercise 34

Find solutions in series for the equations:

1. $3x\frac{d^2y}{dx^2} + 2\frac{dy}{dx} + y = 0.$

2. $2x\frac{d^2y}{dx^2} + \frac{dy}{dx} + y = 0.$

3. $\frac{d^2y}{dx^2} + y = 0.$

(*Note.* If the indicial equation has two solutions, $c = c_1$ and $c = c_2$, differing by an integer, and if a_1 then becomes indeterminate when $c = c_1$, this value of c will give the general solution; substituting $c = c_2$ merely gives a solution which is a multiple of one of the other separate solutions. In such cases a_0 is one arbitrary constant and a_1 another.)

4. Obtain the general solution in series of the differential equation

$$4x\frac{d^2y}{dx^2} + (4x + 3)\frac{dy}{dx} + y = 0.$$

Give the range of values of x for which each series is valid. [L.U.]

5. Obtain the general solution, in series of ascending powers of x, of the equation

$$(1 - x^2)\frac{d^2y}{dx^2} - 2x\frac{dy}{dx} + 6y = 0.$$ [L.U.]

6. Obtain the general solution in series of the equation

$$(1 - x^2)\frac{d^2y}{dx^2} - 2x\frac{dy}{dx} + 12y = 0,$$

showing that one of the series is a polynomial in x.

(*Note.* Questions 5 and 6 are special cases of *Legendre's equation*, another important type of differential equation. Its general form is

$$(1 - x^2)\frac{d^2y}{dx^2} - 2x\frac{dy}{dx} + n(n + 1)y = 0,$$

where n is zero or any positive integer. One of the two series in the general solution is always finite and is a polynomial in x. These give rise to *Legendre polynomials*, $P_n(x)$ being the usual symbol for the Legendre polynomial of degree n.)

7. $$\frac{d^2y}{dx^2} + \frac{1}{x}\frac{dy}{dx} + b(k^2 - x^2)y = 0,$$

where k, b are constants, is a type of equation which occurs in the theory of ionic diffusion, when studying the conduction of electricity through gases. Show that a solution can be found of the form

$$y = A\left\{1 - \frac{k^2bx^2}{4} + \frac{1}{16}\left(b + \frac{k^4b^2}{4}\right)x^4 - \ldots\right\}.$$

8. Using the series $1 + \sum_{r=1}^{\infty}\frac{(-1)^r\left(\frac{x}{2}\right)^{2r}}{(r!)^2}$ for the expansion of

$J_0(x)$, find the values of $J_0(x)$, correct to 2 decimal places, for the values $x = 0, 1, 2, 3$ and 4. Hence, with the aid of the following table, plot a graph of $J_0(x)$ from $x = 0$ to $x = 10$, verifying that it oscillates with decreasing amplitude:

x	0·4	0·8	1·4	1·8	2·405	3·5	4·5	5	5·52	6	7	8
$J_0(x)$	0·96	0·85	0·57	3·4	0	−0·38	−0·32	−0·18	0	0·15	0·30	0·17

x	9	10
$J_0(x)$	−0·09	−0·25

Answers

1. $$y = A\left\{1 - \frac{x}{1.2} + \frac{x^2}{(1.2)(2.5)} - \frac{x^3}{(1.2)(2.5)(3.8)} + \ldots\right\}$$
$$+ Bx^{\frac{1}{3}}\left\{1 - \frac{x}{1.4} + \frac{x^2}{(1.4)(2.7)} - \frac{x^3}{(1.4)(2.7)(3.10)} + \ldots\right\}.$$

2. $$y = A\left\{1 - x + \frac{x^2}{2.3} - \frac{x^3}{(2.3)(3.5)} + \frac{x^4}{(2.3)(3.5)(4.7)} - \ldots\right\}$$
$$+ Bx^{\frac{1}{2}}\left\{1 - \frac{x}{1.3} + \frac{x^2}{(1.3)(2.5)} - \frac{x^3}{(1.3)(2.5)(3.7)} + \ldots\right\}.$$

3. $$y = A\left(1 - \frac{x^2}{2!} + \frac{x^4}{4!} - \ldots\right) + B\left(x - \frac{x^3}{3!} + \frac{x^5}{5!} - \ldots\right),$$

i.e. $y = A\cos x + B\sin x$. [The original equation is the well-known s.h.m. equation and, of course, can be solved easily as a second-order linear equation with constant coefficients.]

4. $$y = A\left\{1 - \frac{x}{1.3} + \frac{5x^2}{(1.3)(2.7)} - \frac{5.9x^3}{(1.3)(2.7)(3.11)} + \ldots\right\}$$
$$+ Bx^{\frac{1}{4}}\left\{1 - \frac{2x}{1.5} + \frac{2^2.3x^2}{(1.5)(2.9)} - \frac{2^3.3.5x^3}{(1.5)(2.9)(3.13)} + \ldots\right\}.$$

For the first series,

$$a_{r+1} = \frac{-(4r+1)a_r}{(r+1)(4r+3)} \qquad (r = 0, 1, 2, \ldots).$$

For the second series,

$$a_{r+1} = \frac{-2.(2r+1)a_r}{(r+1)(4r+5)} \quad (r = 0, 1, 2, \ldots).$$

In both cases

$$\lim_{r\to\infty} \left| \frac{a_{r+1}}{a_r} \right| = 0;$$

thus both series are convergent for all values of x.

5. $y = A(1-3x^2) + Bx\left\{1 - \frac{1.4}{2.3}x^2 - \frac{(1.4)(1.6)x^4}{(2.3)(4.5)} - \frac{(1.4)(1.6)(3.8)x^6}{(2.3)(4.5)(6.7)} - \ldots\right\}.$

Note. $c = 0$ gives both series, taking a_0 and a_1 as the arbitrary constants.

$$a_{r+2} = \frac{(r-2)(r+3)}{(r+2)(r+1)} a_r \quad (r = 0, 1, 2, 3, \ldots).$$

6. $y = A\left\{1 - \frac{3.4}{1.2}x^2 + \frac{(3.4)(1.6)}{(1.2)(3.4)}x^4 + \frac{(3.4)(1.6)(1.8)}{(1.2)(3.4)(5.6)}x^6 + \ldots\right\} + Bx\left(1 - \frac{5}{3}x^2\right).$

Note. With $c = 0$,

$$a_{r+2} = \frac{(r-3)(r+4)a_r}{(r+1)(r+2)} \quad (r = 0, 1, 2, \ldots).$$

7. [The equations satisfied by the coefficients are: $a_0c^2 = 0$, the indicial equation, giving $c = 0$ only. $a_1(c+1)^2 = 0$;

$$a_2(c+2)^2 + a_0k^2b = 0, \quad a_3(c+3)^2 + k^2ba_1 = 0;$$

$$a_{r+2}(c+r+2)^2 + k^2ba_r - ba_{r-2} = 0 \quad \text{for} \quad r \geqslant 2.$$

With $c = 0$, these give:

$$a_1 = 0; \quad a_3.3^2 + k^2ba_1 = 0, \quad \text{therefore} \quad a_3 = 0;$$

$$a_2.2^2 + a_0k^2b = 0, \quad \text{therefore} \quad a_2 = -\frac{k^2ba_0}{2};$$

$a_{r+2}(r+2)^2 + k^2ba_r - ba_{r-2} = 0$, for $r \geqslant 2$ (general equation).

It follows from the general equation that, as $a_1 = a_3 = 0$, all odd coefficients are zero.

Knowing a_2 in terms of a_0, all even coefficients are easily found from the general equation, e.g.

$$a_4 . 4^2 + k^2 b a_2 - b a_0 = 0, \text{ putting } r = 2.]$$

8. $J_0(0) = 1$, $J_0(1) \simeq 0{\cdot}76(5)$, $J_0(2) \simeq 0{\cdot}22$, $J_0(3) \simeq -0{\cdot}26$, $J_0(4) \simeq -0{\cdot}40$.

10·2. The Laplace transformation method

10·21. The Laplace transforms of some standard functions

The Laplace transformation is a device which can be used to solve linear differential equations when initial conditions are known. It is suitable for use in problems on the behaviour of electrical systems which are 'started' at some given instant, taken as $t = 0$, with known initial conditions. The behaviour of the system for $t > 0$ can then be determined. The use of Laplace transforms reduces the solution to algebraic manipulation; in this connexion a knowledge of the technique of splitting into partial fractions is essential. The transforms of standard functions have to be used, but these are usually available in the form of tables (see pp. 254, 255).

Let $y(t)$ be a known function of t for $t > 0$; then the Laplace transform of $y(t)$ is defined as

$$\bar{y}(s) = \int_0^\infty e^{-st} . y(t)\, dt. \tag{1}$$

Several points should be noted:

(i) $\bar{y}(s)$ is the symbol to be used for the Laplace transform of $y(t)$. The symbols are often abbreviated to $\bar{y}$ and y. Some books use the symbols $L(y)$ or $\tilde{y}$ for the transform. Occasionally $L(y)$ will be used in this book when the 'bar' notation is inconvenient.

(ii) It is stressed that, as the integral in (1) is a definite integral, t will not occur in the evaluated integral, which will be a function of s only.

(iii) In (1) above s is a parameter, that is, a constant of no particular given value. As the definition involves an integral with an infinite limit, it is assumed inherent in the definition that s is a positive number large enough to make the integral converge to a definite value. For example, if $y(t)$ were e^{3t}, s would be assumed to be a number greater than 3.

(iv) The original variable need not be t. Thus the Laplace transform of a function of x, e.g. $y(x)$, is given by

$$\bar{y}(s) = \int_0^\infty e^{-sx} \,.\, y(x)\, dx.$$

In electrical work, however, the independent variable is usually time.

Some Laplace transforms will now be evaluated, using t as the independent variable.

(a) $y = c$ (*constant*)

$$\bar{y} = \int_0^\infty e^{-st} c\, dt = -\frac{c}{s}\left[e^{-st}\right]_0^\infty = -\frac{c}{s}(0 - 1) = \frac{c}{s}.$$

Thus $$\underline{\bar{y} = \frac{c}{s}}. \tag{2}$$

Here it sufficient to consider s as any number greater than zero.

(b) $y = e^{ct}$ (c *constant*)

$$\bar{y} = \int_0^\infty e^{-st} e^{ct}\, dt = \int_0^\infty e^{-(s-c)t}\, dt$$

$$= \frac{1}{-(s-c)}\left[e^{-(s-c)t}\right]_0^\infty.$$

As s is considered large enough to make the integral convergent, s can be taken greater than c.

Thus $$\bar{y} = \frac{-1}{(s-c)}(0 - 1),$$

$$\underline{\bar{y} = \frac{1}{(s-c)}}. \tag{3}$$

(*Note.* If c is complex, it is sufficient to consider s greater than the real part of c.)

(c) $y = \cos \omega t$ *or* $\sin \omega t$ (ω *real*). These are treated as particular cases of (b) above.

Thus $\cos \omega t = \frac{1}{2}(e^{j\omega t} + e^{-j\omega t})$ and $\sin \omega t = \dfrac{1}{2j}(e^{j\omega t} - e^{-j\omega t})$.

For $\cos \omega t$,

$$\bar{y} = \frac{1}{2}\left\{\frac{1}{(s - j\omega)} + \frac{1}{(s + j\omega)}\right\},$$

using (3) with $c = j\omega$ and $c = -j\omega$.

This gives

$$\bar{y} = \frac{1}{2}\left(\frac{2s}{s^2 + \omega^2}\right) = \underline{\frac{s}{s^2 + \omega^2}}. \qquad (4)$$

Similarly if $y = \sin \omega t$, $\underline{\bar{y} = \dfrac{\omega}{(s^2 + \omega^2)}}$. (5)

(d) *cosh at or sinh at*. Taking $\cosh at = \frac{1}{2}(e^{at} + e^{-at})$ and $\sinh at = \frac{1}{2}(e^{at} - e^{-at})$, the use of (3) above gives

$$\underline{\bar{y} = \frac{s}{(s^2 - a^2)}}, \quad \text{for } \cosh at, \qquad (6)$$

and

$$\underline{\bar{y} = \frac{a}{(s^2 - a^2)}}, \quad \text{for } \sinh at. \qquad (7)$$

(e) $y = t^n$, *n a positive integer*. Taking first the case of $y = t$,

$$\bar{y} = \int_0^\infty e^{-st} t \, dt = \left[-\frac{e^{-st} t}{s}\right]_0^\infty + \int_0^\infty \frac{e^{-st}}{s} dt, \quad \text{by parts.}$$

Since $\lim\limits_{t \to \infty} te^{-st} = 0$ ($s > 0$) (see Exercise 2, Question 9), this gives

$$\bar{y} = \int_0^\infty \frac{e^{-st}}{s} dt = -\frac{1}{s^2}\left[e^{-st}\right]_0^\infty = -\frac{1}{s^2}(0 - 1) = \frac{1}{s^2},$$

i.e.

$$\int_0^\infty te^{-st} \, dt = \frac{1}{s^2}. \qquad (8)$$

Differentiating under the integral sign with respect to s:

$$\frac{d}{ds}\left\{\int_0^\infty te^{-st}\,dt\right\} = \frac{-2}{s^3},$$

i.e. $$\int_0^\infty \frac{\partial}{\partial s}(te^{-st})\,dt = \frac{-2}{s^3} \quad \text{(see § 3·31)},$$

giving $$\int_0^\infty t^2e^{-st}\,dt = \frac{2}{s^3}. \tag{9}$$

Differentiating again with respect to p:

$$\int_0^\infty t^3e^{-st}\,dt = \frac{3\,.\,2}{s^4} = \frac{3!}{s^4}. \tag{10}$$

Continuing this process, in general,

$$\int_0^\infty t^ne^{-st}\,dt = \frac{n!}{s^{n+1}} \quad (n = 0,\ 1,\ 2,\ \ldots). \tag{11}$$

That is, the Laplace transform of $y = t^n$ is given by

$$\bar{y} = \frac{n!}{s^{n+1}}. \tag{12}$$

(f) $y = \dfrac{t}{2\omega}\, sin\ \omega t$. Equation (4) above effectively states that

$$\int_0^\infty e^{-st}\cos\omega t\,dt = \frac{s}{s^2+\omega^2}.$$

Differentiating under the integral sign with respect to ω:

$$\int_0^\infty -\,te^{-st}\sin\omega t\,dt = -\frac{s\,.\,2\omega}{(s^2+\omega^2)^2}.$$

$$\therefore\ \int_0^\infty e^{-st}\frac{t}{2\omega}\sin\omega t\,dt = \frac{s}{(s^2+\omega^2)^2},$$

i.e. $$\bar{y} = \frac{s}{(s^2+\omega^2)^2}. \tag{13}$$

(g) $y = \dfrac{1}{2\omega^3}(sin\ \omega t - \omega t\ cos\ \omega t)$. Equation (5) above gives

$$\int_0^\infty e^{-st}\sin\omega t\,dt = \frac{\omega}{s^2+\omega^2}.$$

Differentiating with respect to ω:

$$\int_0^\infty te^{-st}\cos\omega t\,dt = \frac{1}{s^2+\omega^2} - \frac{2\omega^2}{(s^2+\omega^2)^2},$$

i.e.

$$\int_0^\infty e^{-st}\,\omega t\cos\omega t\,dt = \frac{\omega}{s^2+\omega^2} - \frac{2\omega^3}{(s^2+\omega^2)^2}.$$

Using (5) again:

$$\int_0^\infty e^{-st}\,\omega t\cos\omega t\,dt = \int_0^\infty e^{-st}\sin\omega t\,dt - \frac{2\omega^3}{(s^2+\omega^2)^2}.$$

$$\therefore \int_0^\infty e^{-st}\frac{(\sin\omega t - \omega t\cos\omega t)}{2\omega^3}\,dt = \frac{1}{(s^2+\omega^2)^2}.$$

Thus

$$\bar{y} = \frac{1}{(s^2+\omega^2)^2}. \tag{14}$$

The results in (2), (3), (4), (5), (6), (7), (12), (13) and (14) are collected in Table I below. Results 8 and 9 arise in cases of resonance.

TABLE I

$y(t)$	$\bar{y}(s)$	
c (constant)	$\frac{c}{s}$	1
t^n	$\frac{n!}{s^{n+1}}$ $(n = 0, 1, 2, 3 \ldots)$	2
e^{ct}	$\frac{1}{(s-c)}$	3
$\sin\omega t$	$\frac{\omega}{(s^2+\omega^2)}$	4
$\cos\omega t$	$\frac{s}{(s^2+\omega^2)}$	5
$\sinh at$	$\frac{a}{(s^2-a^2)}$	6
$\cosh at$	$\frac{s}{(s^2-a^2)}$	7
$\frac{t}{2\omega}\sin\omega t$	$\frac{s}{(s^2+\omega^2)^2}$	8
$\frac{1}{2\omega^3}(\sin\omega t - \omega t\cos\omega t)$	$\frac{1}{(s^2+\omega^2)^2}$	9

10·22. Generalization of standard results

The transforms given in Table I can be extended by the use of the following theorem:

If $\bar{y}(s)$ is the Laplace transform of $y(t)$, then $\bar{y}(s+k)$ is the Laplace transform of $e^{-kt}.y(t)$, where k is any number, real or complex.

This is proved easily, as the Laplace transform of $e^{-kt}y(t)$ is given by

$$\int_0^\infty e^{-st}\{e^{-kt}y(t)\}\,dt = \int_0^\infty e^{-(s+k)t}y(t)\,dt = \bar{y}(s+k).$$

Using this theorem on transforms 2, 4, 5, 6 and 7 of Table I, the results shown in Table II are obtained:

TABLE II

$y(t)$	$\bar{y}(s)$	
$t^n e^{-ct}$	$\dfrac{n!}{(s+c)^{n+1}}$ $(n = 0, 1, 2, 3, \ldots)$	1
$e^{-ct}\sin\omega t$	$\dfrac{\omega}{(s+c)^2+\omega^2}$	2
$e^{-ct}\cos\omega t$	$\dfrac{(s+c)}{(s+c)^2+\omega^2}$	3
$e^{-ct}\sinh at$	$\dfrac{a}{(s+c)^2-a^2}$	4
$e^{-ct}\cosh at$	$\dfrac{(s+c)}{(s+c)^2-a^2}$	5

10·23. To find $y(t)$ given $\bar{y}(s)$

The final step in solving a differential equation by the Laplace transformation method is to deduce the function $y(t)$ given its transform $\bar{y}(s)$.

The procedure is to arrange $\bar{y}(s)$ into sums or differences of the standard types given in Tables I and II and thus read off $y(t)$. It can be shown that the $y(t)$ found in this way is the *only* function which has $\bar{y}(s)$ for its transform.

Partial fractions are often used to arrange $\bar{y}(s)$ into standard forms.

EXAMPLES

(1) If $\bar{y} = \dfrac{1}{s+3}$, it follows from Table I, 3 that

$$\underline{y = e^{-3t}.}$$

(2) If $\bar{y} = \dfrac{3s + 7}{s^2 + 9} = 3\dfrac{s}{s^2 + 9} + \dfrac{7}{3}\dfrac{3}{s^2 + 9}$,

then from Table I, 4 and 5, it follows that

$$\underline{y = 3 \cos 3t + \tfrac{7}{3} \sin 3t.}$$

(3) If the denominator is a quadratic in s the square is completed and Table II is used.

If
$$\bar{y} = \frac{s - 2}{s^2 + 6s + 16} = \frac{s - 2}{(s + 3)^2 + 7}$$

$$= \frac{(s + 3) - 5}{(s + 3)^2 + (\sqrt{7})^2},$$

i.e.
$$\bar{y} = \frac{(s + 3)}{(s + 3)^2 + (\sqrt{7})^2} - \frac{5}{\sqrt{7}} \frac{\sqrt{7}}{(s + 3)^2 + (\sqrt{7})^2}.$$

Using Table II, 2 and 3,

$$\underline{y = e^{-3t} \cos \sqrt{7}t - \frac{5}{\sqrt{7}} e^{-3t} \sin \sqrt{7}t.}$$

Note the method of expressing the numerator of $\bar{y}$ in terms of $(s + 3)$.

(4) Finding $y(t)$ given $\bar{y}(s)$ is often called finding the *inverse* of $\bar{y}(s)$.

Find the inverse of

$$\frac{s + 1}{s^2 + 5s + 6}.$$

Here the denominator factorizes:

$$\frac{s + 1}{s^2 + 5s + 6} = \frac{s + 1}{(s + 2)(s + 3)} = \frac{2}{s + 3} - \frac{1}{s + 2},$$

from the 'cover-up' rule (see vol. 1, p. 64).

From Table I, 3, the inverse is

$$\underline{2e^{-3t} - e^{-2t}.}$$

(5) If $$\bar{y} = \frac{s-1}{s(s^2+6s+13)},$$

then $$\bar{y} = -\frac{1}{13}\frac{1}{s} + \frac{As+B}{s^2+6s+13},$$

i.e. $$s - 1 \equiv -\tfrac{1}{13}(s^2+6s+13) + (As+B)s.$$

On equating coefficients:

$$0 = -\tfrac{1}{13} + A, \quad 1 = -\tfrac{6}{13} + B,$$

giving $$A = \tfrac{1}{13}, \quad B = \tfrac{19}{13}.$$

Thus $$\bar{y} = -\frac{1}{13}\frac{1}{s} + \frac{1}{13}\frac{(s+19)}{s^2+6s+13}$$

$$= -\frac{1}{13}\frac{1}{s} + \frac{1}{13}\left\{\frac{(s+3)+16}{(s+3)^2+4}\right\}.$$

From Table I, 1 and Table II, 2 and 3,

$$\underline{y = -\tfrac{1}{13} + \tfrac{1}{13}\{e^{-3t}\cos 2t + 8e^{-3t}\sin 2t\}}.$$

(6) Find the inverse of $\dfrac{1}{(s^4-16)}$.

$$\frac{1}{(s^4-16)} = \frac{1}{(s^2-4)(s^2+4)} = \frac{1}{8}\left\{\frac{1}{(s^2-2^2)} - \frac{1}{(s^2+2^2)}\right\},$$

treating s^2 as a single element.

From Table I, 6 and 4, the inverse is

$$\tfrac{1}{8}\{\tfrac{1}{2}\sinh 2t - \tfrac{1}{2}\sin 2t\}$$

$$= \underline{\tfrac{1}{16}(\sinh 2t - \sin 2t)}.$$

(7) If $\bar{y} = \dfrac{1}{(s-1)^2(s^2+1)}$,

then $$\bar{y} = \frac{A}{(s-1)} + \frac{B}{(s-1)^2} + \frac{Cs+D}{(s^2+1)}.$$

Using normal partial fraction methods:

$$A = -\tfrac{1}{2}, \quad B = \tfrac{1}{2}, \quad C = \tfrac{1}{2}, \quad D = 0.$$

Thus $\bar{y} = -\frac{1}{2}\frac{1}{(s-1)} + \frac{1}{2}\frac{1}{(s-1)^2} + \frac{1}{2}\frac{s}{(s^2+1)}$.

Using Table I, 3 and 5 and Table II, 1,

$$y = -\tfrac{1}{2}e^t + \frac{1}{2}\frac{te^t}{1!} + \tfrac{1}{2}\cos t,$$

i.e.

$$\underline{y = \tfrac{1}{2}e^t(t-1) + \tfrac{1}{2}\cos t.}$$

(8) Show that the Laplace transform of $\cos(\omega t - \phi)$ is

$$\frac{s\cos\phi + \omega\sin\phi}{(s^2+\omega^2)}.$$

$$\cos(\omega t - \phi) = \cos\omega t\cos\phi + \sin\omega t\sin\phi.$$

The result then follows at once from Table I, 4 and 5.

Exercise 35

Find the Laplace transforms of:

1. $\sin(3t - \frac{1}{6}\pi)$.
2. $e^{-3t}\cos 2t$.
3. $\cos 2t - \cosh 4t$.
4. $e^{-t}(\cos t + 5\sin t)$.
5. $\sin 3t - 3t\cos 3t$.
6. $t^4e^{3t} - e^{-2t}\sin t$.

Find the inverses of:

7. (i) $\dfrac{3}{(s+3)(s-2)}$; (ii) $\dfrac{7}{s^2-5s+6}$;

(iii) $\dfrac{1}{(s+3)^2+3^2}$; (iv) $\dfrac{s+5}{(s+5)^2+2^2}$;

(v) $\dfrac{s}{(s^2+9)^2}$; (vi) $\dfrac{5}{(s^2+16)^2}$;

(vii) $\dfrac{7!}{(s-3)^8}$; (viii) $\dfrac{5}{s}$;

(ix) $\dfrac{s}{s^2-4} + \dfrac{1}{s^2+49}$; (x) $\dfrac{5}{(s-3)} + \dfrac{7s}{s^2+25}$.

8. (i) $\dfrac{7s^2+2s}{(s^2+4)(s^2-9)}$; (ii) $\dfrac{3s-4}{(s^2+9)(s-2)}$.

9. $\dfrac{1}{s^2(s^2+1)(s^2+4)}$.

Answers

1. $\sin(3t-\frac{1}{6}\pi) = \sin 3t \cos\frac{1}{6}\pi - \cos 3t \sin\frac{1}{6}\pi$

$$= \frac{\sqrt{3}}{2}\sin 3t - \frac{1}{2}\cos 3t.$$

Its transform is therefore

$$\frac{3\sqrt{3}}{2(s^2+9)} - \frac{s}{2(s^2+9)}.$$

2. $\dfrac{s+3}{(s+3)^2+4} = \dfrac{s+3}{s^2+6s+13}$.

3. $\dfrac{s}{s^2+4} - \dfrac{s}{s^2-16}$.

4. $\dfrac{(s+1)}{(s+1)^2+1} + \dfrac{5}{(s+1)^2+1} = \dfrac{s+6}{s^2+2s+2}$.

5. $\dfrac{54}{(s^2+9)^2}$. (Use formula 9 of Table I.)

6. $\dfrac{4!}{(s-3)^5} - \dfrac{1}{(s+2)^2+1} = \dfrac{24}{(s-3)^5} - \dfrac{1}{s^2+4s+5}$.

7. (i) $\frac{3}{5}(e^{2t}-e^{-3t})$; (ii) $7(e^{3t}-e^{2t})$;
(iii) $\frac{1}{3}e^{-3t}\sin 3t$; (iv) $e^{-5t}\cos 2t$;
(v) $\frac{1}{6}t\sin 3t$; (vi) $\frac{5}{128}(\sin 4t - 4t\cos 4t)$;
(vii) t^7e^{3t}; (viii) 5;
(ix) $\cosh 2t + \frac{1}{7}\sin 7t$; (x) $5e^{3t} + 7\cos 5t$.

8. (i) The partial fractions are

$$\frac{23}{26(s-3)} - \frac{19}{26(s+3)} - \frac{(2s-28)}{13(s^2+4)}.$$

The inverse is $\frac{23}{26}e^{3t} - \frac{19}{26}e^{-3t} - \frac{2}{13}\cos 2t + \frac{14}{13}\sin 2t$.

(ii) The partial fractions are

$$\frac{2}{13(s-2)} - \frac{(2s-35)}{13(s^2+9)}.$$

The inverse is $\frac{2}{13}e^{2t} - \frac{2}{13}\cos 3t + \frac{35}{39}\sin 3t$.

9. Treat as a function of s^2 and use 'cover-up' rule, giving s^2 values of 0, -1 and -4. The partial fractions are

$$\frac{1}{4s^2} - \frac{1}{3(s^2+1)} + \frac{1}{12(s^2+4)}.$$

The inverse is $\frac{1}{4}t - \frac{1}{3}\sin t + \frac{1}{24}\sin 2t$.

(*Note.* Problems concerning resistanceless circuits often give transforms such as this which can be treated in terms of s^2 only.)

10·3. Transforms of derivatives and solutions of equations

So far there has been no real indication as to *why* Laplace transforms are useful in solving linear differential equations with constant coefficients. On investigating the transforms of derivatives however, the reason becomes apparent.

It may be noted that the results which follow can be proved rigorously provided $y(t)$ and its derivatives are continuous and tend to definite values as $t \to 0$. In what follows D will signify d/dt. From definition, the Laplace transform of $dy/dt = Dy$ is given by

$$(\overline{Dy}) = \int_0^\infty e^{-st}\frac{dy}{dt}\,dt = [ye^{-st}]_0^\infty + s\int_0^\infty ye^{-st}\,dt \quad \text{(by parts).}$$

Let y_0 be the value of y as $t \to 0$. Now $ye^{-st} \to 0$ as $t \to \infty$, thus

$$(\overline{Dy}) = [0 - y_0] + s\bar{y},$$

i.e.

$$\underline{(\overline{Dy}) = s\bar{y} - y_0.} \quad (1)$$

Again

$$(\overline{D^2y}) = \int_0^\infty e^{-st}\left(\frac{d^2y}{dt^2}\right)dt = \left[e^{-st}\frac{dy}{dt}\right]_0^\infty$$

$$+ s\int_0^\infty e^{-st}\left(\frac{dy}{dt}\right)dt \quad \text{(by parts)}$$

$$= \left[e^{-st}\frac{dy}{dt}\right]_0^\infty + s(\overline{Dy}).$$

Let y_1 be the value of dy/dt as $t \to 0$. Using (1) above, and the fact that $(dy/dt)e^{-st} \to 0$ as $t \to \infty$ gives

$$(\overline{D^2y}) = [0 - y_1] + s(s\bar{y} - y_0),$$

i.e.
$$\underline{(\overline{D^2y}) = s^2\bar{y} - sy_0 - y_1.} \tag{2}$$

Continuing in this way, using y_r to denote the value of d^ry/dt^r as $t \to 0$,

$$\underline{(\overline{D^3y}) = s^3\bar{y} - s^2y_0 - sy_1 - y_2,} \tag{3}$$

and in general

$$\underline{(\overline{D^ny}) = s^n\bar{y} - s^{n-1}y_0 - s^{n-2}y_1 - \dots - sy_{n-2} - y_{n-1}.} \tag{4}$$

Now suppose that the solution of the equation

$$(D^2 + aD + b)y = f(t)$$

is required for $t > 0$, and it is given that y_0 and y_1 are the values of y and Dy at $t = 0$. Taking the Laplace transforms of each side of the equation, from (1) and (2) above:

$$(s^2\bar{y} - sy_0 - y_1) + a(s\bar{y} - y_0) + b\bar{y} = L\{f(t)\},$$

where $L\{f(t)\}$ denotes the transform of $f(t)$.

Solving for $\bar{y}$:

$$\underline{(s^2 + as + b)\bar{y} = L\{f(t)\} + ay_0 + (sy_0 + y_1).} \tag{5}$$

This equation is called the *subsidiary equation* corresponding to the given differential equation and its given initial conditions. For first- and second-order equations it soon becomes easy to write down this subsidiary equation at sight.

Equation (5) gives

$$\bar{y} = \frac{L\{f(t)\} + ay_0 + (sy_0 + y_1)}{(s^2 + as + b)}. \tag{6}$$

Now y_0, y_1 are given and $L\{f(t)\}$ can be evaluated using Tables I or II. Thus $\bar{y}$ can be found as a rational function of s. By use of partial fractions, or otherwise, this can be arranged as a sum of fractions of the type occurring in the tables; then y is found as the inverse of $\bar{y}$.

Equations of higher order can be treated in a similar way, as can also simultaneous linear differential equations.

The following worked examples will be found instructive.

EXAMPLE 1

Solve $(D^2 + 9)y = 0$ with $y = 4$, $dy/dt = 0$ at $t = 0$.

Transforming both sides:

$$(s^2\bar{y} - sy_0 - y_1) + 9\bar{y} = 0.$$

This gives
$$\bar{y} = \frac{sy_0 + y_1}{s^2 + 9}.$$

Now $y_0 = 4$ and $y_1 = 0$, therefore

$$\bar{y} = \frac{4s}{s^2 + 9}.$$

Thus
$$\underline{y = 4 \cos 3t.}$$

EXAMPLE 2

$(D^2 - 5D + 6)y = 3$, given $y = 2$, $Dy = 1$ at $t = 0$.

Transforming both sides:

$$(s^2\bar{y} - sy_0 - y_1) - 5(s\bar{y} - y_0) + 6\bar{y} = \frac{3}{s}.$$

This gives

$$(s^2 - 5s + 6)\bar{y} = \frac{3}{s} + sy_0 + y_1 - 5y_0.$$

Now $y_0 = 2$ and $y_1 = 1$, thus

$$(s^2 - 5s + 6)\bar{y} = \frac{3}{s} + 2s - 9$$

and
$$\bar{y} = \frac{2s^2 - 9s + 3}{s(s^2 - 5s + 6)} = \frac{2s^2 - 9s + 3}{s(s - 2)(s - 3)},$$

i.e.
$$\bar{y} = \frac{1}{2s} + \frac{7}{2(s - 2)} - \frac{2}{(s - 3)}.$$

$$\underline{\therefore\ y = \tfrac{1}{2} + \tfrac{7}{2}e^{2t} - 2e^{3t}.}$$

Note. It will give the student confidence in the use of this method if he verifies, in the case of these worked examples, that the solutions obtained *do* fit the given initial conditions and the given equation.

EXAMPLE 3

$(D + 4)^2y = \sin 2t$, given that $y = Dy = 0$ at $t = 0$.

Transforming both sides:

$$(s^2\bar{y} - sy_0 - y_1) + 8(s\bar{y} - y_0) + 16\bar{y} = L(\sin 2t) = \frac{2}{s^2 + 4},$$

$$(s^2 + 8s + 16)\bar{y} = sy_0 + y_1 + 8y_0 + \frac{2}{s^2 + 4},$$

$$(s^2 + 8s + 16)\bar{y} = 0 + 0 + 0 + \frac{2}{(s^2 + 4)},$$

i.e.
$$\bar{y} = \frac{2}{(s + 4)^2 (s^2 + 4)}$$

$$= \frac{1}{25(s + 4)} + \frac{1}{10(s + 4)^2} - \frac{(2s - 3)}{50(s^2 + 4)}.$$

$$\therefore\ y = \tfrac{1}{25} e^{-4t} + \tfrac{1}{10} te^{-4t} - \tfrac{1}{25} \cos 2t + \tfrac{3}{100} \sin 2t.$$

EXAMPLE 4

$(D + 2)y = t^3e^{-2t}$ given that $y = 1$ at $t = 0$.

Transforming both sides:

$$(s\bar{y} - y_0) + 2\bar{y} = L\{t^3e^{-2t}\} = \frac{3!}{(s + 2)^4} = \frac{6}{(s + 2)^4}.$$

Thus

$$(s + 2)\bar{y} = 1 + \frac{6}{(s + 2)^4},$$

$$\bar{y} = \frac{1}{(s + 2)} + \frac{6}{(s + 2)^5}.$$

$$\therefore\ y = e^{-2t} + 6\,\frac{t^4e^{-2t}}{4!},$$

$$y = e^{-2t}(1 + \tfrac{1}{4}t^4).$$

EXAMPLE 5

$(D^2 + n^2)y = \sin nt$, given that $y = y_0$, $Dy = y_1$ at $t = 0$.

$$(s^2\bar{y} - sy_0 - y_1) + n^2\bar{y} = L(\sin nt) = \frac{n}{s^2 + n^2},$$

giving
$$(s^2 + n^2)\bar{y} = sy_0 + y_1 + \frac{n}{s^2 + n^2}.$$

$$\therefore\ \bar{y} = y_0 \frac{s}{(s^2 + n^2)} + \frac{y_1}{(s^2 + n^2)} + \frac{n}{(s^2 + n^2)^2}.$$

Thus

$$y = y_0 \cos nt + \frac{y_1}{n} \sin nt + n \frac{1}{2n^3} (\sin nt - nt \cos nt),$$

i.e. $y = \dfrac{1}{2n^2} (1 + 2ny_1) \sin nt + \left(y_0 - \dfrac{t}{2n}\right) \cos nt.$

EXAMPLE 6

Solve the equation $(D^2 + 2D + 2)y = e^{-t}$.

This is a case where no initial conditions are given. A useful method is to write down the complementary function by the ordinary D-rules and find a particular integral using Laplace transforms and *assuming zero initial conditions*.

Thus:

A.E. $$m^2 + 2m + 2 = 0,$$

$$m = -1 \pm j.$$

C.F. $$y = e^{-t}(A \cos t + B \sin t). \tag{1}$$

P.I. Assuming zero initial conditions:

$$(s^2 + 2s + 2)\bar{y} = L(e^{-t}) = \frac{1}{(s + 1)}.$$

$$\therefore\ \bar{y} = \frac{1}{(s + 1)(s^2 + 2s + 2)}$$

$$= \frac{1}{(s + 1)} - \frac{(s + 1)}{s^2 + 2s + 2}$$

$$= \frac{1}{(s + 1)} - \frac{(s + 1)}{\{(s + 1)^2 + 1\}}.$$

Thus $$y = e^{-t} - e^{-t}\cos t. \tag{2}$$

From (1) and (2), the general solution is

$$y = e^{-t} + (A-1)e^{-t}\cos t + Be^{-t}\sin t,$$

As A is arbitrary, this may be written

$$\underline{y = e^{-t}(1 + B\sin t + C\cos t).}$$

It should now be obvious that speed in the use of the Laplace transformation method is dependent on acquiring skill, through practice, in arranging functions of s into fractions whose inverses can be quickly looked up in standard tables.

EXAMPLE 7. *Simultaneous differential equations*

Solve
$$(D^2 + 4)x - 2Dy = 2,$$
$$Dx + (D^2 + 4)y = 0,$$

given that $x = 1$, $y = Dx = Dy = 0$, at $t = 0$.

No change in method is required.
The transforms are:

$$(s^2\bar{x} - sx_0 - x_1) + 4\bar{x} - 2(s\bar{y} - y_0) = \frac{2}{s},$$

$$(s\bar{x} - x_0) + (s^2\bar{y} - sy_0 - y_1) + 4\bar{y} = 0.$$

These give the subsidiary equations

$$(s^2 + 4)\bar{x} - 2s\bar{y} = \frac{2}{s} + sx_0 + x_1 - 2y_0,$$

$$s\bar{x} + (s^2 + 4)\bar{y} = x_0 + sy_0 + y_1.$$

Now $x_0 = 1$, $y_0 = x_1 = y_1 = 0$.

Thus

$$(s^2 + 4)\bar{x} - 2s\bar{y} = \frac{2}{s} + s,$$

$$s\bar{x} + (s^2 + 4)\bar{y} = 1.$$

Solving as ordinary simultaneous equations:

$$\bar{x}\{(s^2+4)^2+2s^2\} = \left(\frac{2}{s}+s\right)(s^2+4)+2s,$$

$$\bar{y}\{(s^2+4)^2+2s^2\} = (s^2+4)-s\left(\frac{2}{s}+s\right).$$

These give

$$\bar{x} = \frac{s^4+8s^2+8}{s(s^4+10s^2+16)}, \quad \bar{y} = \frac{2}{(s^4+10s^2+16)},$$

i.e. $$\bar{x} = \frac{s^4+8s^2+8}{s(s^2+2)(s^2+8)}, \quad \bar{y} = \frac{2}{(s^2+2)(s^2+8)}.$$

In partial fractions:

$$\bar{x} = \frac{1}{2s}+\frac{s}{3(s^2+2)}+\frac{s}{6(s^2+8)},$$

$$\bar{y} = \frac{1}{3(s^2+2)}-\frac{1}{3(s^2+8)}.$$

Thus $$\underline{x = \tfrac{1}{2}+\tfrac{1}{3}\cos\sqrt{2}t+\tfrac{1}{6}\cos 2\sqrt{2}t,}$$

$$\underline{y = \frac{1}{3\sqrt{2}}\sin\sqrt{2}t-\frac{1}{6\sqrt{2}}\sin 2\sqrt{2}t.}$$

Exercise 36

1. Solve $(D+4)y = 1$, with $y = 2$ at $t = 0$.

2. Solve $(D^2+5D+6)y = 3$, with $y = 2$, $Dy = 0$ at $t = 0$.

3. Solve $(D+1)^2y = \sin t$, with $y = 3$, $Dy = 1$ at $t = 0$.

4. Solve $(D^2+n^2)y = \sin \omega t$, where $\omega \neq n$ and $y = Dy = 0$ at $t = 0$.

5. Solve Questions 5 and 14 of Exercise 27 using the Laplace transformation method.

6. Solve $(D^2+2D+4)y = 1+t^2$, with $y = 4$, $Dy = 2$ at $t = 0$.

7. If $h(s) = \int_0^\infty e^{-st} f(t)\, dt$, find $h(s)$ if (i) $f(t) = e^{-at}$; (ii) $f(t) = t$. Show that if $f(t)$ is finite at infinity

$$\int_0^\infty e^{-st} f'(t)\, dt = s h(s) - f(0).$$

If $f'(t) + 2f(t) = e^{-t}$ and $f(0) = 0$, show that

$$h(s) = \frac{1}{(s+1)} - \frac{1}{(s+2)}$$

and by inverting the result (i) determine $f(t)$ in this case. [L.U.]

8. Solve the simultaneous differential equations

$$\frac{dx}{dt} + \frac{dy}{dt} + 2x + y = e^{-3t},$$

$$\frac{dy}{dt} + 5x + 3y = 5e^{-2t},$$

given that when $t = 0$, $x = -1$ and $y = 4$. [L.U.]

9. If $L\frac{dx}{dt} - Ry = E$, $RCy + q = 0$, $x + y = \frac{dq}{dt}$, where L, R, C, E are constants, find the differential equation relating q and t and show that q oscillates only if $L < 4CR^2$. If this condition is satisfied and $q = x = 0$ at $t = 0$, show that

$$q = EC - \frac{E}{2Rp} e^{-t/2RC} \{\sin pt + 2RC\, p \cos pt\},$$

where $$p^2 = \frac{1}{LC} - \frac{1}{4R^2C^2}.$$ [L.U.]

10. Solve the equations

$$(D+2)x + Dy = 0,$$

$$(D-2)x + 4Dy = e^{-2t},$$

given that $x = 2$, $y = 0$, when $t = 0$.

Answers

1. $y = \frac{1}{4} + \frac{7}{4}e^{-4t}$.
2. $y = \frac{1}{2} + \frac{9}{2}e^{-2t} - 3e^{-3t}$.
3. $y = \frac{9}{2}te^{-t} + \frac{7}{2}e^{-t} - \frac{1}{2}\cos t$.

4. $y = \dfrac{\omega}{n(\omega^2 - n^2)} \sin nt - \dfrac{1}{(\omega^2 - n^2)} \sin \omega t.$

5. As for Exercise 27.

6. $y = \frac{3}{4}e^{-t}\left(5 \cos \sqrt{3}t + \dfrac{8}{\sqrt{3}} \sin \sqrt{3}t\right) + \frac{1}{4}(1 - t + t^2).$

7. (i) $\dfrac{1}{(s + a)}$, (ii) $\dfrac{1}{s^2}$.

$f(t) = e^{-t} - e^{-2t}.$

[This question consists mainly of standard bookwork, using functional notation throughout.]

8. $x = 3 \sin t = 2 \cos t + e^{-2t};$

$y = -\frac{7}{2} \sin t + \frac{9}{2} \cos t - \frac{1}{2}e^{-3t}.$

9. The differential equation is

$$(RCLD^2 + LD + R)q = ERC.$$

10. $x = \frac{9}{4}e^{-\frac{10}{3}t} - \frac{1}{4}e^{-2t}; \quad y = \frac{9}{10}(1^z - e^{-\frac{10}{3}t}).$

10·4. Applications to electrical circuits

The examples which follow will show the application of Laplace transformations to electrical circuits. Before proceeding with the examples however, a simple theorem will be proved which is often useful for circuits including capacitors.

The theorem states: *if $\bar{y}(s)$ is the Laplace transform of $y(t)$ then the transform of* $\int_0^t y(t')\,dt'$ is $\dfrac{1}{s}\bar{y}(s)$.

The proof is as follows:

Given $$\bar{y}(s) = \int_0^\infty e^{-st}y(t)\,dt,$$

then $$L\left\{\int_0^t y(t')\,dt'\right\} = \int_0^\infty e^{-st}\left\{\int_0^t y(t')\,dt'\right\}dt$$

$$= \left[-\frac{1}{s}e^{-st}\left\{\int_0^t y(t')\,dt'\right\}\right]_0^\infty$$

$$+\frac{1}{s}\int_0^\infty e^{-st}y(t)\,dt. \qquad (1)$$

Integration by parts has been used, and also the fact that

$$\frac{d}{dt}\left\{\int_0^t y(t')\,dt'\right\} = y(t).$$

As $\int_0^t y(t')\,dt' = 0$ when $t = 0$ and also $e^{-st} \to 0$ as $t \to \infty$, the first term of (1) (in the square brackets) is zero.

Thus $\quad L\left\{\int_0^t y(t')\,dt'\right\} = \frac{1}{s}\int_0^\infty e^{-st}y(t)\,dt = \frac{1}{s}\bar{y}(s).$

Once the theorem has been proved and there is no risk of confusion, there is no need to distinguish between t and t' as $\int_0^t y(t')\,dt' = \int_0^t y(t)\,dt$, the integral being a function of t.

The main use of this theorem in electrical work arises in the case of a capacitor whose charge q is solely due to the flow of a current i for time t. In this case $q = \int_0^t i\,dt$, and the voltage drop is given by $\frac{q}{C} = \frac{1}{C}\int_0^t i\,dt$. The Laplace transform is then given by $\frac{\bar{q}}{C} = \frac{\bar{\imath}}{sC}$, using the fact that $\frac{\bar{\imath}}{sC}$ is the transform of

$$\frac{1}{C}\int_0^t i\,dt. \tag{2}$$

Example 1 shows a case of this transform being used.

When the charge on the capacitor is *not* solely due to a current i flowing for a time t, it is better to use $i = dq/dt$ and work in terms of q. Example 2 illustrates this.

A further example of the use of the theorem is:

As $\frac{\omega}{s^2 + \omega^2}$ is the transform of $\sin\omega t$, then

$\frac{\omega}{s(s^2 + \omega^2)}$ is the transform of $\int_0^t \sin\omega t\,dt = \frac{1}{\omega}(1 - \cos\omega t)$.

EXAMPLE 1

A constant voltage E_0 is applied at $t = 0$ to a series L-C-R circuit with $i = 0 = q$ at $t = 0$.

The equation is

$$L\frac{di}{dt} + Ri + \frac{q}{C} = E_0.$$

The charge q is solely due to the current i acting for time t, therefore, using (1) above,

$$L\frac{di}{dt} + Ri + \frac{\int_0^t idt}{C} = E_0$$

gives as its transform

$$L(p\bar{\imath} - i_0) + R\bar{\imath} + \frac{\bar{\imath}}{Cp} = \frac{E_0}{p},$$

As $i_0 = 0$, this gives

$$\left(Lp + R + \frac{1}{Cp}\right)\bar{\imath} = \frac{E_0}{p}. \tag{1}$$

$$\therefore\ \bar{\imath} = \frac{E_0}{\left(Lp^2 + Rp + \frac{1}{C}\right)} = \frac{E_0}{L\left(p^2 + \frac{R}{L}p + \frac{1}{CL}\right)}.$$

Completing the square in the denominator and writing

$$\underline{k = \frac{R}{2L},} \quad \underline{n^2 = \frac{1}{LC} - \frac{R^2}{4L^2},} \tag{2}$$

$$\bar{\imath} = \frac{E_0}{L}\frac{1}{\{(p+k)^2 + n^2\}}. \tag{3}$$

From the tables of transforms three cases arise:

$$\left.\begin{aligned} &\text{(i)} \quad n^2 > 0: \quad i = \frac{E_0}{nL}e^{-kt}\sin nt, \\ &\text{(ii)} \quad n^2 = 0: \quad i = \frac{E_0}{L}te^{-kt}, \\ &\text{(iii)} \quad n^2 < 0: \quad i = \frac{E_0}{aL}e^{-kt}\sinh at, \end{aligned}\right\}$$

where $a^2 = -n^2 = \dfrac{R^2}{4L^2} - \dfrac{1}{LC}$.

Note. If i is current and V voltage and the transformed equation is of the form $\bar{\imath}.Z(s) = \bar{V}$, then $Z(s)$ is often called the *generalized impedance.*

In the example above this would be $\left(Ls + R + \dfrac{1}{Cs}\right)$.

EXAMPLE 2

A capacitor of capacitance C, charged to a voltage E_0, is discharged at $t = 0$ through an inductive reactance L, R (see fig. 49).

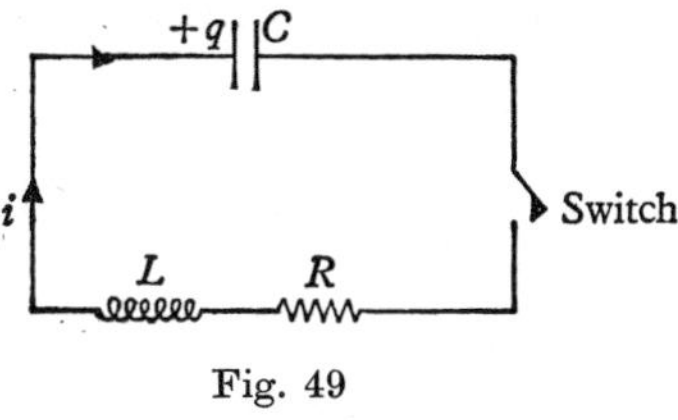

Fig. 49

Here the charge on the capacitor is *not* solely due to current i, and it is best in this case to work from the fact that $i = dq/dt$.

The initial conditions are $q_0 = CE_0$, $i_0 = 0$.

The equations are

$$L\frac{di}{di} + Ri + \frac{q}{C} = 0,$$

$$i = \frac{dq}{dt}.$$

Their transforms are

$$L(s\bar{\imath} - i_0) + R\bar{\imath} + \frac{\bar{q}}{C} = 0,$$

$$\bar{\imath} = (s\bar{q} - q_0).$$

Using the values of q_0 and i_0:

$$(Ls + R)\,\bar{\imath} + \frac{\bar{q}}{C} = 0, \tag{1}$$

$$\bar{\imath} = s\bar{q} - CE_0. \tag{2}$$

Solving for $\bar{q}$:

$$(Ls + R)\,(s\bar{q} - CE_0) + \frac{\bar{q}}{C} = 0,$$

$$\left(Ls^2 + Rs + \frac{1}{C}\right)\bar{q} = CE_0(Ls + R).$$

$$\therefore\ \bar{q} = \frac{CE_0(Ls + R)}{L\{(s + k)^2 + n^2\}},$$

where once again

$$k = \frac{R}{2L}, \quad n_2 = \frac{1}{LC} - \frac{R^2}{4L^2}. \tag{3}$$

Taking the case where $n^2 > 0$,

$$\bar{q} = \frac{CE_0(s+k)}{\{(s+k)^2+n^2\}} + \frac{CE_0\left(\dfrac{R}{L}-k\right)}{\{(s+k)^2+n^2\}}.$$

Using the fact that $R/L = 2k$:

$$\bar{q} = CE_0\left[\frac{s+k}{\{(s+k)^2+n^2\}} + \frac{k}{\{(s+k)^2+n^2\}}\right].$$

$$\therefore\ q = CE_0 e^{-kt}\left(\cos nt + \frac{k}{n}\sin nt\right). \tag{4}$$

Note. If i were needed it could now be found as dq/dt.

Alternatively, (1) and (2) above could be solved originally for $\bar{\imath}$ instead of $\bar{q}$, giving

$$\left(Ls + R + \frac{1}{Cs}\right)\bar{\imath} = -\frac{E_0}{s}. \tag{5}$$

This equation is interesting when compared with equation (1) of Example 1. They are the same except for the negative sign in equation (5) above. This arises from the fact that the current starts flowing *away* from the positive side of the condenser.

EXAMPLE 3

The case of a series L-R circuit with a voltage $E\sin\omega t$ applied at $t = 0$, and initial current zero.

The equation is $\quad L\dfrac{di}{dt} + Ri = E\sin\omega t.$

The transform is $\quad L(s\bar{\imath} - i_0) + R\bar{\imath} = \dfrac{E\omega}{s^2+\omega^2}.$

As $i_0 = 0$, $\quad \bar{\imath} = \dfrac{E\omega}{(s^2+\omega^2)(R+sL)}.$

By partial fractions:

$$\bar{i} = E\omega \left\{ \frac{R - Ls}{(R^2 + \omega^2 L^2)(s^2 + \omega^2)} + \frac{L}{(R^2 + \omega^2 L^2)\left(s + \frac{R}{L}\right)} \right\}.$$

$$\therefore\ i = E\omega \left\{ \frac{R}{\omega(R^2 + \omega^2 L^2)} \sin \omega t - \frac{L}{(R^2 + \omega^2 L^2)} \cos \omega t + \frac{L}{(R^2 + \omega^2 L^2)} e^{-(R/L)t} \right\},$$

i.e. $$i = \frac{E}{(R^2 + \omega^2 L^2)} \{R \sin \omega t - \omega L \cos \omega t + \omega L e^{-(R/L)t}\}$$

$$= \frac{E}{(R^2 + \omega^2 L^2)^{\frac{1}{2}}} \sin (\omega t - \phi) + \frac{E\omega L}{(R^2 + \omega^2 L^2)} e^{-(R/L)t},$$

where $\phi = \tan^{-1} \dfrac{\omega L}{R}$.

EXAMPLE 4

Fig. 50 shows two circuits coupled by mutual inductance M. A constant voltage E_0 is applied to the primary at time $t = 0$

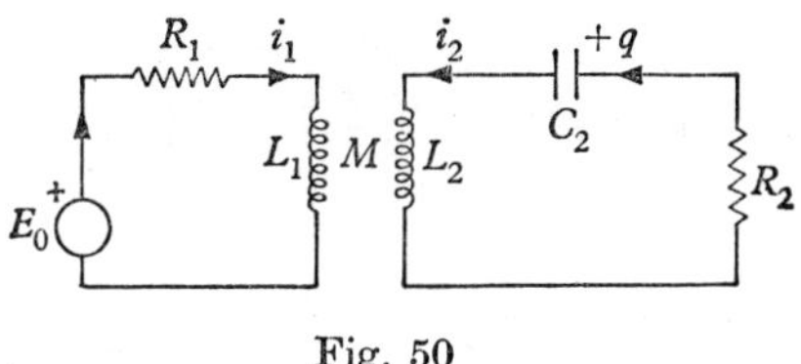

Fig. 50

with $i_1 = i_2 = 0$ as initial conditions. It is required to find the secondary current i_2.

The equations are

$$L_1 \frac{di_1}{dt} + R_1 i_1 + M \frac{di_2}{dt} = E_0,$$

$$L_2 \frac{di_2}{dt} + R_2 i_2 + \frac{1}{C_2} \int_0^t i_2\, dt + M \frac{di_1}{dt} = 0.$$

Taking the transforms and remembering that $i_1 = 0 = i_2$ at $t = 0$:

$$L_1 s\bar{i}_1 + R_1 \bar{i}_1 + Ms\bar{i}_2 = \frac{E_0}{s},$$

$$L_2 s\bar{i}_2 + R_2 \bar{i}_2 + \frac{\bar{i}_2}{sC_2} + Ms\bar{i}_1 = 0,$$

i.e.
$$(L_1 s + R_1)\,\bar{i}_1 + Ms\bar{i}_2 = \frac{E_0}{s}, \tag{1}$$

$$Ms\bar{i}_1 + \left(L_2 s + R_2 + \frac{1}{C_2 s}\right)\bar{i}_2 = 0. \tag{2}$$

Solving (1) and (2) for $\bar{i}_2$:

$$\bar{i}_2 = \frac{ME_0}{\left\{M^2 s^2 - (L_1 s + R_1)\left(L_2 s + R_2 + \frac{1}{C_2 s}\right)\right\}}$$

$$= \frac{ME_0}{\left\{(M^2 - L_1 L_2)\, s^2 - (R_1 L_2 + R_2\, L_1)\, s - \left(R_1 R_2 + \frac{L_1}{C_2}\right) - \frac{R_1}{C_2 s}\right\}}.$$

As $L_1 L_2$ is usually greater than M^2, it is best to write:

$$\bar{i}_2 = \frac{-ME_0}{\left\{(L_1 L_2 - M^2)\, s^2 + (R_1 L_2 + R_2 L_1)\, s + \left(R_1 R_2 + \frac{L_1}{C_2}\right) + \frac{R_1}{C_2 s}\right\}}. \tag{3}$$

Taking the simplest case, when R_1, the primary resistance, is negligible:

$$\bar{i}_2 = \frac{-ME_0}{\left\{(L_1 L_2 - M^2)\, s^2 + R_2 L_1 s + \frac{L_1}{C_2}\right\}}$$

$$= \frac{-ME_0}{(L_1 L_2 - M^2)} \frac{1}{\left[\left\{s + \frac{R_2 L_1}{2(L_1 L_2 - M^2)}\right\}^2 + \left\{\frac{L_1}{C_2(L_1 L_2 - M^2)} - \frac{R_2^2 L_1^2}{4(L_1 L_2 - M^2)^2}\right\}\right]}.$$

Taking $k = \dfrac{R_2L_1}{2(L_1L_2 - M^2)}$ and $\omega^2 = \dfrac{L_1}{C_2(L_1L_2 - M^2)} - k^2$,

then

$$\bar{i}_2 = \frac{-ME_0}{(L_1L_2 - M^2)}\left\{\frac{1}{(s+k)^2 + \omega^2}\right\}.$$

Thus $$i_2 = \frac{-ME_0}{\omega(L_1L_2 - M^2)} e^{-kt} \sin \omega t.$$

In conclusion, it should be remarked that Laplace transforms can be used in more complicated cases. For example (i) when a circuit has a steady voltage applied for a given time interval or (ii) a circuit with a surge (impulsive) voltage applied. To cope with these, special functions called respectively the Heaviside unit function and Dirac's delta function are used. These will not be dealt with in this book. In addition, Laplace transform methods can be adapted to the solution of partial differential equations. Again, the student is referred to more advanced text-books.

Exercise 37

1. A voltage $E \sin(nt + \phi)$ is applied at $t = 0$ to a circuit consisting of an inductance L and capacitance C in series. If $n^2 = 1/LC$ and the initial current and charge are zero, show that the current at any time is

$$\frac{E}{2nL}\{nt \sin(nt + \phi) + \sin\phi \sin nt\}.$$

[Deal with the transform of $E \sin(nt + \phi)$ as in Example 8 of §10·23.]

2. A circuit contains a capacitor of capacitance C, a wire of resistance R, and a coil of self-inductance $L = 2R^2C$ and negligible resistance. The coil is in parallel with the wire and the capacitor is in series with them. At time $t = 0$ the charge on the capacitor is Q_0 and there is no current passing through the coil. Show that the charge of the capacitor at a later time t is $Q_0 e^{-kt}(\cos kt - \sin kt)$, where $k = \dfrac{1}{2RC}$. [C.U.]

3. In the circuit shown in fig. 51 the voltage $E \sin \omega t$ is applied at time $t = 0$. Find the value of i_1 at a later time t if the initial currents are zero.

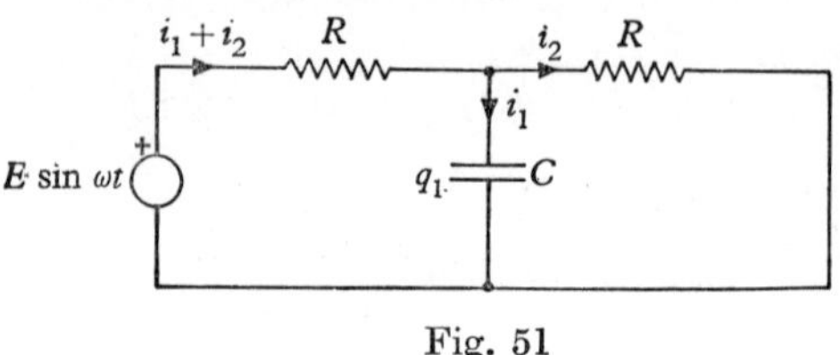

Fig. 51

4. The terminals of a generator producing a variable voltage $V(t)$ are connected through a resistance R in series with a self-inductance L. A capacitance C is in parallel with R. Find the differential equation satisfied by the current J through the resistance. Show that if

$$L < 4R^2C \quad \text{and} \quad V(t) = \begin{cases} 0 & (t < 0), \\ V_0 & (t \geqslant 0), \end{cases}$$

where V_0 is a constant, then for $t > 0$, J is given by

$$J(t) = \frac{V_0}{R}\left[1 - \left(\cos \omega t + \frac{\alpha}{\omega} \sin \omega t\right)e^{-\alpha t}\right],$$

where $$\alpha = \frac{1}{2RC}, \quad \omega^2 = \frac{1}{LC} - \frac{1}{4R^2C^2}.$$ [C.U.]

5. Fig. 52 shows a parallel circuit supplied with a *constant current* of value E/R for $t > 0$. Find the voltage drop across the capacitor at any later time t. The initial values are zero.

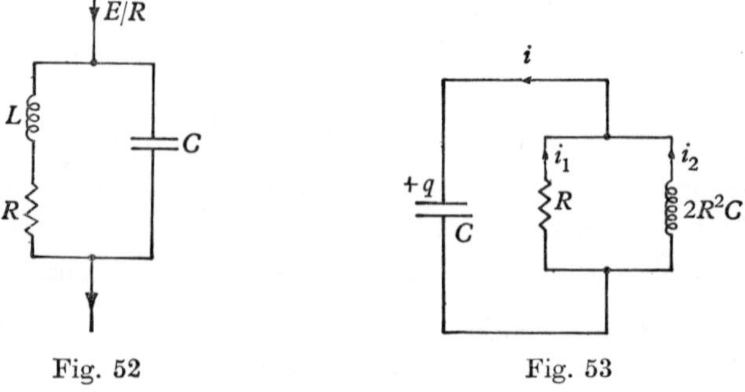

Fig. 52

Fig. 53

6. A circuit consists of an inductive resistance L_1, R_1 in series with a combination of R, L, C in series. $L = L_1 = 1$ H, $R = 8\ \Omega$, $R_1 = 4\ \Omega$ and $C = 0{\cdot}02$ F. The R–C–L combination can be short-circuited by a switch S. At $t = 0$, when a steady current of 10 A, from a battery of 40 V, is flowing through the circuit with S closed, the switch S is opened. Show that, for $t > 0$, the subsequent current is given by

$$i = \tfrac{5}{4}e^{-3t}(4\cos 4t + \sin 4t).$$

Deduce that the current in the circuit changes impulsively from 10 A to 5 A at the instant the switch S is opened.

ANSWERS

1. $$\left[\bar{\imath} = \frac{E}{L}\left\{\frac{sn\cos\phi}{(s^2+n^2)^2} + \frac{\sin\phi}{(s^2+n^2)} - \frac{n^2\sin\phi}{(s^2+n^2)^2}\right\}.\right.$$

Then use Table I, formulae, 8, 4 and 9.]

1. [With the notation of fig. 53, use

$$i = \frac{dq}{dt}, \quad \therefore\ \bar{\imath} = s\bar{q} - Q_0;$$

$$i = i_1 + i_2, \quad \therefore\ \bar{\imath} = \bar{\imath}_1 + \bar{\imath}_2;$$

$$\frac{q}{C} + i_1 R = 0, \quad \therefore\ \frac{\bar{q}}{C} + \bar{\imath}_1 R = 0;$$

$$\frac{q}{C} + 2R^2C\frac{di_2}{dt} = 0, \quad \therefore\ \frac{\bar{q}}{C} + 2R^2C(s\bar{\imath}_2) = 0.$$

Solving for $\bar{q}$ gives $\bar{q} = \dfrac{Q_0 s}{(s^2 + 2ks + 2k^2)}$.]

3. $$i_1 = \frac{E\omega C}{(4 + \omega^2C^2R^2)}\{2\cos\omega t + \omega CR\sin\omega t - 2e^{-2t/CR}\}.$$

4. $(CLRD^2 + LD + R)\,J = V$, giving, with $V = V_0$,

$$J = \frac{V_0}{RCLs\{(s+\alpha)^2 + \omega^2\}} = \frac{V_0}{R}\left\{\frac{1}{s} - \frac{(s+2\alpha)}{(s+\alpha)^2+\omega^2}\right\}.$$

[Note that $1/CL = \alpha^2 + \omega^2$.]

5. $$E - E\,e^{-kt}\left\{\cos nt + \left(\frac{R}{2nL} - \frac{1}{nRC}\right)\sin nt\right\},$$

where $$k = \frac{R}{2L} \quad\text{and}\quad n^2 = \frac{1}{LC} - \frac{R^2}{4L^2}.$$

[The equation is $\dfrac{q}{C} = L\dfrac{d}{dt}\left(\dfrac{E}{R} - i\right) + R\left(\dfrac{E}{R} - i\right)$, where i is the current through the capacitor and q is the charge of the capacitor. When taking transforms, with initial values zero, note that the transform of

$$L\frac{d}{dt}\left(\frac{E}{R} - i\right) \text{ is } Ls\left(\frac{E}{Rs} - \bar{\imath}\right) = \frac{LE}{R} - Ls\bar{\imath},$$

whilst that of $R\left(\dfrac{E}{R} - i\right)$ is $\dfrac{E}{s} - R\bar{\imath}$.

Using $i = dq/dt$ and therefore $\bar{\imath} = s\bar{q}\;(q_0 = 0)$, gives

$$\left(Ls^2 + Rs + \frac{1}{C}\right)\bar{q} = \frac{EL}{R} + \frac{E}{s}.$$

After solving for q, the voltage drop across the capacitor is q/C.]

6. Note that the Laplace transform method will give correct results for the state of a circuit at $t = 0_+$ even when impulsive readjustments of current have taken place, due to switching, for example. Each current, or charge, must be given its own initial value at $t = 0$. In Q6, take *separate* currents i_1, i_2 for the L_1, R_1 and L, R, C combinations; then $(i_1)_0 = 10$ and $(i_2)_0 = 0$. *After* taking transforms and putting in these initial values, use $\bar{\imath}_1 = \bar{\imath}_2$ for $t > 0$.

CHAPTER 11

INTRODUCTION TO PARTIAL DIFFERENTIAL EQUATIONS

11·1. Formation

Partial differential equations can be formed in much the same way as ordinary differential equations, namely, by the elimination of arbitrary constants by means of differentiation. In addition, however, arbitrary *functions* can also be eliminated.

EXAMPLES

(1) Eliminate a and b from $z = ax^2 + by^2$.

$$\frac{\partial z}{\partial x} = 2ax, \quad \text{giving} \quad a = \frac{1}{2x}\frac{\partial z}{\partial x}.$$

$$\frac{\partial z}{\partial y} = 2by, \quad \text{giving} \quad b = \frac{1}{2y}\frac{\partial z}{\partial y}.$$

Thus,
$$\underline{z = \frac{1}{2}\left(x\frac{\partial z}{\partial x} + y\frac{\partial z}{\partial y}\right).}$$

This is a first-order partial differential equation.

(2) Eliminate the arbitrary function f, when

$$z = yf(x/y).$$

Now
$$\frac{\partial z}{\partial x} = yf'\left(\frac{x}{y}\right)\frac{1}{y} = f'(x/y). \tag{1}$$

And
$$\frac{\partial z}{\partial y} = f\left(\frac{x}{y}\right) + yf'(x/y)\left(-\frac{x}{y^2}\right). \tag{2}$$

From the original equation, $f\left(\frac{x}{y}\right) = \frac{z}{y}$, and from equation (1),

$$f'\left(\frac{x}{y}\right) = \frac{\partial z}{\partial x}.$$

Putting these into equation (2) gives

$$\frac{\partial z}{\partial y} = \frac{z}{y} - \frac{x}{y}\frac{\partial z}{\partial x},$$

giving
$$x\frac{\partial z}{\partial x} + y\frac{\partial z}{\partial y} = z.$$

It is obvious that the reverse process of finding a solution, given the partial differential equation, is not easy, especially if a completely general solution is required. Both arbitrary constants and arbitrary functions may arise. In this respect it may be noted that solutions containing the maximum number of arbitrary constants are known as *complete integrals*, whilst solutions containing the maximum number of arbitrary functions are known as *general integrals*. In the majority of cases which arise in engineering only special solutions are required which fit known initial or boundary conditions.

11·2. Solution with variables separated

When it is not easy to find the general solution to a partial differential equation, it is often possible to find a particular type of solution in which the variables are separated.

The starting point of this technique is to assume a solution which is the product of two (or more) functions each containing only one of the independent variables. The following example should make the method clear.

EXAMPLE

If the equation

$$x^2\frac{\partial^2 u}{\partial x^2} + x\frac{\partial u}{\partial x} + \frac{\partial^2 u}{\partial y^2} = 0$$

has a solution of the form $u = XY$, where X, Y are respectively functions of x and y only, find the differential equations satisfied by X and Y and solve them when Y involves real trigonometric functions only.

If $\dfrac{\partial u}{\partial x} = -\cos 2y$ when $x = a$, and u tends to zero as x tends to infinity, find u. [L.U.]

Substituting $u = XY$ into the given equation, remembering that X is a function of x only, and Y a function of y only:

$$x^2 Y \frac{d^2X}{dx^2} + xY \frac{dX}{dx} + X \frac{d^2Y}{dy^2} = 0.$$

$$\therefore \; Y\left(x^2 \frac{d^2X}{dx^2} + x \frac{dX}{dx}\right) = -X \frac{d^2Y}{dy^2},$$

giving

$$\frac{1}{X}\left(x^2 \frac{d^2X}{dx^2} + x \frac{dX}{dx}\right) = -\frac{1}{Y}\frac{d^2Y}{dy^2}. \qquad (1)$$

The left-hand side of (1) is a function of x *only*, and the right-hand side is a function of y *only*, therefore the only possibility for equality is that each side must be a constant, c say.

(*Note.* This argument is essential for this type of solution.)

Thus $$x^2 \frac{d^2X}{dx^2} + x \frac{dX}{dx} = cX \qquad (2)$$

and $$\frac{d^2Y}{dy^2} = -cY,$$

i.e. $$\frac{d^2Y}{dy^2} + cY = 0. \qquad (3)$$

Now it is given that Y involves only real trigonometric functions, therefore c must be positive.

Let $c = p^2$.

Then $$x^2 \frac{d^2X}{dx^2} + x \frac{dX}{dx} - p^2X = 0 \qquad (4)$$

and $$\frac{d^2Y}{dy^2} + p^2Y = 0. \qquad (5)$$

Solving equation (5): $$\underline{Y = A\cos py + B\sin py.} \qquad (6)$$

To solve equation (4), put $x = e^t$, then

$$x^2 \frac{d^2X}{dx^2} + x \frac{dX}{dx} = \frac{d^2X}{dt^2} \quad \text{(see § 7·5).}$$

Thus $$\frac{d^2X}{dt^2} - p^2X = 0,$$

giving $$X = Ce^{pt} + Fe^{-pt}.$$

Thus $$X = Cx^p + Fx^{-p}. \tag{7}$$

From (6) and (7):

$$u = (Cx^p + Fx^{-p})(A \cos py + B \sin py)$$

and $$\frac{\partial u}{\partial x} = p(Cx^{p-1} - Fx^{-p-1})(A \cos py + B \sin py).$$

If $\partial u / \partial x = -\cos 2y$ when $x = a$, then

$$-\cos 2y = p(Ca^{p-1} - Fa^{-p-1})(A \cos py + B \sin py)$$

for all y.

This can be so only if

$$B = 0, \quad p = 2 \tag{8}$$

and $$2A\left(Ca - \frac{F}{a^3}\right) = -1. \tag{9}$$

Thus

$$u = \left(Cx^2 + \frac{F}{x^2}\right)A \cos 2y = \left(ACx^2 + \frac{AF}{x^2}\right) \cos 2y. \tag{10}$$

Now, if $u \to 0$ as $x \to \infty$, from (10), $AC = 0$. $\therefore$ $C = 0$. In this case (9) and (10) become

$$-\frac{2AF}{a^3} = -1 \quad \text{and} \quad u = \frac{AF}{x^2} \cos 2y;$$

i.e. $$AF = \frac{a^3}{2} \quad \text{and} \quad u = \frac{a^3}{2x^2} \cos 2y.$$

11·3. Some common types of equations

Three of the most frequently occurring equations in engineering are given below.

(*a*) *The diffusion equation* (or equation of heat conduction):

$$\frac{\partial^2 y}{\partial x^2} - \frac{1}{c^2}\frac{\partial y}{\partial t} = 0. \tag{1}$$

This equation arises in line transmission, the laminar motion of a viscous fluid, the diffusion of solids and liquids, heat conduction, etc.

(*b*) *The wave equation of one dimension*:

$$\frac{\partial^2 y}{\partial x^2} - \frac{1}{c^2}\frac{\partial^2 y}{\partial t^2} = 0. \tag{2}$$

This equation arises in the study of longitudinal and transverse vibrations and in most wave problems, whether they be electric, fluid or sound waves. It is sometimes called D'Alembert's equation.

(*c*) *Laplace's equation of two dimensions*:

$$\frac{\partial^2 u}{\partial x^2} + \frac{\partial^2 u}{\partial y^2} = 0. \tag{3}$$

This equation arises in the general study of the flow of electric current, heat or fluids and in problems on electrostatics.

The above equations have their corresponding forms in two or three dimensions as the case may be. Many techniques are used in finding special solutions of them to fit given conditions; for example, Laplace transformations, Fourier series, etc.

11·4. A general solution of the wave equation

It is easily verified that $y = f(x + ct) + F(x - ct)$, where f and F refer to completely arbitrary functions, is a solution of the wave equation

$$\frac{\partial^2 y}{\partial x^2} - \frac{1}{c^2}\frac{\partial^2 y}{\partial t^2} = 0 \quad \text{(see Exercise 38, Question 1).}$$

Consider the equation $y = f(x + ct)$.

At $t = 0$, $y = f(x)$, which, if graphed, will give some curve.

At $t = 1$, $y = f(x + c)$; if graphed, this will give the same curve moved a distance c to the *left*.

At $t = 2$, $y = f(x + 2c)$, and the curve is now a distance $2c$ to the left.

It appears therefore that $y = f(x + ct)$ represents, graphically, some curve moving to the left with constant speed c.

Similarly, $y = F(x - ct)$ represents a curve travelling to the *right* with speed c.

The general equation $y = f(x + ct) + F(x - ct)$ represents a combination of motions travelling, one to the left and one to the right, with constant speed c.

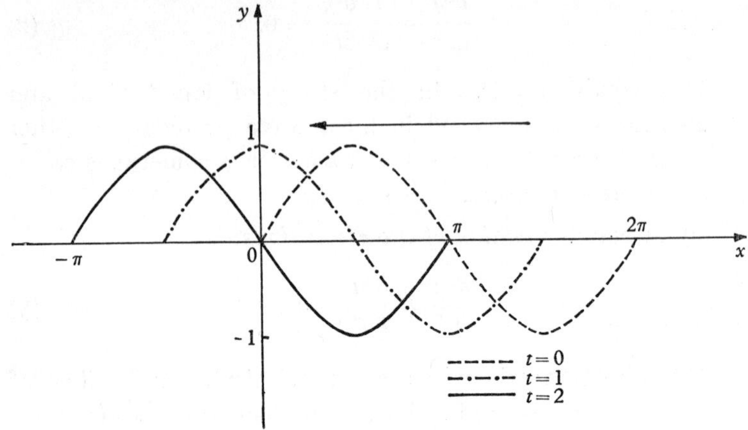

Fig. 54

As an illustration, fig. 54 shows the case of $y = f(x + ct)$ when the function is taken to be a sine function and c is taken as $\frac{1}{2}\pi$. The figure shows the position of the sine wave at $t = 0$, 1 and 2.

If
$$y = \sin(x + \tfrac{1}{2}\pi t),$$
at $t = 0$, $y = \sin x$; at $t = 1$, $y = \sin(x + \frac{1}{2}\pi)$; at $t = 2$, $y = \sin(x + \pi)$.

Two waves travelling in opposite directions can give rise to *stationary* waves. As an example of this, in the equation

$$y = f(x + ct) + F(x - ct)$$

let both of the functions f and F be sine waves. The equation then becomes

$$y = a\sin(x + ct) + a\sin(x - ct)$$
$$= a.2\sin x\cos ct \quad \text{(using the product formula)},$$

i.e. $\underline{y = (2a\cos ct)\sin x.}$

This wave travels neither to the right nor left; variation in t merely alters the amplitude $2a\cos ct$. At distances $x = 0, \pm\pi$,

$\pm 2\pi$, etc., its value is permanently zero, no matter what the value of t may be. These points are called nodes. Midway between the nodes, at points $x = \pm \frac{1}{2}\pi$, $\pm \frac{3}{2}\pi$, etc., the amplitude varies between $\pm 2a$.

These places are often called loops (or antinodes).

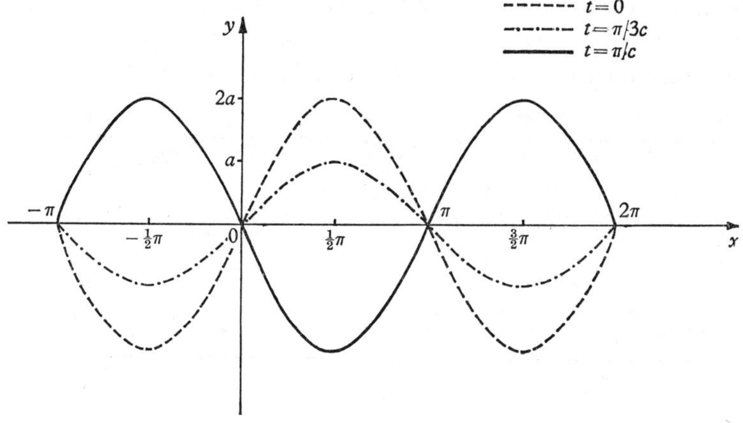

Fig. 55

Fig. 55 shows a picture of the standing wave for three values of t.

$$t = 0,\ y = 2a \sin x;\quad t = \pi/3c,\ y = a \sin x;$$

$$t = \pi/c,\ y = -2a \sin x.$$

Treating $y = f(x + ct) + F(x - ct)$ as a general solution of the wave equation $\frac{\partial^2 y}{\partial x^2} - \frac{1}{c^2}\frac{\partial^2 y}{\partial t^2}$ may seem to give very little exact information except for the general idea that it represents two 'phenomena' travelling in opposite directions with constant speed c. However, when it is remembered that an arbitrary function can be replaced, over a definite range, by trigonometrical series using Fourier series, some of the 'arbitrariness' disappears. In actual practice the difficulties encountered are usually (i) to choose the right type of series, and (ii) having done so, to determine the coefficients.

For examples of the use of Fourier series in solving partial differential equations, the student should refer to §§ 12·52 and 12·53.

The rest of this chapter will be devoted to a study of a special case of particular interest to line engineers.

11·5. Transmission along a uniform cable

In this section, transmission of electric current along a uniform cable in which both capacitance and inductance are present will be considered. This, then, is the general case, whereas only a special case was considered in § 9·23.

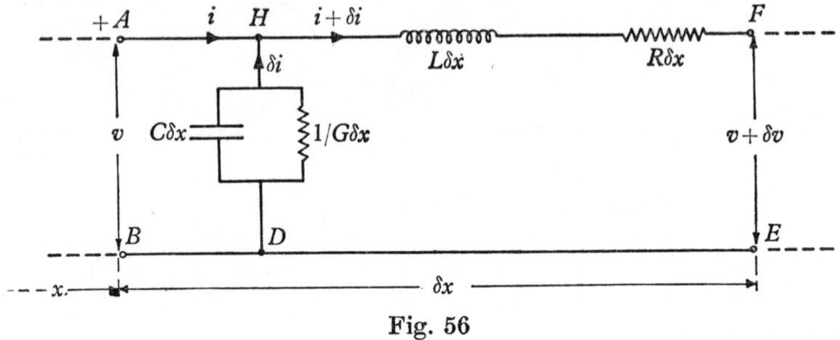

Fig. 56

Fig. 56 shows an enlarged drawing of the equivalent circuit for a length from x to $x + \delta x$ of a uniform transmission line. The term 'uniform' will be taken as meaning that the line has uniform resistance, capacitance, inductance and leakance. It may make the interpretation of the circuit easier if the student imagines AHF to be the line of the cable and BDE to be at zero potential (earthed). The inductance L and resistance R are then in series along the cable and the capacitance C and leakage conductance G are across from the cable to earth.

Let L, R, C and G be respectively the inductance, resistance, capacitance and leakage conductance/unit length of the cable.

Let i, v be the current and voltage at distance x, and $i + \delta i$, $v + \delta v$ the current and voltage at distance $x + \delta x$.

The voltage *drop* across the portion AHF is $-\delta v$.

The back e.m.f. across this portion is $(L\,\delta x)\dfrac{\partial i}{\partial t} + (R\,\delta x)\,i$.

$$\therefore -\delta v = (L\,\delta x)\frac{\partial i}{\partial t} + (R\,\delta x)\,i.$$

($\partial i/\partial t$ is used as i is a function of both x and t, as is also the voltage v. Second-order quantities are omitted.)

Dividing by δx, and taking the limit as $\delta x \to 0$,

$$-\frac{\partial v}{\partial x} = L\frac{\partial i}{\partial t} + Ri. \tag{1}$$

Considering the part of the circuit HD: the charge on the capacitor is $(C\delta x)\,v$; the current flowing through it is thus

$$\frac{\partial}{\partial t}(C\,\delta x)v = (C\,\delta x)\frac{\partial v}{\partial t}.$$

The current flowing through the resistance $\dfrac{1}{G\,\delta x}$, due to leakage conductance, is

$$v\bigg/\frac{1}{G\,\delta x} = v(G\,\delta x).$$

Thus $$(C\,\delta x)\frac{\partial v}{\partial t} + v(G\,\delta x) = -\,\delta i.$$

Dividing by δx and taking the limit as $\delta x \to 0$,

$$-\frac{\partial i}{\partial x} = C\frac{\partial v}{\partial t} + Gv. \tag{2}$$

(1) and (2) are a pair of simultaneous partial differential equations for v and i. Either v or i can be eliminated by normal means. For example, from (2)

$$-\frac{\partial^2 i}{\partial x\,\partial t} = C\frac{\partial^2 v}{\partial t^2} + G\frac{\partial v}{\partial t},$$

and from (1)

$$L\frac{\partial^2 i}{\partial x\,\partial t} = -\frac{\partial^2 v}{\partial x^2} - R\frac{\partial i}{\partial x}.$$

Thus

$$-L\frac{\partial^2 i}{\partial x\,\partial t} = \frac{\partial^2 v}{\partial x^2} + R\frac{\partial i}{\partial x} = LC\frac{\partial^2 v}{\partial t^2} + LG\frac{\partial v}{\partial t}.$$

Substitution for $\partial i/\partial x$ from (2) now gives

$$\frac{\partial^2 v}{\partial x^2} - R\left(C\frac{\partial v}{\partial t} + Gv\right) = LC\frac{\partial^2 v}{\partial t^2} + LG\frac{\partial v}{\partial t},$$

i.e. $$\frac{\partial^2 v}{\partial x^2} - LC\frac{\partial^2 v}{\partial t^2} - (LG + RC)\frac{\partial v}{\partial t} - RGv = 0. \tag{3}$$

In a similar manner an equation for i can be found. It is obviously a difficult task to find a general solution for such an equation, and no such general solution will be attempted here. It may be noted that equations of types (a) and (b) of § 11·3 can be obtained as special cases of equation (3).

(i) If $R = 0 = G$ (a 'lossless' line), the equation reduces to

$$\frac{\partial^2 v}{\partial x^2} - LC\frac{\partial^2 v}{\partial t^2} = 0,$$

a wave-type equation.

(ii) If $L = 0 = G$ (an ideal submarine cable), the equation reduces to

$$\frac{\partial^2 v}{\partial x^2} - RC\frac{\partial v}{\partial t} = 0,$$

a diffusion-type equation.

Ignoring any transients, the steady-state solution of equations (1) and (2) can be determined by assuming that v and i are the imaginary (or real) parts of $Ve^{j\omega t}$ and $Ie^{j\omega t}$, where V and I are (complex) functions of x only and $\omega/2\pi$ is the frequency of the operating voltage. This assumption is feasible, as when the steady state has been reached it is reasonable to suppose that the voltage and current are dependent only on the distance along the line, the only time effect being due to the frequency of the operating voltage.

Substituting $v = Ve^{j\omega t}$, $i = Ie^{j\omega t}$ into equations (1) and (2) gives

$$-\frac{dV}{dx}e^{j\omega t} = LI(j\omega e^{j\omega t}) + RIe^{j\omega t}$$

and $$-\frac{dI}{dx}e^{j\omega t} = CV(j\omega e^{j\omega t}) + GVe^{j\omega t}.$$

(dV/dx and dI/dx, not $\partial V/\partial x$, $\partial I/\partial x$, as V and I are functions of x only.)

On cancelling $e^{j\omega t}$, these equations give

$$-\frac{dV}{dx} = (R + j\omega L)I \tag{4}$$

and

$$-\frac{dI}{dx} = (G + j\omega C)V. \tag{5}$$

Thus

$$\frac{d^2V}{dx^2} = -(R + j\omega L)\frac{dI}{dx} = (R + j\omega L)(G + j\omega C)V,$$

i.e.

$$\frac{d^2V}{dx^2} - (R + j\omega L)(G + j\omega C)V = 0. \tag{6}$$

This equation could have been obtained directly from equation (3) by the substitution of $v = Ve^{j\omega t}$, but the product $(R + j\omega L)(G + j\omega C)$ would not then have appeared already factorized.

Now put $\gamma = \alpha + j\beta = \{(R + j\omega L)(G + j\omega C)\}^{\frac{1}{2}}$. (7)

γ is called the *propagation constant* of the line, α the *attenuation constant*, and β the *phase constant*. Reasons for these names will become apparent as the student reads further.

In practice the square root in equation (7) is taken in any numerical case so that α is positive.

$$\frac{d^2V}{dx^2} - \gamma^2 V = 0. \tag{8}$$

The solution to this equation is

$$V = Ae^{\gamma x} + Be^{-\gamma x} = Ae^{(\alpha+j\beta)x} + Be^{-(\alpha+j\beta)x}. \tag{9}$$

Thus

$$Ve^{j\omega t} = \{Ae^{(\alpha+j\beta)x} + Be^{-(\alpha+j\beta)x}\}e^{j\omega t}. \tag{10}$$

From equations (4) and (9):

$$I = \frac{-1}{(R + j\omega L)}\frac{dV}{dx} = \frac{-1}{(R + j\omega L)}(\gamma Ae^{\gamma x} - \gamma Be^{-\gamma x})$$

$$= \frac{-\gamma}{(R + j\omega L)}(Ae^{\gamma x} - Be^{-\gamma x}).$$

Using equation (7):

$$I = -\left(\frac{G + j\omega C}{R + j\omega L}\right)^{\frac{1}{2}}\{Ae^{(\alpha+j\beta)x} - Be^{-(\alpha+j\beta)x}\}. \tag{11}$$

Thus

$$Ie^{j\omega t} = -\left(\frac{G + j\omega C}{R + j\omega L}\right)^{\frac{1}{2}}\{Ae^{(\alpha+j\beta)x} - Be^{-(\alpha+j\beta)x}\}e^{j\omega t}. \tag{12}$$

The calculation of A and B will depend on the given initial and boundary conditions. As an example, take the case of an infinitely long line for which $v = E_0 \sin \omega t$ at $x = 0$. This means that $V = E_0$ at $x = 0$ and implies that v and i are to be taken as the *imaginary* parts of $Ve^{j\omega t}$ and $Ie^{j\omega t}$. From equation (9) it is clear that in this case $A = 0$, as the voltage cannot be infinite when x is infinite.

Thus $$Ve^{j\omega t} = Be^{j\omega t}e^{-(\alpha+j\beta)x}.$$

At $x = 0$, $V = E_0$. $\quad \therefore B = E_0$.

This gives $$Ve^{j\omega t} = E_0 e^{j\omega t} e^{-(\alpha+j\beta)x},$$

i.e. $$\underline{Ve^{j\omega t} = E_0 e^{-\alpha x} e^{j(\omega t-\beta x)}.} \tag{13}$$

Now v is the imaginary part of $Ve^{j\omega t}$.

Therefore $$\underline{v = E_0 e^{-\alpha x} \sin(\omega t - \beta x).} \tag{14}$$

It should now be clear why α is called the attenuation constant and β the phase constant.

Putting $A = 0$, $B = E_0$ into equation (12) gives

$$Ie^{j\omega t} = \left(\frac{G + j\omega C}{R + j\omega L}\right)^{\frac{1}{2}} E_0 e^{-\alpha x} e^{j(\omega t-\beta x)}. \tag{15}$$

Now $\left(\dfrac{R + j\omega L}{G + j\omega C}\right)^{\frac{1}{2}}$ is a complex number and can be expressed in polar form as $|Z|e^{j\phi}$, where $|Z|$ is its modulus and ϕ its argument.

Writing $$|Z|e^{j\phi} = \left(\frac{R + j\omega L}{G + j\omega C}\right)^{\frac{1}{2}}, \tag{16}$$

$$Ie^{j\omega t} = \frac{E_0}{|Z|e^{j\phi}} e^{-\alpha x} e^{j(\omega t-\beta x)},$$

giving $$\underline{Ie^{j\omega t} = \frac{E_0}{|Z|} e^{-\alpha x} e^{j(\omega t-\beta x-\phi)}.} \tag{17}$$

Now i is the imaginary part of $Ie^{j\omega t}$.

$$\therefore \underline{i = \frac{E_0}{|Z|} e^{-\alpha x} \sin(\omega t - \beta x - \phi).} \tag{18}$$

Equations (14) and (18) give the steady-state values of v and i in the case considered. They are both waves travelling with velocity ω/β, but having their amplitudes damped, or attenuated, by the factor $e^{-\alpha x}$.

$\left(\dfrac{R + j\omega L}{G + j\omega C}\right)^{\frac{1}{2}}$ is called the *characteristic impedance* of the line and is often written Z_0. It is the ratio of V to I when $x = 0$, as can easily be seen from equations (13) and (15).

The values of V and I at $x = 0$ are often called the sending-end voltage and current, denoted by E_s and I_s. Thus $Z_0 = E_s/I_s$, for an infinite line.

11·51. Practical considerations

11·511. Calculation of V and I

In many cases the line engineer is satisfied to find the values of V and I, that is, the complex values of the voltage and current at any distance along the line. He may not be concerned with the actual wave equation involving time as well as distance.

In this case equations (9) and (11) are fundamental. They may be written

$$V = Ae^{\gamma x} + Be^{-\gamma x}, \tag{1}$$

$$I = -\frac{1}{Z_0}(Ae^{\gamma x} - Be^{-\gamma x}), \tag{2}$$

where $\gamma = \alpha + j\beta = \{(R + j\omega L)(G + j\omega C)\}^{\frac{1}{2}}$ is the propagation constant, and $Z_0 = \left(\dfrac{R + j\omega L}{G + j\omega C}\right)^{\frac{1}{2}}$ is the characteristic impedance. The constants A and B can be found from given conditions at either end of the line.

Two typical cases are given below.

(*a*) *A line of infinite length.* The voltage must be finite as $x \to \infty$, therefore $A = 0$. (Alternatively it can be argued that the current must be zero as $x \to \infty$, and this also gives $A = 0$.)

Thus
$$I = \frac{B}{Z_0}e^{-\gamma x}.$$

If I_s is the sending-end current (at $x = 0$), then

$$I_s = \frac{B}{Z_0}, \quad \text{i.e.} \quad B = Z_0 I_s.$$

Equations (1) and (2) now become

$$V = I_s Z_0 e^{-\gamma x} \quad \text{and} \quad I = I_s e^{-\gamma x}. \tag{3}$$

If E_s is the sending-end voltage (at $x = 0$) $E_s = I_s Z_0$.

Thus
$$\frac{E_s}{I_s} = Z_0. \tag{4}$$

The characteristic impedance Z_0 is in fact often defined as the ratio of the sending-end voltage and current for an infinite line.

It follows that
$$V = E_s e^{-\gamma x}. \tag{5}$$

From (3),
$$I = I_s e^{-\gamma x} = I_s e^{-(\alpha + j\beta)x}.$$

In practice x is usually measured in kilometres.
Let I_1 be the current at 1 kilometre, then

$$I_s = I_1 e^{(\alpha + j\beta)} = I_1 e^{\alpha} e^{j\beta}.$$

Using polar complex number notation:

$$\frac{I_s}{I_1} = e^{\alpha} \angle \beta. \tag{6}$$

Thus
$$\left|\frac{I_s}{I_1}\right| = e^{\alpha}$$

and
$$\alpha = \log_e \left|\frac{I_s}{I_1}\right|. \tag{7}$$

Equation (7) can be used to find α, the attenuation constant. When calculated from this formula it is given in nepers per kilometre. For practical work the neper is often converted to decibels. In this respect it may be noted that the attenuation in decibels is equal to $20 \log_{10} e \times$ the attenuation in nepers; this is approximately $8{\cdot}685 \times$ the attenuation in nepers. Alternatively, the attenuation may be measured in decibels as $10 \log_{10} |P_s/P_1|$, where P is the power. Only when the resistive components of the impedances at the input and output of the line are equal can one readily be converted to the other, for then $P_s/P_1 = (I_s/I_1)^2$.

β is usually measured as the phase constant per kilometre. It is measured in radians per kilometre for use in the formulae given here. For practical work it is often given in degrees, but these are easily converted into radians before use in the formulae.

For a distance m kilometres:

$$\frac{I_s}{I_m} = e^{m\gamma} \quad \text{(from (3))}$$
$$= e^{m\alpha + jm\beta} = e^{m\alpha} \angle m\beta.$$

Thus the attenuations for m kilometres is $m\alpha$ nepers and the phase shift is $m\beta$ radians.

(b) *Any line. Conditions known at either end.* Equations (1) and (2) above are now best written as hyperbolic functions, using $e^{\gamma x} = \cosh \gamma x + \sinh \gamma x$ and $e^{-\gamma x} = \cosh \gamma x - \sinh \gamma x$.

Thus $$V = C \cosh \gamma x + D \sinh \gamma x, \tag{8}$$

$$I = -\frac{1}{Z_0}(D \cosh \gamma x + C \sinh \gamma x) \tag{9}$$

(where $C = A + B$ and $D = A - B$).

Let E_s, I_s be the known voltage and current at the sending end, $x = 0$.

From (8) and (9):

$$E_s = C \quad \text{and} \quad I_s = -\frac{D}{Z_0}.$$

Thus $$V = E_s \cosh \gamma x - Z_0 I_s \sinh \gamma x \tag{10}$$

and $$I = I_s \cosh \gamma x - \frac{E_s}{Z_0} \sinh \gamma x. \tag{11}$$

In a similar manner, if E_R, I_R are the known receiving-end voltage and current, at $x = l$, it is easily shown that

$$V = E_R \cosh \gamma (l - x) + Z_0 I_R \sinh \gamma (l - x), \tag{12}$$

$$I = \frac{E_R}{Z_0} \sinh \gamma (l - x) + I_R \cosh \gamma (l - x). \tag{13}$$

Note. As γ is a complex number, in working numerical examples use is often made of such expressions as

$$\sinh(\alpha + j\beta)l = \sinh \alpha l . \cos \beta l + j \cosh \alpha l \sin \beta l,$$

etc. (see § 5·4).

(c) *Finite line terminated by the characteristic impedance.* From (1) and (2) above, at the receiving end

$$V = Ae^{\gamma l} + Be^{-\gamma l} \quad \text{and} \quad I = -\frac{1}{Z_0}(Ae^{\gamma l} - Be^{-\gamma l}).$$

Now the ratio V/I at the receiving end gives the impedance at the receiving end. Thus, if this is Z_0,

$$-\frac{(Ae^{\gamma l} + Be^{-\gamma l})Z_0}{(Ae^{\gamma l} - Be^{-\gamma l})} = Z_0,$$

giving

$$Ae^{\gamma l} + Be^{-\gamma l} = -Ae^{\gamma l} + Be^{-\gamma l},$$

i.e.

$$2Ae^{\gamma l} = 0,$$

therefore $A = 0$ and

$$V = Be^{-\gamma x}, \quad I = \frac{B}{Z_0} e^{-\gamma x}. \tag{14}$$

Reference to paragraph (a) above shows that these equations are the same as for an infinite line. Thus a line terminated by its characteristic impedance behaves as if it were a line of infinite length.

11·512. Determination of Z_0 and γ

γ and Z_0 can both be calculated by measuring the values of the impedance of a known length, l, of a line on open-circuit and on short-circuit.

(a) *Open-circuit at distant end.* The current I_R at the distant end is zero, that is,

$$I = 0 \text{ at } x = l.$$

From equation (11) of § 11·511,

$$0 = I_s \cosh \gamma l - \frac{E_s}{Z_0} \sinh \gamma l,$$

giving

$$\frac{E_s}{I_s} = Z_0 \coth \gamma l.$$

But E_s/I_s is the input impedance of the line, which for this open-circuit case will be written Z_{oc}.

Thus $$Z_{oc} = Z_0 \coth \gamma l. \tag{1}$$

(*Note.* As $l \to \infty$, $\coth \gamma l \to 1$, and $Z_{oc} \to Z_0$.)

(*b*) *Short-circuit at distant end.* In this case the voltage, E_R, at the distant end is zero, i.e.

$$V = 0 \text{ at } x = l.$$

Equation (10) of § 11·511 then gives

$$0 = E_s \cosh \gamma l - Z_0 I_s \sinh \gamma l.$$

$$\therefore \quad \frac{E_s}{I_s} = Z_0 \tanh \gamma l.$$

But E_s/I_s is the input impedance, which for this short-circuit case will be written Z_{sc}.

Thus $$Z_{sc} = Z_0 \tanh \gamma l. \tag{2}$$

(*Note.* As $l \to \infty$, $\tanh \gamma l \to 1$ and $Z_{sc} \to Z_0$.)

Multiplying equations (1) and (2):

$$Z_{oc} Z_{sc} = Z_0^2. \tag{3}$$

Dividing equations (1) and (2):

$$\frac{Z_{sc}}{Z_{oc}} = \tanh^2 \gamma l,$$

i.e. $$\tanh \gamma l = \sqrt{\frac{Z_{sc}}{Z_{oc}}}. \tag{4}$$

If Z_{oc} and Z_{sc} are measured, Z_0 and γ can thus be determined.

11·52. Worked examples

(1) At 1500 Hz a certain type of cable has an attenuation constant of 2·5 db. per kilometre and a phase constant of 0·31 rad per kilometre. If 2 volts at 1500 Hz are applied at the sending end, and the line is terminated in its characteristic impedance, what is the voltage at a point 10 km down the line?

From §11·511 (c), a line terminated by its characteristic impedance behaves as an infinite line. Thus, using §11·511, equation (5), the required voltage is given by

$$V = 2e^{-\gamma 10} = 2e^{-10\alpha} \angle -10\beta,$$

where α is in nepers per kilometre and β in radians per kilometre.

$$\text{Now } 2{\cdot}5 \text{ db. per kilometre} = \frac{2{\cdot}5}{8{\cdot}685} \text{ nepers per kilometre}$$

$$\simeq 0{\cdot}288 \text{ neper per kilometre.}$$

Thus $\qquad \alpha = 0{\cdot}288.$

Also $\qquad \beta = 0{\cdot}31.$

$$\therefore V = 2e^{-2{\cdot}88} \angle -3{\cdot}1 \simeq 0{\cdot}112 \angle -3{\cdot}1.$$

The voltage 10 km down the line is 0·112 volt lagging by 3·1 rad, or 177° 36′, behind the sending-end voltage.

(2) At 1200 Hz a certain open-wire type of line has a characteristic impedance, Z_0, of $500 \angle -10°$ and a propagation constant $\gamma = 0{\cdot}01 + j0{\cdot}06$. When 4 volts are applied to the sending end, a current of 5 mA flows. What will be the current at the receiving end 50 km away?

$I_s = 0{\cdot}005$, $E_s = 4$, $x = 50$, $Z_0 = 500 \angle -10°$.

From equation (11), §11·511.

$$I = 0{\cdot}005 \cosh 50(0{\cdot}01 + j0{\cdot}06)$$

$$- \frac{4}{500 \angle -10°} \sinh 50(0{\cdot}01 + j0{\cdot}06)$$

$$= 0{\cdot}005 \cosh (0{\cdot}5 + j3) - 0{\cdot}008 \angle 10° \sinh (0{\cdot}5 + j3)$$

$$= 0{\cdot}005 (\cosh 0{\cdot}5 \cos 3 + j \sinh 0{\cdot}5 \sin 3)$$

$$- 0{\cdot}008 \angle 10° (\sinh 0{\cdot}5 \cos 3 + j \cosh 0{\cdot}5 \sin 3)$$

$$\simeq 0{\cdot}005 (-1{\cdot}116 + j0{\cdot}0735)$$

$$- 0{\cdot}008 \angle 10° (-0{\cdot}516 + j0{\cdot}159)$$

$$\simeq 0{\cdot}001 [(-5{\cdot}58 + j0{\cdot}3675)$$

$$- 8\{(0{\cdot}985 + j0{\cdot}174)(-0{\cdot}516 + j0{\cdot}159)\}]$$

$$\simeq 0{\cdot}001[(-5{\cdot}58 + j0{\cdot}3675) - 8(-0{\cdot}536 + j0{\cdot}067)]$$

$$\simeq 0{\cdot}001 [-1{\cdot}29 - j0{\cdot}169]$$

$$\simeq 0{\cdot}0013 \angle -172\tfrac{1}{2}°.$$

The current at the receiving end is therefore 1·3 mA, lagging behind the sending-end current by $172\frac{1}{2}°$.

Note. It is not surprising that tables and diagrams are to be found in text-books on line communication, from which the numerical values of quantities such as cosh $(0{\cdot}5 + j3)$ can be read off.

(3) Explain what is meant by (*a*) the characteristic impedance and (*b*) the attenuating constant of a transmission line. A telephone line, 10 km. long, is termined by its characteristic impedance. At a frequency of 800 Hz the attenuation of the line is 6 db. and the p.d. at the receiving end lags 44° behind that at the sending end. If the sending-end voltage is $10\angle 0°$ volts, find (*a*) the value of the receiving-end voltage and (*b*) the value of the voltage half-way along the line. Also find the propagation constant per loop kilometre and the velocity of propagation. [L.U.]

$$l = 10, \quad E_s = 10\angle 0°.$$

Let α = attenuation in nepers, then

$$\alpha = \frac{0{\cdot}6}{20\log_{10}e} = \frac{0{\cdot}6}{20}\log_e 10. \tag{1}$$

(*a*) As the line is terminated by its characteristic impedance, it behaves as an infinite line.

Thus $$V_x = E_s e^{-\gamma x}.$$

At $x = 10$, $$V_R = 10e^{-10\gamma}.$$

As the receiving-end voltage lags by 44°:

$$E_R\angle -44° = 10e^{-10(\alpha+j\beta)},$$

i.e. $$E_R\angle -44° = 10e^{-10\alpha}e^{-j10\beta}$$

$$= 10e^{-0{\cdot}3\log_e 10}e^{-j10\beta}\text{ (from (1))}$$

$$= \frac{10}{10^{0{\cdot}3}}\angle -10\beta \simeq 5{\cdot}012\angle -10\beta.$$

Thus, $\underline{E_R \simeq 5{\cdot}012\text{ volts and }\beta = 4{\cdot}4°\text{ per kilometre.}}$ (2)

(b) When $x = 5$, $V_5 = 10e^{-5\gamma}$

$$= 10e^{-5\alpha}e^{-j5\beta}$$

$$= 10e^{-\frac{3}{20}\log_e 10}e^{-j22^\circ} \quad \text{(from (1) and (2))}$$

$$= \frac{10}{10^{\frac{3}{20}}} \angle -22^\circ$$

$$\simeq 7{\cdot}08 \angle -22^\circ.$$

The voltage at 5 *km is* 7·08 *volts lagging by* 22°.

(c) $\gamma = \alpha + j\beta$

$$= \frac{0{\cdot}3}{10}\log_e 10 + j4{\cdot}4\frac{\pi}{180},$$

$$\gamma \simeq 0{\cdot}0691 + j0{\cdot}0768.$$

The velocity of propagation is $\dfrac{\omega}{\beta} = \dfrac{2\pi\cdot 800}{0{\cdot}0768}$

$$\simeq 65{,}450\,\text{km/s}.$$

(4) The voltage at any point on a uniform transmission line can be expressed as

$$V = V_r \cosh P(l - x) + I_r Z_0 \sinh P(l - x).$$

Define all the quantities in this expression, and deduce from it the ratio of the sending-end voltage to receiving-end voltage in terms of the terminating impedance and the characteristics of the line.

Calculate the ratio for a 20 km length of telephone line having a propagation constant of $0{\cdot}06 \angle 45^\circ$ per loop kilometre and a characteristic impedance of $500 \overline{\angle} 30^\circ\,\Omega$ when it is terminated by an impedance of $200 \overline{\angle} 30^\circ\,\Omega$. [L.U.]

V is the (complex) voltage at distance x.

V_r, I_r are the receiving-end voltage and current.

P is the propagation constant for the line and is given by

$$P = \{(R + j\omega L)(G + j\omega C)\}^{\frac{1}{2}},$$

where R, L, G, C, ω have their usual meanings (see §11·5). l is the length of the line. Z_0 is the characteristic impedance of the line and is given by

$$Z_0 = \left(\frac{R + j\omega L}{G + j\omega C}\right)^{\frac{1}{2}}.$$

The terminating impedance is given by $V_r/I_r = Z_r$ say. Let V_s = sending-end voltage, at $x = 0$, then

$$V_s = V_r \cosh Pl + I_r Z_0 \sinh Pl.$$

$$\therefore \frac{V_s}{V_r} = \cosh Pl + \frac{I_r}{V_r} Z_0 \sinh Pl,$$

i.e.
$$\frac{V_s}{V_r} = \cosh Pl + \frac{Z_0}{Z_r} \sinh Pl. \tag{1}$$

If $P = 0{\cdot}06 \angle 45°$, $Z_0 = 500 \overline{\angle} 30°$, $l = 20$ and $Z_r = 200 \overline{\angle} 30°$, then

$$\frac{V_s}{V_r} = \cosh (1{\cdot}2 \angle 45°) + \frac{500 \overline{\angle} 30°}{200 \overline{\angle} 30°} \sinh (1{\cdot}2 \angle 45°)$$

$$\simeq \cosh (0{\cdot}8485 + j0{\cdot}8485) + 2{\cdot}5 \sinh (0{\cdot}8485 + j0{\cdot}8485)$$

$$\simeq (0{\cdot}9137 + j0{\cdot}7158) + 2{\cdot}5(0{\cdot}6306 + j1{\cdot}037)$$

$$\simeq 2{\cdot}49 + j3{\cdot}31$$

$$\simeq 4{\cdot}14 \angle 53°.$$

The magnitude of the required ratio is 4·14.

Exercise 38

1. Show that the wave equation $\dfrac{\partial^2 y}{\partial x^2} - \dfrac{1}{c^2}\dfrac{\partial^2 y}{\partial t^2} = 0$ is satisfied by $y = f(x + ct) + F(x - ct)$, where f and F denote arbitrary functions.

2. Show that $z = \sin \dfrac{r\pi x}{a} \sinh \dfrac{r\pi(b - y)}{a}$, where r is any integer and a, b are constants, satisfies Laplace's equation $\dfrac{\partial^2 z}{\partial x^2} + \dfrac{\partial^2 z}{\partial y^2} = 0$.

3. The equation $\dfrac{\partial^2 u}{\partial z^2} - \dfrac{1}{\nu}\dfrac{\partial u}{\partial t} = 0$ occurs in the study of the linear motion of a fluid. Show that it is satisfied by

$$u = \exp\left\{-\sqrt{\left(\frac{\omega}{2\nu}\right)} z\right\} \sin\left\{\omega t - \sqrt{\left(\frac{\omega}{2\nu}\right)} z\right\}.$$

4. Show that the equation $\dfrac{w}{g}\dfrac{\partial^2 y}{\partial t^2} = -EI\dfrac{\partial^4 y}{\partial x^4}$ is satisfied by
$y = (A\cos kx + B\sin kx + C\cosh kx + D\sinh kx)\sin(nt + \alpha)$,
where $k^4 = \dfrac{wn^2}{gEI}$ and A, B, C, D are arbitrary constants.

5. Prove that $V = \dfrac{1}{r}\{\phi(ct + r) + \psi(ct - r)\}$ satisfies

$$\frac{\partial^2 V}{\partial t^2} = \frac{c^2}{r^2}\frac{\partial}{\partial r}\left(r^2\frac{\partial V}{\partial r}\right).$$ [L.U.]

6. Prove that

$$x^2\frac{\partial^2 z}{\partial x^2} + 2xy\frac{\partial^2 z}{\partial x\,\partial y} + y^2\frac{\partial^2 z}{\partial y^2} + x\frac{\partial z}{\partial x} + y\frac{\partial z}{\partial y} = z$$

is satisfied by $z = x\phi\left(\dfrac{y}{x}\right) + \dfrac{1}{y}\psi\left(\dfrac{y}{x}\right)$, where ϕ and ψ are arbitrary functions. [L.U.]

7. Find X, as a function of x only, such that $v = X\cos ct$ is a solution of the equation $\dfrac{\partial^2 v}{\partial x^2} = \dfrac{\partial^2 v}{\partial t^2} + n^2 v$, for the cases (*a*) $c < n$, (*b*) $c > n$, (*c*) $c = n$, where c and n are constants. Determine the solution when $n = 2{\cdot}5$, given that $v = \cos 1{\cdot}5t$ when $x = 0$ and when $x = 0{\cdot}5$. Verify that the greatest value of v when $x = 1$ is $e + e^{-1} - 1$. [L.U.]

8. If $V = x^n f(y/x)$, where f is any function, show that $x\dfrac{\partial V}{\partial x} + y\dfrac{\partial V}{\partial y} = nV$. If $\dfrac{\partial^2 V}{\partial x\,\partial y} = 0$ and u denotes $\dfrac{y}{x}$, show that $(n-1)\dfrac{df}{du} = u\dfrac{d^2 f}{du^2}$ and hence show that $V = ax^n + by^n$, where a and b are constants. [L.U.]

9. If $V = r^n\sin\theta$ is to satisfy the equation

$$r\frac{\partial}{\partial r}\left(r\frac{\partial V}{\partial r}\right) + \frac{\partial^2 V}{\partial \theta^2} = 0,$$

show that n must have one of the values ± 1. Find a solution of this equation which vanishes when $\theta = 0$ for all values of r, and which vanishes when $r = a$ for all values of θ, and for which $\partial V/\partial r = 1$ when $r = a$, $\theta = \frac{1}{2}\pi$. [I.E.E.]

10. v is a function of x and t satisfying the equation

$$\frac{\partial^2 v}{\partial x^2} + 3\frac{\partial v}{\partial x} = \frac{1}{4}\frac{\partial^2 v}{\partial t^2}.$$

If $v = V\cos 5t$, where V is a function of x only, obtain and solve the differential equation giving V as a function of x. If, at $t = 0$, $v = 4\cos 5t$ and $\partial v/\partial x = 0$ for all values of t, show that $v = 5e^{-\frac{3}{2}x}.\cos(2x - \text{arc tan } 3/4).\cos 5t$. [I.E.E.]

11. A finite length, l, of transmission line is terminated by an impedance Z_R. Find the input impedance in terms of l, Z_R, γ and Z_0, where γ and Z_0 are respectively the propagation constant and characteristic impedance of the line. What does the value of this input impedance approach as l tends to infinity?

12. A line, 20 km long, has the following constants: $Z_0 = 600\angle 0°$, $\alpha = 0{\cdot}15$ neper per kilometre, $\beta = 0{\cdot}01$ radian per kilometre.

Find the receiving-end current if 30 mA is the sending-end current and the receiving end is short-circuited.

For the same line constants and length of line, find the sending-end current and voltage if the receiving-end current is 5 mA and the receiving-end voltage is 0·1 volt.

13. Calculate the characteristic impedance, propagation constant and velocity of wave propagation at a frequency of 400 kHz of a uniform transmission line which has the following constants:

$$L = 0{\cdot}5\,\text{mH/km}, \quad C = 0{\cdot}08\,\mu F/\text{km}.$$

Resistance and leakance negligible.

Derive the necessary formulae from the differential equations for a uniform line which has negligible resistance and leakance. [L.U.]

14. A telephone line, 10 km long, has the following constant per loop kilometre:

$R = 196\,\Omega$, $L = 7{\cdot}1\,\text{mH}$, $C = 0{\cdot}09\,\mu\text{F}$, G negligible.

The line is short-circuited at the far end and a p.d. of 10 volts at a frequency of $\dfrac{5000}{2\pi}$ Hz is applied at the sending end. Assuming the general equations

$$V_x = V_s \cosh Px - I_s Z_0 \sinh Px$$

and

$$I_x = I_s \cosh Px - \frac{V_s}{Z_0} \sinh Px,$$

derive an expression for the current at the far end, and calculate its value and its phase relation to the applied voltage. [L.U.]

15. Prove that the attenuation per unit length of a uniform two-wire line is given approximately by the expression $\alpha = \dfrac{R}{2}\sqrt{\dfrac{C}{L}} + \dfrac{G}{2}\sqrt{\dfrac{L}{C}}$, provided that the inductive reactance is large compared with the line resistance and that the insulation resistance of the line is high. Hence deduce what value of L/C is needed if the attenuation of a line with given resistance and leakance is to be a minimum. [L.U.]

ANSWERS

7. (a) $X = A \cosh px + B \sinh px$, where $p^2 = n^2 - c^2$.

(b) $X = C \cos \omega x + D \sin \omega x$, where $\omega^2 = c^2 - n^2$.

(c) $X = Ex + F$.

$$v = \left\{\cosh 2x + \frac{(2 - e - e^{-1})}{(e - e^{-1})} \sinh 2x\right\} \cos 1{\cdot}5t.$$

v has its largest value when $\cos 1{\cdot}5t = 1$.

8. [Note that

$$\frac{\partial^2 V}{\partial x\, \partial y} = nx^{n-2}f'\left(\frac{y}{x}\right) - x^{n-2}f'\left(\frac{y}{x}\right) - yx^{n-3}f''\left(\frac{y}{x}\right).$$

If $\dfrac{\partial^2 V}{\partial x\, \partial y} = 0$, this gives, with $u = \dfrac{y}{x}$,

$$x^{n-2}\left\{(n-1)\frac{df}{du} - u\frac{d^2 f}{du^2}\right\} = 0.$$

Also note that $(n-1)\dfrac{df}{du} = u\dfrac{d^2f}{du^2}$ can be written

$$\frac{d^2f}{du^2}\bigg/\frac{df}{du} = \frac{(n-1)}{u}.$$

When integrating both sides use

$$\int\left(\frac{d^2f}{du^2}\bigg/\frac{df}{du}\right)du = \log_e\left(\frac{df}{du}\right).\Bigg]$$

9. $V = \frac{1}{2}r\sin\theta - \dfrac{a^2}{2r}\sin\theta.$

10. $\dfrac{d^2V}{dx^2} + 3\dfrac{dV}{dx} + \dfrac{25}{4}V = 0;\quad V = e^{-\frac{3}{2}x}(A\cos 2x + B\sin 2x).$

11. $\dfrac{Z_0(Z_R + Z_0\tanh\gamma l)}{(Z_0 + Z_R\tanh\gamma l)};\quad Z_0.$

12. [Use equations (13) and (12) of §11·511.]
(i) 3 mA lagging by $11\frac{1}{2}°$.
(ii) $I_s \simeq 52$ mA leading I_R by $11\frac{1}{2}°$,
$E_s \simeq 31$ volts leading E_R by $11\frac{1}{2}°$.

13. $Z_0 = 79{\cdot}06\angle 0°\,\Omega.$
$\gamma = 15{\cdot}9\angle 90°$ per kilometre; about $1{\cdot}58 \times 10^5$ km/s.

14. $I_R = \dfrac{V_s}{Z_0\sinh Pl}.$

Approx. 4·4 mA lagging by $90\frac{1}{2}°$.

15. $\dfrac{L}{C} = \dfrac{R}{G}.$

CHAPTER 12

FURTHER FOURIER SERIES

12·1. Foreword

Some colleges do not include harmonic analysis in the first year (A 1) course. In this case the chapter on this subject in volume I of this book should be taken before reading further. In any case, it would be advisable for the student to revise the work on Fourier series before reading the following chapter.

12·2. Half-range series

If it is desired to represent a function defined only in the range 0 to T by a Fourier series of period $2T$, there is a choice between a sine and a cosine series.

As an example, take the function defined as

$$f(t) = t \quad (0 < t < T).$$

The range $-T$ to 0 can be completed to give either an odd function (sine series) or an even function (cosine series). Sketches of the two Fourier series which would be obtained are shown in fig. 57.

The two series for the particular function defined above will now be worked out in detail. Other functions can be treated in a similar manner.

(*a*) $f(t) = t$ $(0 < t < T)$. *Expansion as a sine series.* To obtain a sine series the function is deliberately defined in the range 0 to $-T$ to make it an odd function.

Thus
$$f(t) = t \quad (0 \leqslant t < T),$$
$$f(t) = t \quad (-T < t \leqslant 0).$$

As the function is odd, in the completed range,

$$a_0 = 0 \quad \text{and} \quad a_n = 0.$$

$b_n = 2 \times$ mean value of $f(t) . \sin \dfrac{n\pi t}{T}$ in the range $-T$ to T, i.e.

$$b_n = \frac{1}{T}\left\{\int_{-T}^{0} t \sin\frac{n\pi t}{T}\,dt + \int_{0}^{T} t \sin\frac{n\pi t}{T}\,dt\right\},$$

giving $$b_n = \frac{1}{T}\int_{-T}^{T} t \sin\frac{n\pi t}{T}\,dt.$$

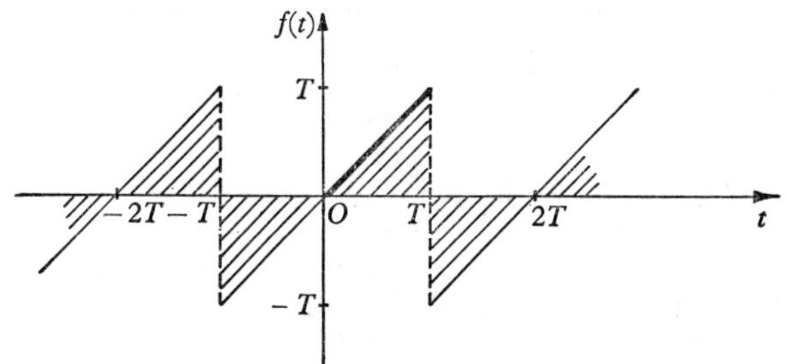

(i) Range completed to make the series an odd (sine) function

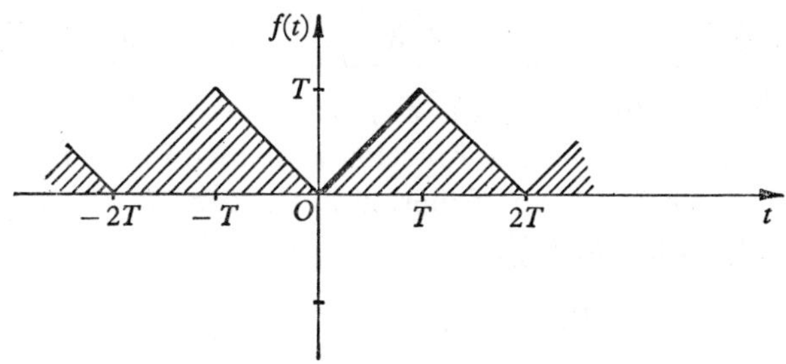

(ii) Range completed to make the series an even (cosine) function

Fig. 57

The integrand is the product of two odd functions $\left(t \text{ and } \sin\frac{n\pi t}{T}\right)$ and is therefore even.

Thus $$b_n = \frac{2}{T}\int_{0}^{T} t \sin\frac{n\pi t}{T}\,dt \quad \text{(see §2·62).} \tag{1}$$

This is quite general. If $f(t)$ is to be expanded as a sine series in the half-range 0 to T, then

$$b_n = \frac{2}{T}\int_{0}^{T} f(t) \sin\frac{n\pi t}{T}\,dt. \tag{2}$$

Proceeding with equation (1):

$$b_n = \frac{2}{T}\left[-\frac{T}{n\pi}t\cos\frac{n\pi t}{T}\right]_0^T - \frac{2}{T}\int_0^T\left(-\frac{T}{n\pi}\cos\frac{n\pi t}{T}\right)dt$$

$$= -\frac{2T}{n\pi}\cos n\pi + \frac{2}{n\pi}\left[\frac{T}{n\pi}\sin\frac{n\pi t}{T}\right]_0^T.$$

Thus $$b_n = -\frac{2T}{n\pi}\cos n\pi.$$

(i) *n odd.* $$b_n = \frac{2T}{n\pi}.$$

(ii) *n even:* $$b_n = -\frac{2T}{n\pi}.$$

The sine series is therefore

$$t = \frac{2T}{\pi}\left(\sin\frac{\pi t}{T} - \frac{1}{2}\sin\frac{2\pi t}{T} + \frac{1}{3}\sin\frac{3\pi t}{T} - \frac{1}{4}\sin\frac{4\pi t}{T} + \dots\right). \quad (3)$$

(*b*) $f(t) = t$ $(0 < t < T)$. *Expansion as a cosine series.* To obtain a cosine series the function is defined in the range $-T$ to 0 so that it is an even function in the completed range $-T$ to $+T$ (see fig. 57 (ii)).

Thus $$f(t) = t \quad (0 \leqslant t < T),$$

$$f(t) = -t \quad (-T < t \leqslant 0).$$

As the function is now even, $b_n = 0$.

$$a_0 = \frac{1}{T}\left\{\int_{-T}^0 -t\,dt + \int_0^T t\,dt\right\}$$

$$= \frac{2}{T}\int_0^T t\,dt = \frac{2}{T}\frac{T^2}{2} = T,$$

i.e. $$a_0 = T. \quad (4)$$

$$a_n = \frac{1}{T}\left\{\int_{-T}^0 -t\cos\frac{n\pi t}{T}\,dt + \int_0^T t\cos\frac{n\pi t}{T}\,dt\right\}.$$

Now the integrand is even over the range $-T$ to T.

$$\therefore a_n = \frac{2}{T}\int_0^T t\cos\frac{n\pi t}{T}\,dt \quad \text{(see § 2.62)}. \quad (5)$$

In general, if $f(t)$ is to be expanded as a cosine series in the half-range 0 to T, then

$$a_n = \frac{2}{T}\int_0^T f(t)\cos\frac{n\pi t}{T}\,dt. \tag{6}$$

Proceeding with equation (5):

$$a_n = \frac{2}{T}\left[\frac{T}{n\pi}\,t\sin\frac{n\pi t}{T}\right]_0^T - \frac{2}{T}\int_0^T \frac{T}{n\pi}\sin\frac{n\pi t}{T}\,dt$$

$$= 0 - \frac{2}{n\pi}\left[-\frac{T}{n\pi}\cos\frac{n\pi t}{T}\right]_0^T,$$

$$a_n = \frac{2T}{n^2\pi^2}(\cos n\pi - 1).$$

(i) *n odd*: $a_n = -\dfrac{4T}{n^2\pi^2}$, as $\cos n\pi$ is then -1.

(ii) *n even*: $a_n = 0$, as then $\cos n\pi$ is 1.

Thus in the half-range 0 to T

$$t = \frac{T}{2} - \frac{4T}{\pi^2}\left(\frac{1}{1^2}\cos\frac{\pi t}{T} + \frac{1}{3^2}\cos\frac{3\pi t}{T} + \frac{1}{5^2}\cos\frac{5\pi t}{T} + \dots\right). \tag{7}$$

Note. If the range is 0 to π (i.e. $T = \pi$), the working is a little easier as $\sin \pi t/T$ is then $\sin t$ and $\cos \pi t/T$ is $\cos t$.

12·3. Differentiation and integration of Fourier series

In general a Fourier series can always be integrated term by term, but it cannot be differentiated term by term unless special conditions are satisfied. Briefly, these conditions are

(i) $f(t)$ is continuous in the range $-T$ to $+T$.

(ii) $f(t)$ is periodic.

(iii) $f'(t)$ must satisfy the conditions given for $f(t)$ in volume I, § 15·15.

As an example of integration, take the sine series for $f(t) = t$ in the range 0 to T, given in § 12·2 above:

$$t = \frac{2T}{\pi}\left(\sin\frac{\pi t}{T} - \frac{1}{2}\sin\frac{2\pi t}{T} + \frac{1}{3}\sin\frac{3\pi t}{T} - \frac{1}{4}\sin\frac{4\pi t}{T} + \dots\right).$$

Integrating:

$$\frac{t^2}{2} = A - \frac{2T^2}{\pi^2}\left(\frac{1}{1^2}\cos\frac{\pi t}{T} - \frac{1}{2^2}\cos\frac{2\pi t}{T} + \frac{1}{3^2}\cos\frac{3\pi t}{T} - \dots\right). \quad (1)$$

Unless the series obtained by integrating vanishes for a simple value of t (e.g. $t = 0$ or $t = \pi$ in the case of a sine series), it is better to calculate the value of the arbitrary constant A as being the mean value of the function (in this case $\frac{1}{2}t^2$) in the range 0 to T.

Thus
$$A = \frac{1}{T}\int_0^T \frac{t^2}{2}\,dt$$
$$= \frac{1}{T}\frac{T^3}{6} = \frac{T^2}{6}.$$

Thus

$$\frac{t^2}{2} = \frac{T^2}{6} - \frac{2T^2}{\pi^2}\left(\frac{1}{1^2}\cos\frac{\pi t}{T} - \frac{1}{2^2}\cos\frac{2\pi t}{T} + \frac{1}{3^2}\cos\frac{3\pi t}{T} - \dots\right)$$

or

$$t^2 = \frac{T^2}{3} - \frac{4T^2}{\pi^2}\left(\frac{1}{1^2}\cos\frac{\pi t}{T} - \frac{1}{2^2}\cos\frac{2\pi t}{T} + \frac{1}{3^2}\cos\frac{3\pi t}{T} - \dots\right).$$

12·4. Worked example

If $f(x)$ can be expanded in the Fourier Series

$$\tfrac{1}{2}a_0 + \sum_{n=1}^{\infty}(a_n \cos nx + b_n \sin nx),$$

show that

$$a_n + jb_n = \frac{1}{\pi}\int_0^{2\pi} f(x)e^{jnx}\,dx.$$

Hence, or otherwise, expand the function $x(x - 1)(x - 2)$ in a full-range Fourier series, valid in the interval (0, 2), giving the general term.

Illustrate your result by drawing a graph of the function represented by the series for values of x outside this range. [L.U.]

This exercise has been specially chosen in order to show how a complex form of Fourier series can be used to calculate a_n and b_n simultaneously.

A slightly more general case of the first part of this exercise will be worked here.

Suppose that $f(x)$ can be expanded in the Fourier series

$$\tfrac{1}{2}a_0 + \sum_{n=1}^{\infty}\left(a_n \cos\frac{n}{c}\pi x + b_n \sin\frac{n}{c}\pi x\right),$$

where c is a constant and the range is 0 to $2c$.

From previous work:

$$a_n = \frac{1}{c}\int_0^{2c} f(x)\cos\frac{n}{c}\pi x\,dx$$

and $$b_n = \frac{1}{c}\int_0^{2c} f(x)\sin\frac{n}{c}\pi x\,dx.$$

$$\therefore\ a_n + jb_n = \frac{1}{c}\int_0^{2c} f(x)\left\{\cos\frac{n}{c}\pi x + j\sin\frac{n}{c}\pi x\right\}dx,$$

i.e. $$\underline{a_n + jb_n = \frac{1}{c}\int_0^{2c} f(x)e^{j\frac{n}{c}\pi x}\,dx.} \qquad (1)$$

Putting $c = \pi$, and the range therefore 0 to 2π:

$$a_n + jb_n = \frac{1}{\pi}\int_0^{2\pi} f(x)e^{jnx}\,dx,$$

which is the required result.

Taking $f(x) = x(x-1)(x-2)$ and the range 0 to 2 (i.e. $c = 1$):

$$a_0 = \frac{1}{c}\int_0^{2c} f(x)\,dx = \int_0^2 x(x-1)(x-2)dx$$

$$= \int_0^2 (x^3 - 3x^2 + 2x)\,dx$$

$$= \left[\frac{x^4}{4} - x^3 + x^2\right]_0^2 \qquad = 0\,.$$

Thus $$\underline{a_0 = 0.} \qquad (2)$$

$$a_n + jb_n = \frac{1}{c}\int_0^{2c} f(x)e^{j\frac{n}{c}\pi x}\,dx, \quad \text{from (1).}$$

$$= \int_0^2 f(x)e^{jn\pi x}\,dx, \quad \text{when } c = 1$$

$$= \int_0^2 (x^3 - 3x^2 + 2x)e^{jn\pi x}\,dx$$

$$= \left[\frac{1}{jn\pi}(x^3 - 3x^2 + 2x)e^{jn\pi x}\right]_0^2 - \frac{1}{jn\pi}\int_0^2 (3x^2 - 6x + 2)e^{jn\pi x}\,dx$$

$$= 0 + \frac{j}{n\pi}\int_0^2 (3x^2 - 6x + 2)e^{jn\pi x}\,dx$$

$$= \frac{j}{n\pi}\left[\frac{1}{jn\pi}(3x^2 - 6x + 2)e^{jn\pi x}\right]_0^2 - \frac{1}{n^2\pi^2}\int_0^2 (6x - 6)e^{jn\pi x}\,dx$$

$$= \frac{1}{n^2\pi^2}(2e^{jn2\pi} - 2) - \frac{6}{n^2\pi^2}\int_0^2 (x - 1)e^{jn\pi x}\,dx$$

$$= 0 - \frac{6}{n^2\pi^2}\int_0^2 (x - 1)e^{jn\pi x}\,dx, \quad (\text{as } e^{jn2\pi} = 1)$$

$$= -\frac{6}{n^2\pi^2}\left[\frac{1}{jn\pi}(x - 1)e^{jn\pi x}\right]_0^2 + \frac{6}{jn^3\pi^3}\int_0^2 e^{jn\pi x}\,dx$$

$$= \frac{j6}{n^3\pi^3}(e^{jn2\pi} + 1) + \frac{6}{jn^3\pi^3}\left[\frac{1}{jn\pi}e^{jn\pi x}\right]_0^2$$

$$= \frac{j12}{n^3\pi^3} - \frac{6}{n^4\pi^4}(e^{jn2\pi} - 1)$$

$$= \frac{j12}{n^3\pi^3} - 0.$$

Thus $$a_n + jb_n = \frac{j12}{n^3\pi^3}.$$

Equating real and imaginary parts:

$$\underline{a_n = 0, \quad b_n = \frac{12}{n^3\pi^3}}. \tag{3}$$

Hence

$$x(x-1)(x-2) = \frac{12}{\pi^3}\left(\frac{1}{1^3}\sin \pi x + \frac{1}{2^3}\sin 2\pi x + \dots + \frac{1}{n^3}\sin n\pi x + \dots\right),$$

in the range (0, 2).

Fig. 58 below shows the Fourier series in dotted line and the function $x(x-1)(x-2)$ in full line.

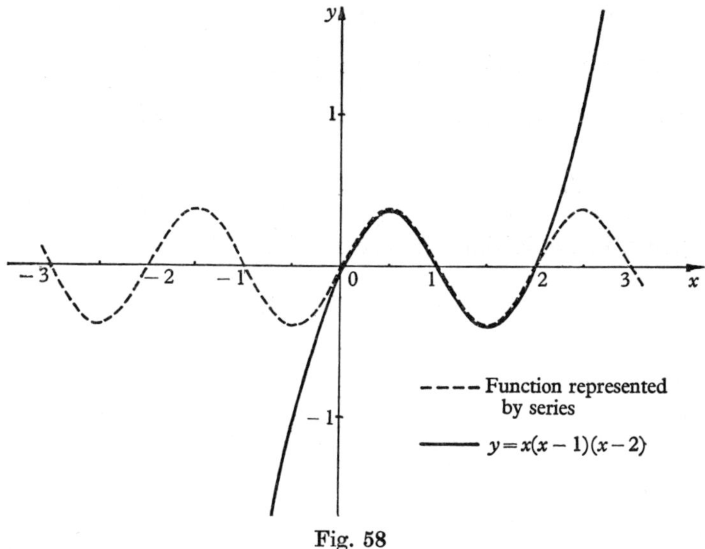

Fig. 58

12·5. Some further applications of Fourier series

12·51. Foreword

A general indication of the uses of Fourier series in engineering problems was given in volume I, § 15·3. In the following sections two detailed examples will be given. These examples are by no means exhaustive, but give an idea of the scope and usefulness of Fourier series.

12·52. Electrical example

In a telegraph cable in which L and G are negligible, it was shown in § 11·5 that the partial differential equation giving the voltage at distance x and time t is

$$\frac{\partial^2 v}{\partial x^2} - RC\frac{\partial v}{\partial t} = 0. \qquad (1)$$

Initially, let there be a constant voltage E_0 applied at $x = 0$, which steadily falls to zero at the end of the line ($x = l$) when the steady state has been reached. If the sending end is now suddenly earthed (i.e. its voltage suddenly reduced to zero), it is required to find the voltage v at any subsequent time.

In the steady state, $\partial v/\partial t = 0$ (the voltage is not changing with the time), therefore from (1):

$$\frac{\partial^2 v}{\partial x^2} = 0$$

and integrating twice $\quad v = Ax + B.$

Now $\quad v = E_0$ at $x = 0$ and $v = 0$ at $x = l.$

$$\therefore\ B = E_0 \quad \text{and} \quad A = -\frac{B}{l} = -\frac{E_0}{l}.$$

Thus, in the initial steady condition,

$$v = \frac{E_0}{l}(l - x). \tag{2}$$

After this the voltage is to be zero at both ends.

This suggests that a solution of (1) should now be sought in the form

$$v = T \sin \frac{n\pi x}{l}, \tag{3}$$

where T is a function of time only.

This does fit the facts that for all t, $v = 0$ at $x = 0$ and $x = l$ (as $\sin n\pi = 0$).

Substitution into (1) gives

$$-\frac{n^2\pi^2}{l^2} T \sin \frac{n\pi x}{l} - RC\frac{dT}{dt} \sin \frac{n\pi x}{l} = 0.$$

Thus $$\frac{dT}{dt} = -\frac{n^2\pi^2 T}{CRl^2},$$

i.e. $$\frac{dT}{T} = -\frac{n^2\pi^2}{CRl^2}\,dt = -n^2\lambda\,dt,$$

where $$\lambda = \frac{\pi^2}{CRl^2}. \tag{4}$$

Integrating:

$$\log_e BT = -n^2\lambda t, \quad B \text{ arbitrary},$$

giving

$$T = Ae^{-n^2\lambda t}, \quad A \text{ arbitrary}.$$

Putting this into (3), $v = Ae^{-n^2\lambda t}\sin\dfrac{n\pi x}{l}$.

A general type of solution is therefore of the form

$$v = \sum_{n=1}^{\infty} A_n e^{-n^2\lambda t}\sin\frac{n\pi x}{l}. \tag{5}$$

The initial value of this at time $t = 0$ is

$$v = \sum_{n=1}^{\infty} A_n \sin\frac{n\pi x}{l}. \tag{6}$$

But this initial value is known. It is given by equation (2) with the additional fact that at $x = 0$, $v = 0$ owing to the sudden earthing of the sending end. The value of v, then, at $t = 0$ is as shown in fig. 59 below.

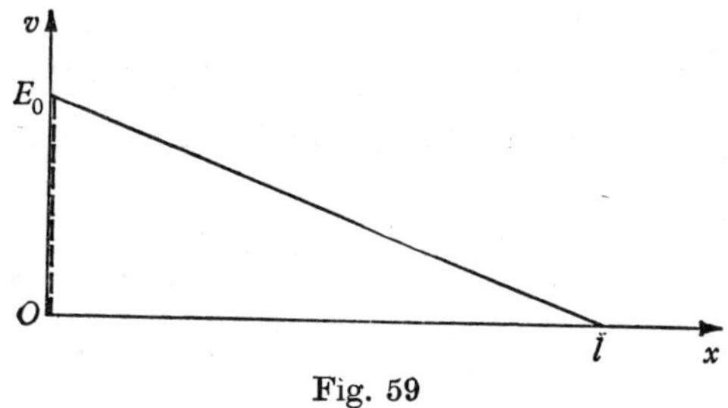

Fig. 59

The dotted line signifies that at $x = 0$ the voltage is suddenly zero (i.e. $x = 0$ is a point of discontinuity in the figure).

Consider the function defined by

$$v = E_0\frac{(l-x)}{l} \quad (0 < x \leqslant l),$$

$$v = 0 \quad \text{at} \quad x = 0.$$

It can be expanded as a Fourier half-range sine series:

$$v = \sum_{n=1}^{\infty} b_n \sin\frac{n\pi x}{l}.$$

(Note that the sine series satisfies $v = 0$ at $x = 0$ and $x = l$.)

Now

$$b_n = \frac{2}{l}\int_0^l \frac{E_0(l-x)}{l}\sin\frac{n\pi x}{l}\,dx$$

$$= \frac{2}{l}\left[-\frac{E_0(l-x)}{l}\frac{l}{n\pi}\cos\frac{n\pi x}{l}\right]_0^l - \frac{2E_0}{ln\pi}\int_0^l \cos\frac{n\pi x}{l}dx$$

$$= \frac{2E_0}{n\pi} - \frac{2E_0}{ln\pi}\left[\frac{l}{n\pi}\sin\frac{n\pi x}{l}\right]_0^l$$

$$= \frac{2E_0}{n\pi}.$$

$$\therefore\ v = \sum_{n=1}^{\infty}\frac{2E_0}{n\pi}\sin\frac{n\pi x}{l} \quad \text{at } t = 0. \tag{7}$$

Comparing (6) and (7):

$$A_n = \frac{2E_0}{n\pi}.$$

Putting this into (5), a general solution for the voltage at any subsequent time to the earthing of the sending end is given by

$$v = \sum_{n=1}^{\infty}\frac{2E_0}{\pi}\frac{1}{n}e^{-n^2\lambda t}\sin\frac{n\pi x}{l}.$$

Expanded, this is

$$v = \frac{2E_0}{\pi}\left(e^{-\lambda t}\sin\frac{\pi x}{l} + \frac{1}{2}e^{-4\lambda t}\sin\frac{2\pi x}{l} + \dots + \frac{1}{n}e^{-n^2\lambda t}\sin\frac{n\pi x}{l} + \dots\right).$$

12·53. An example of a solution of Laplace's equation, using Fourier series

The student may have noticed that the examples in the previous sections have covered two of the three common types of partial differential equations given in § 11·3. This final example will deal with the third type, Laplace's equation (of two dimensions).

The previous example has also illustrated the usual method of applying Fourier series to solve a partial differential equation which has given boundary conditions. This consists of first trying to write down a simple solution of the equation involving trigonometrical functions and which satisfies some of the boundary conditions; a more general solution is then assumed to be an infinite series of such terms with arbitrary coefficients; finally, these coefficients are evaluated by forming a Fourier series which is determined by the remaining boundary conditions.

This same method will now be applied to solve the following problem:

Find a solution for the equation

$$\frac{\partial^2 z}{\partial x^2} + \frac{\partial^2 z}{\partial y^2} = 0, \tag{1}$$

which is bounded by the rectangle

$$0 < x < a, \quad 0 < y < b, \tag{2}$$

and which also satisfies the conditions

$$z = 0 \quad \text{when} \quad x = 0 \quad (0 < y < b), \tag{3}$$

$$z = 0 \quad \text{when} \quad x = a \quad (0 < y < b), \tag{4}$$

$$z = 0 \quad \text{when} \quad y = b \quad (0 < x < a), \tag{5}$$

$$z = x \quad \text{when} \quad y = 0 \quad (0 < x < a). \tag{6}$$

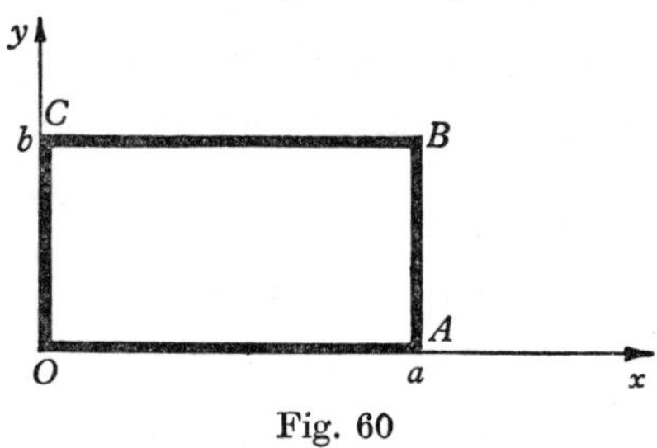

Fig. 60

In geometrical language (see fig. 60) this means that a function, $z = f(x, y)$, is being sought which both satisfies equation (1) and is also such that $z = 0$ along OC, $z = 0$ along AB, $z = 0$ along CB and $z = x$ along OA. The function, when found, will give the value for z anywhere within the rectangle $OABC$.

As a trial solution, let $z = XY$, where X and Y are respectively functions of x and y only (cf. § 11·2).

Substituting in (1) gives

$$Y\frac{d^2X}{dx^2} + X\frac{d^2Y}{dy^2} = 0,$$

giving
$$\frac{1}{Y}\frac{d^2Y}{dy^2} = -\frac{1}{X}\frac{d^2X}{dx^2}.$$

As the left-hand side is a function of y only, and the right-hand side is a function of x only, they must both be constant. Let this constant be p^2, then

$$\frac{1}{Y}\frac{d^2Y}{dy^2} = p^2, \quad \text{giving} \quad \frac{d^2Y}{dy^2} - p^2Y = 0$$

and
$$\frac{1}{X}\frac{d^2X}{dx^2} = -p^2, \quad \text{giving} \quad \frac{d^2X}{dx^2} + p^2X = 0.$$

The solutions to these equations can be written

$$Y = A\sinh p(y+\alpha), \quad X = B\sin p(x+\beta),$$

where A, B, α, β are arbitrary constants.

A solution of (1) is therefore

$$z = C\sin p(x+\beta)\sinh p(y+\alpha), \tag{7}$$

where C, α, β are arbitrary constants.

Using conditions (3), (4) and (5) in turn:

(i) $0 = C\sin p\beta \sinh p(y+\alpha)$ for $0 < y < b$.

Neglecting the trivial solution $C = 0$, $\sin p\beta = 0$ and therefore β can be taken as zero. Thus

$$z = C\sin px\sinh p(y+\alpha).$$

(ii) $\quad 0 = C\sin pa\sinh p(y+\alpha) \quad$ for $\quad 0 < y < b$.

$$\therefore\ 0 = \sin pa, \quad \text{and} \quad pa = n\pi \quad (n = 1, 2, 3, \ldots).$$

Thus $\quad z = C\sin\frac{n\pi x}{a}\sinh\frac{n\pi}{a}(y+\alpha).$

(iii) $$0 = C \sin\frac{n\pi x}{a}\sinh\frac{n\pi}{a}(b+\alpha) \quad \text{for} \quad 0 < x < a.$$

$$\therefore \sinh\frac{n\pi}{a}(b+\alpha) = 0,$$

giving $\alpha = -b$.

Thus $$z = C\sin\frac{n\pi x}{a}\sinh\frac{n\pi}{a}(y-b).$$

Now C is arbitrary and the range of y is from 0 to b; it is therefore more convenient to write this as

$$z = E\sin\frac{n\pi x}{a}\sinh\frac{n\pi}{a}(b-y), \tag{8}$$

where E is arbitrary.

With more experience, solutions such as that given in (8) above can in fact be guessed on inspection of the boundary conditions.

A more general solution, which will satisfy equations (1), (3), (4) and (5), can be written as

$$\underline{z = \sum_{n=1}^{\infty} A_n \sin\frac{n\pi x}{a}\sinh\frac{n\pi}{a}(b-y).} \tag{9}$$

The remaining boundary condition, given by (6), will now be used.

When $y = 0$, equation (9) reduces to

$$z = \sum_{n=1}^{\infty} A_n \sinh\frac{n\pi b}{a}\sin\frac{n\pi x}{a}. \tag{10}$$

As this must equal x, in the range $0 < x < a$ (see (6)), a Fourier half-range sine series for x, of period $2a$, is determined, and compared with (10).

Let $$x = \sum_{n=1}^{\infty} b_n \sin\frac{n\pi x}{a} \quad (0 < x < a). \tag{11}$$

Then $$b_n = \frac{2}{a}\int_0^a x\sin\frac{n\pi x}{a}\,dx$$

$$= \frac{2}{a}\left[-\frac{a}{n\pi}x\cos\frac{n\pi x}{a}\right]_0^a + \frac{2}{n\pi}\int_0^a \cos\frac{n\pi x}{a}\,dx$$

$$= -\frac{2a}{n\pi}\cos n\pi + \frac{2}{n\pi}\left[\frac{a}{n\pi}\sin\frac{n\pi x}{a}\right]_0^a,$$

i.e. $$b_n = -\frac{2a}{n\pi}\cos n\pi.$$

$$\text{(i) } n \textit{ odd}: \ b_n = \frac{2a}{n\pi}.$$

$$\text{(ii) } n \textit{ even}: \ b_n = -\frac{2a}{n\pi}.$$

Thus $$b_n = (-1)^{n+1}\frac{2a}{n\pi}$$

for all positive integers n.

From (11), when $y = 0$,

$$z = \sum_{n=1}^{\infty} (-1)^{n+1}\frac{2a}{n\pi}\sin\frac{n\pi x}{a}. \tag{12}$$

Comparing coefficients in (10) and (12)

$$A_n \sinh\frac{n\pi b}{a} = (-1)^{n+1}\frac{2a}{n\pi}.$$

$$\therefore A_n = (-1)^{n+1}\frac{2a}{n\pi\sinh\frac{n\pi b}{a}}. \tag{13}$$

From (9) and (13) a general solution to fit the given conditions is

$$z = \sum_{n=1}^{\infty} (-1)^{n+1}\frac{2a}{n\pi}\,\frac{\sin\frac{n\pi x}{a}\sinh\frac{n\pi}{a}(b-y)}{\sinh\frac{n\pi b}{a}}.$$

Note. Laplace's equation in three dimensions is

$$\frac{\partial^2 V}{\partial x^2} + \frac{\partial^2 V}{\partial y^2} + \frac{\partial^2 V}{\partial z^2} = 0.$$

It is satisfied by the potential V (whether gravitational, electrical or magnetic) of a system in free space. It is extremely important in advanced theory.

EXERCISE 39

N.B. These will include miscellaneous revision exercises on Fourier Series.

1. A function $f(x)$ is such that $f(x) = f(x + 2\pi)$ and $f(x) = \frac{1}{4}x^2$ for $-\pi \leqslant x \leqslant \pi$. Sketch the graph of $f(x)$ from $x = -2\pi$ to $x = 2\pi$ and state what terms are absent from the Fourier expansion of $f(x)$ in the range $-\pi$ to π of x.

Obtain the general term in this series and deduce the value of $\sum_{n=1}^{\infty} n^{-2}$. [L.U.]

2. A function is defined by $f(x) = \pi x - x^2$ for $0 \leqslant x \leqslant \pi$. Expand the function in (i) a series of sines only, (ii) a series of cosines only of integral multiples of x, each series being valid for $0 \leqslant x \leqslant \pi$. Sketch the graph of the sum of each series for $-\pi \leqslant x \leqslant \pi$. [L.U.]

3. A full wave rectifier delivers a current $i = I \left| \sin \omega t \right|$. Determine the direct component and the frequency and amplitude of the fundamental oscillatory component of this current. [I.E.E.]

4. A function of x, of period 2π, is zero for $0 \leqslant x \leqslant \frac{1}{2}\pi$ and equal to $\cos x$ for $\frac{1}{2}\pi \leqslant x \leqslant \pi$. Expand the function in a series of cosines of multiples of x, valid when $0 \leqslant x \leqslant \pi$. Sketch the graph of the sum of the series for $0 < x < 3\pi$. [L.U·]

5. Integrate the two series obtained in Exercise 2 and evaluate the arbitrary constants. State the functions which each series represents within the range 0 to π.

6. (i) The function $f(\theta)$ is an odd function of θ, with a period of 2π, which is defined by the equations:

$$f(\theta) = c \text{ (constant)} \quad (0 < \theta < \tfrac{1}{2}\pi),$$

$$f(\theta) = 0 \quad (\tfrac{1}{2}\pi < \theta < \pi).$$

Express $f(\theta)$ in a Fourier sine series.

(ii) Draw the graph of $\int_0^\theta f(\theta)\, d\theta$ in the range from $-\pi$ to π.

(iii) From the series obtained above, by integrating twice in succession from 0 to θ and then putting $\theta = 0$ and $\theta = \pi$ to find the constants of integration, two new series, a cosine series and a sine series, can be deduced. Show graphically what functions these two series represent. [L.U.]

7. If $f(x) = -f(-x)$ in the range $-\pi < x < \pi$, show that the Fourier series for $f(x)$ in this range contains no cosine terms.

If the above function is defined by $f(x) = x$ when $0 \leqslant x \leqslant \frac{1}{2}\pi$, and $f(x) = \pi - x$ when $\frac{1}{2}\pi \leqslant x \leqslant \pi$, show that in the range $-\pi \leqslant x \leqslant \pi$,

$$f(x) = \frac{4}{\pi} \sum_{n=0}^{\infty} (-1)^n \frac{\sin(2n+1)x}{(2n+1)^2}. \qquad \text{[L.U.]}$$

8. Solve the equation

$$\frac{\partial^2 v}{\partial x^2} - \frac{1}{k}\frac{\partial v}{\partial t} = 0,$$

given that $v = 0$ when $x = l$, $\dfrac{\partial v}{\partial x} = 0$ when $x = 0$, and $v = c$ (constant) when $t = 0$.

It is suggested that a solution of the type $v = T\cos px$, where T is a function of t only, will best fit the given conditions.

9. Data for a half-cycle of a symmetrical magnetizing current wave for an iron-cored inductor are given in the following table. Plot the wave to scale, and determine by Fourier analysis (e.g. a tabular method using mid-ordinates) approximate values for the amplitude and phase of the fundamental component of the wave.

Electrical degrees	0	30	45	60	75	82·5	90	97·5	105	120	135	150	180
Amperes	1·0	1·3	1·55	1·95	2·65	3·35	4·0	3·15	2·1	0·6	− 0·1	− 0·5	− 1·0

[L.U.]

10. (i) Express $x \sin x$ as a Fourier series of the form

$$\tfrac{1}{2}a_0 + \sum_{n=1}^{\infty} (a_n \cos nx + b_n \sin nx)$$

in the range $-\pi \leqslant x \leqslant \pi$.

(ii) What is the function represented by the series in the range $\pi \leqslant x \leqslant 3\pi$?

(iii) By taking a particular value of x deduce that

$$\frac{1}{1.3} - \frac{1}{3.5} + \frac{1}{5.7} - \dots = \tfrac{1}{4}(\pi - 2).$$

[C.U.]

Answers

1. $f(x)$ is an even function, therefore there are no sine terms.

$$\frac{x^2}{4} = \frac{\pi^2}{12} + \sum_{n=1}^{\infty} \frac{1}{n^2} \cos n\pi \cos nx.$$

Putting $x = \pi$ gives $\sum_{n=1}^{\infty} n^{-2} = \frac{1}{6}\pi^2$.

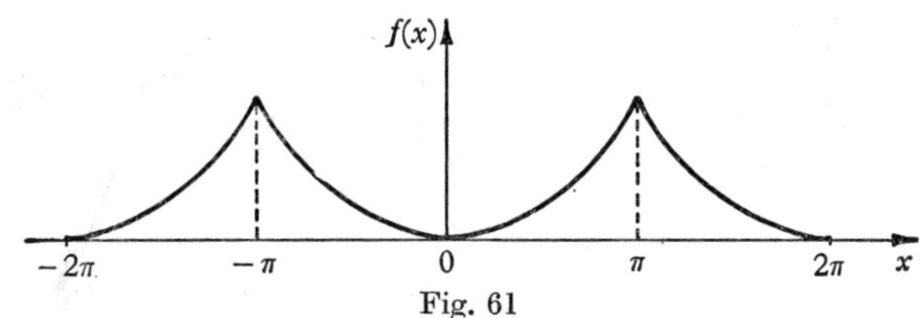

Fig. 61

2. (i) $$\frac{8}{\pi}\left(\frac{\sin x}{1^3} + \frac{\sin 3x}{3^3} + \frac{\sin 5x}{5^3} + \ldots\right);$$

(ii) $$\frac{\pi^2}{6} - 4\left(\frac{\cos 2x}{2^2} + \frac{\cos 4x}{4^2} + \frac{\cos 6x}{6^2} + \ldots\right).$$

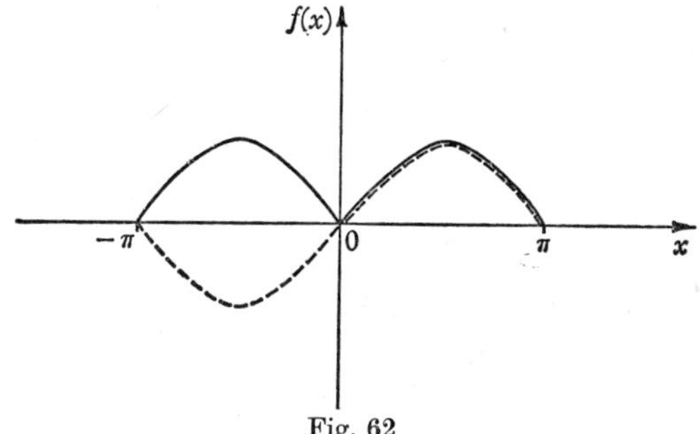

Fig. 62

3. $\frac{2I}{\pi}$; $-\frac{4I}{3\pi} \cos 2\,\omega t.$

4. $$-\frac{1}{\pi} + \frac{1}{2}\cos x - \frac{2}{\pi}\left(\frac{1}{1\,.\,3}\cos 2x - \frac{1}{3\,.\,5}\cos 4x + \frac{1}{5\,.\,7}\cos 6x - \ldots\right).$$

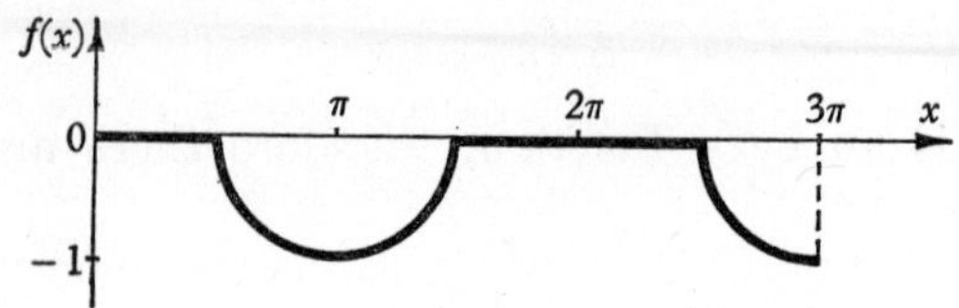

Fig. 63

5. (i) $$\frac{\pi x^2}{2} - \frac{x^3}{3} = \frac{\pi^3}{12} - \frac{8}{\pi}\left(\frac{\cos x}{1^4} + \frac{\cos 3x}{3^4} + \frac{\cos 5x}{5^4} + \dots\right);$$

(ii) $$\frac{\pi x^2}{2} - \frac{\pi^2 x}{6} - \frac{x^3}{3} = -4\left(\frac{\sin 2x}{2^3} + \frac{\sin 4x}{4^3} + \frac{\sin 6x}{6^3} + \dots\right).$$

6. (i) $$f(\theta) = \frac{2c}{\pi}(\sin\theta + \tfrac{1}{3}\sin 3\theta + \tfrac{1}{5}\sin 5\theta + \dots) + \frac{4c}{\pi}(\tfrac{1}{2}\sin 2\theta + \tfrac{1}{6}\sin 6\theta + \tfrac{1}{10}\sin 10\theta + \dots).$$

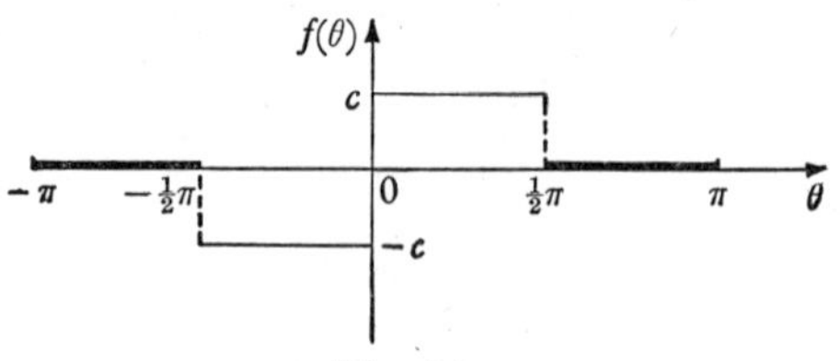

Fig. 64

The graph of $f(\theta)$ is shown in fig. 64.

$$f(\theta) = c, \quad 0 < \theta < \tfrac{1}{2}\pi$$
$$f(\theta) = 0, \quad \tfrac{1}{2}\pi < \theta < \pi.$$

(ii) Let $F(\theta) = \int_0^\theta f(\theta)\,d\theta$. The graph of $F(\theta)$, in the range $-\pi$ to $+\pi$, is shown in fig. 65.

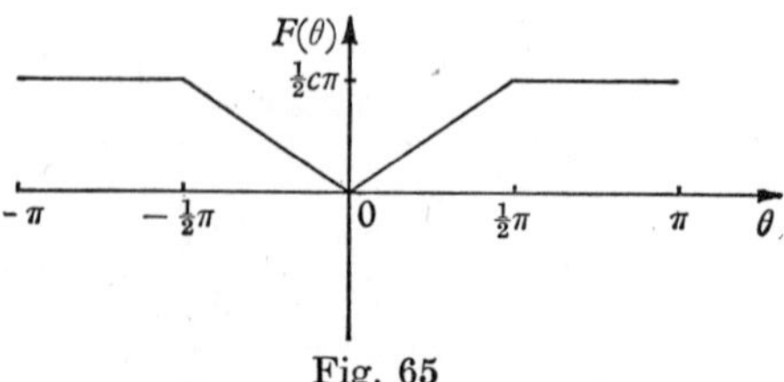

Fig. 65

(Notice that as $f(\theta)$ is odd, $F(\theta)$ is even.)

$$F(\theta) = c\theta, \qquad 0 \leqslant \theta \leqslant \tfrac{1}{2}\pi$$
$$F(\theta) = \tfrac{1}{2}c\pi, \quad \tfrac{1}{2}\pi \leqslant \theta \leqslant \pi.$$

(These two graphs can be compared with figs. 89 and 90 in volume I, which show shearing force and bending moment for a beam with a point load. In both cases a point of discontinuity of the first graph becomes a point of continuity on the second graph, but the *slope* at this point is discontinuous.)

(iii) Integration of the series in (i) above gives

$$F(\theta) = -\frac{2c}{\pi}\left(\frac{1}{1^2}\cdot\cos\theta + \frac{1}{3^2}\cos 3\theta + \frac{1}{5^2}\cos 5\theta + \ldots\right)$$
$$-\frac{4c}{\pi}\left(\frac{1}{2^2}\cos 2\theta + \frac{1}{6^2}\cos 6\theta + \frac{1}{10^2}\cos 10\theta + \ldots\right) + A.$$

Putting $\theta = 0$ (in which case $F(\theta) = 0$) and $\theta = \pi$ (in which case $F(\theta) = \frac{1}{2}c\pi$) gives $A = \frac{3}{8}c\pi$. In this connexion note that

$$\left(\frac{1}{2^2} + \frac{1}{6^2} + \frac{1}{10^2} + \ldots\right) = \frac{1}{4}\left(\frac{1}{1^2} + \frac{1}{3^2} + \frac{1}{5^2} + \ldots\right).$$

A quicker way of finding A is to use the method given in § 12·3, i.e. A is the mean value of $F(\theta)$ from 0 to π.

Thus the cosine series obtainable is for a function $\phi(\theta)$ given by

$$\phi(\theta) = c\theta - \tfrac{3}{8}c\pi \;\; (0 \leqslant \theta \leqslant \tfrac{1}{2}\pi);$$
$$\phi(\theta) = \tfrac{1}{2}c\pi - \tfrac{3}{8}c\pi = \tfrac{1}{8}c\pi \;\; (\tfrac{1}{2}\pi \leqslant \theta \leqslant \pi),$$

and the cosine series for $\phi(\theta)$ is

$$-\frac{2c}{\pi}\left(\frac{1}{1^2}\cos\theta + \frac{1}{3^2}\cos 3\theta + \frac{1}{5^2}\cos 5\theta + \ldots\right)$$
$$-\frac{4c}{\pi}\left(\frac{1}{2^2}\cos 2\theta + \frac{1}{6^2}\cos 6\theta + \frac{1}{10^2}\cos 10\theta + \ldots\right).$$

The graph of $\phi(\theta)$ is shown in fig. 66.

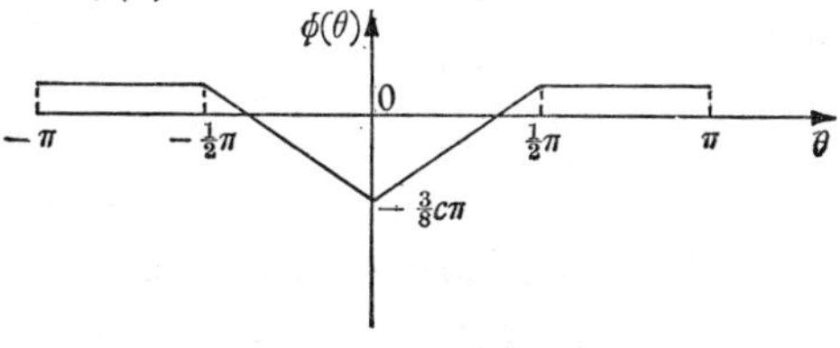

Fig. 66

Integrating the cosine series above gives a sine series:

$$-\frac{2c}{\pi}\left(\frac{1}{1^3}\sin\theta+\frac{1}{3^3}\sin 3\theta+\frac{1}{5^3}\sin 5\theta+\dots\right)$$
$$-\frac{4c}{\pi}\left(\frac{1}{2^3}\sin 2\theta+\frac{1}{6^3}\sin 6\theta+\frac{1}{10^3}\sin 10\theta+\dots\right)+B.$$

This will represent a function, say $\psi(\theta)$, such that

$$\psi(\theta)=\tfrac{1}{2}c\theta^2-\tfrac{3}{8}c\pi\theta \quad (0\leqslant\theta\leqslant\tfrac{1}{2}\pi);$$
$$\psi(\theta)=\tfrac{1}{8}c\pi\theta-\tfrac{1}{8}\pi^2c \quad (\tfrac{1}{2}\pi\leqslant\theta\leqslant\pi).$$

This can be obtained by integrating $\phi(\theta)$ of (iii), bearing in mind that at $\theta=\frac{1}{2}\pi$ both $\psi(\theta)$ and $\psi'(\theta)$ have the same values taken in either sub-interval 0 to $\frac{1}{2}\pi$ or $\frac{1}{2}\pi$ to π.

Putting $\theta=0$ shows that $B=0$.

The graph of $\psi(\theta)$ is shown in fig. 67.

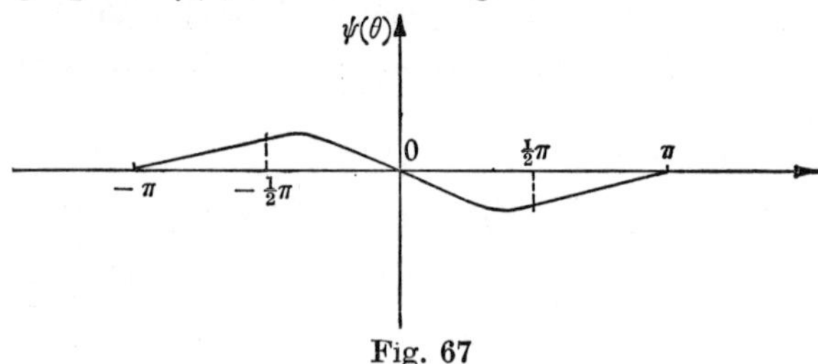

Fig. 67

8. $v=\frac{4c}{\pi}\sum_{n=0}^{\infty}\frac{(-1)^n}{(2n+1)}\exp\left\{-\frac{(2n+1)^2\pi^2kt}{4l^2}\right\}\cos\frac{(2n+1)\pi x}{2l}.$

9. Fundamental is approximately
$3\sin(x+14°30')$; of amplitude 3 and phase $14°30'$.

10. (i) $$1-2\left\{\tfrac{1}{4}\cos x+\frac{1}{1.3}\cos 2x-\frac{1}{2.4}\cos 3x+\frac{1}{3.5}\cos 4x-\dots\right\}.$$

Note that

$$b_n=0 \text{ and } a_n=\frac{2(-1)^{n+1}}{(n-1)(n+1)} \text{ for } n=2, 3, 4, \dots.$$

a_1 is an exception, and has to be calculated separately.

(ii) The function represented by the series in the range $\pi\leqslant x\leqslant 3\pi$ is $(x-2\pi)\sin x$.

(iii) [Taking $x=\frac{1}{2}\pi$ gives the required result.]

Chapter 13

DETERMINANTS. SOLUTION OF SIMULTANEOUS LINEAR EQUATIONS

13·1. Introduction

At this stage there are two main uses of determinants for engineers: their use in solving simultaneous linear equations, and their application to coordinate geometry. Apart from purely theoretical work, there is a growing feeling that their use in solving simultaneous equations may decline. In practice, for equations with simple numerical coefficients, it is just as quick to use the ordinary elimination process. For cases involving cumbersome coefficients, methods giving answers correct to any desired degree of accuracy are coming into use. One of these, the relaxation method, is given later.

13·2. Definitions

$\begin{vmatrix} 7 & 5 \\ 2 & 3 \end{vmatrix}$ is called a determinant of the second order. The sign for a determinant is the vertical pair of lines shown above. This signifies that a definite sequence of operations must be performed on the numbers enclosed. The determinant shown above has two *rows* (7 5 and 2 3) and two *columns* (7 2 and 5 3). All determinants have the same number of rows as columns. One with two rows and two columns is called a determinant of the second *order*; one with three rows and three columns is of the third order, and so on. The individual numbers are called the *elements* of the determinant. The top left-hand element is called the *leading* element, the number 7 in the case above.

Consider the third-order determinant shown below:

$$\begin{vmatrix} 7 & 5 & 9 \\ 2 & 4 & 11 \\ 3 & 12 & 10 \end{vmatrix}$$

The numbers 7 4 10 form the leading *diagonal*. If the column and row containing any element is imagined to be ruled through, the determinant left is called the *minor* of that element. In the case shown above $\begin{vmatrix} 5 & 9 \\ 12 & 10 \end{vmatrix}$ is the minor for the element 2.

13·3. Expansion of determinants

Every determinant has one specific value, which is found by performing on it a definite sequence of operations.

Determinants were invented to deal concisely with the theory of simultaneous linear equations. The rules for their evaluation are based on this original intention.

$$\begin{vmatrix} a_1 & b_1 \\ a_2 & b_2 \end{vmatrix} \equiv a_1b_2 - a_2b_1,$$

e.g. $$\begin{vmatrix} 7 & 5 \\ 2 & 3 \end{vmatrix} = 7 . 3 - 2 . 5 = 21 - 10 = 11.$$

The dotted lines indicate the 'cross-multiplication' rule.

To expand a determinant of higher order any row (or column) is first selected. The elements in this row (or column) are then multiplied by plus or minus their corresponding minors, and added together algebraically. This rule holds for determinants of any order, though mostly third-order determinants will be used here. The plus or minus sign to be attached to each minor, when evaluating, may be found by the following rule: Step from the leading element along the top row and go down the column containing the element whose minor is desired, until this element is reached. If the number of steps is even, the sign to be attached to the minor is positive; if odd, the sign is negative.

As an example, take the general third-order determinant

$$\Delta = \begin{vmatrix} a_1 & b_1 & c_1 \\ a_2 & b_2 & c_2 \\ a_3 & b_3 & c_3 \end{vmatrix}$$

When evaluating, the sign for the minor of b_3 is negative, there being three steps.

The required signs build up to the pattern given below:

$$\begin{vmatrix} + & - & + & - & + & - & + \\ - & + & - & + & - & + & - \\ + & - & + & - & + & - & + \\ - & + & - & + & \cdots & \cdots & \cdots \\ + & - & + & \cdots & \cdots & \cdots & \cdots \\ - & + & \cdots & \cdots & \cdots & \cdots & \cdots \\ \cdots & \cdots & \cdots & \cdots & \cdots & \cdots & \cdots \end{vmatrix}$$

The expansion of the third-order determinant given above is:

(i) *Selecting the first column*:

$$\Delta = a_1 \begin{vmatrix} b_2 & c_2 \\ b_3 & c_3 \end{vmatrix} - a_2 \begin{vmatrix} b_1 & c_1 \\ b_3 & c_3 \end{vmatrix} + a_3 \begin{vmatrix} b_1 & c_1 \\ b_2 & c_2 \end{vmatrix}$$

$$= a_1(b_2c_3 - b_3c_2) - a_2(b_1c_3 - b_3c_1) + a_3(b_1c_2 - b_2c_1)$$

$$= a_1b_2c_3 - a_1b_3c_2 - a_2b_1c_3 + a_2b_3c_1 + a_3b_1c_2 - a_3b_2c_1. \quad (1)$$

(ii) *Selecting the second row*:

$$\Delta = -a_2 \begin{vmatrix} b_1 & c_1 \\ b_3 & c_3 \end{vmatrix} + b_2 \begin{vmatrix} a_1 & c_1 \\ a_3 & c_3 \end{vmatrix} - c_2 \begin{vmatrix} a_1 & b_1 \\ a_3 & b_3 \end{vmatrix}$$

$$= -a_2(b_1c_3 - b_3c_1) + b_2(a_1c_3 - a_3c_1) - c_2(a_1b_3 - a_3b_1)$$

$$= -a_2b_1c_3 + a_2b_3c_1 + a_1b_2c_3 - a_3b_2c_1 - a_1b_3c_2 + a_3b_1c_2.$$

The student can easily check that these two values are identical. He should also work out the expansions selecting the other rows or columns, and verify that they all give the same result.

It will have been noticed that the Greek capital delta, Δ, is often used to signify a determinant.

13·31. Cofactors

If the proper plus or minus sign is attached to the minor of an element, it is called the cofactor of that element. For example, the cofactor of b_1 is

$$-\begin{vmatrix} a_2 & c_2 \\ a_3 & c_3 \end{vmatrix}.$$

If capital letters are written to stand for the cofactors of the elements, the expansion of a determinant can be written quite shortly. For example,

$$\Delta = a_1A_1 + b_1B_1 + c_1C_1.$$

13·32. Rule of Sarrus

In the case of a third-order determinant only, the six terms of the expansion can be written down at sight from the following (alternative) rule:

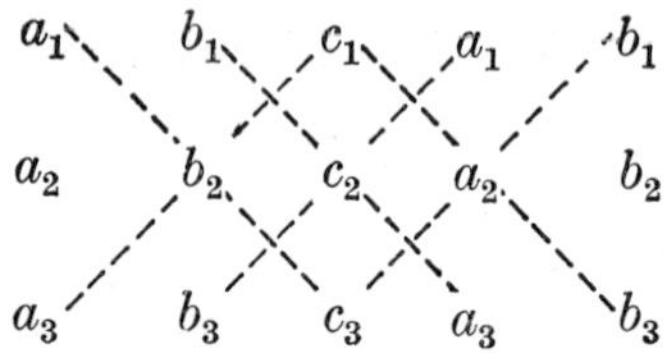

Repeat the first two columns. Draw diagonals, starting from the leading top left-hand element and going downwards from left to right. The products along these three diagonals give the positive terms of the expansion, viz. $a_1b_2c_3 + a_3b_1c_2 + a_2b_3c_1$.

Now start from the bottom left-hand element and draw the three diagonals going upwards from left to right. The products along these three diagonals give the negative terms for the expansion, viz. $- a_3b_2c_1 - a_1b_3c_2 - a_2b_1c_3$.

13·33. Worked examples

(1) Evaluate $\begin{vmatrix} 7 & 5 & 9 \\ 2 & 0 & 11 \\ 3 & 1 & 8 \end{vmatrix}$.

(i) Choosing the second column because this is the easiest, having one of the elements zero:

$$\begin{aligned} \Delta &= -5\begin{vmatrix} 2 & 11 \\ 3 & 8 \end{vmatrix} + 0\begin{vmatrix} 7 & 9 \\ 3 & 8 \end{vmatrix} - 1\begin{vmatrix} 7 & 9 \\ 2 & 11 \end{vmatrix} \\ &= -5(16 - 33) - (77 - 18) \\ &= -5(-17) - 59 \\ &= 85 - 59 \\ &= \underline{26.} \end{aligned}$$

(ii) Alternatively, using the Rule of Sarrus:

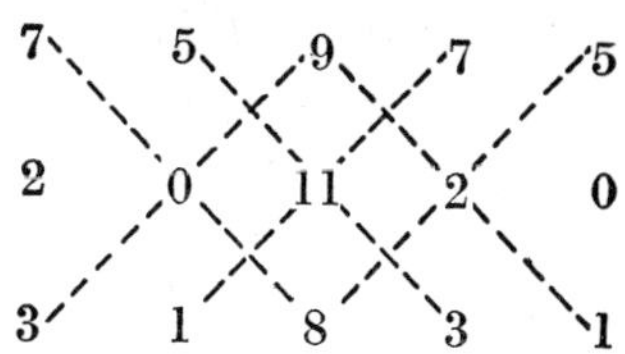

$$\Delta = 7.0.8 + 5.11.3 + 9.2.1 - 3.0.9 - 1.11.7 - 8.2.5$$
$$= 165 + 18 - 77 - 80$$
$$= 183 - 157$$
$$= \underline{26.}$$

(2) Solve the equation

$$\begin{vmatrix} x+2 & 2x-1 & 3 \\ 2x+1 & x+2 & 1 \\ 3x & 1 & 3 \end{vmatrix} = 0.$$

Expanding by the third column:

$$3\begin{vmatrix} 2x+1 & x+2 \\ 3x & 1 \end{vmatrix} - 1\begin{vmatrix} x+2 & 2x-1 \\ 3x & 1 \end{vmatrix} + 3\begin{vmatrix} x+2 & 2x-1 \\ 2x+1 & x+2 \end{vmatrix} = 0.$$

$$\therefore\ 3(2x + 1 - 3x^2 - 6x) - (x + 2 - 6x^2 + 3x) + 3(x^2 + 4x + 4 - 4x^2 + 1) = 0.$$

$$\therefore\ 3(-3x^2 - 4x + 1) - (-6x^2 + 4x + 2) + 3(-3x^2 + 4x + 5) = 0.$$

$$\therefore\ -12x^2 - 4x + 16 = 0.$$

$$\therefore\ 3x^2 + x - 4 = 0.$$

$$\therefore\ (3x + 4)(x - 1) = 0.$$

$$\therefore\ \underline{x = 1 \quad \text{or} \quad -\tfrac{4}{3}.}$$

(3) Expand $\begin{vmatrix} a & h & g \\ h & b & f \\ g & f & c \end{vmatrix}$.

$$\Delta = a\begin{vmatrix} b & f \\ f & c \end{vmatrix} - h\begin{vmatrix} h & g \\ f & c \end{vmatrix} + g\begin{vmatrix} h & g \\ b & f \end{vmatrix}$$

$$= a(bc - f^2) - h(ch - fg) + g(fh - bg)$$

$$= abc - af^2 - ch^2 + fgh + fgh - bg^2$$

$$\underline{= abc - af^2 - bg^2 - ch^2 + 2fgh.}$$

Exercise 40

1. Evaluate:

(i) $\begin{vmatrix} 5 & 3 \\ 4 & 9 \end{vmatrix}$. (ii) $\begin{vmatrix} 6 & 7 \\ 2 & 9 \end{vmatrix}$. (iii) $\begin{vmatrix} 15 & 11 \\ 17 & 2 \end{vmatrix}$. (iv) $\begin{vmatrix} 35 & 9 \\ 42 & 10 \end{vmatrix}$.

2. Give the minors of the ringed elements of the following determinants:

(i) $\begin{vmatrix} 5 & 7 & 9 \\ 2 & 3 & 4 \\ 8 & \enclose{circle}{7} & 5 \end{vmatrix}$. (ii) $\begin{vmatrix} a & b & c \\ d & e & f \\ g & h & \enclose{circle}{k} \end{vmatrix}$. (iii) $\begin{vmatrix} x & y & 1 \\ a & \enclose{circle}{b} & 1 \\ l & m & 1 \end{vmatrix}$.

3. Give the cofactors of the ringed elements of the following determinants:

(i) $\begin{vmatrix} 1 & 2 & 3 \\ 4 & 5 & \enclose{circle}{6} \\ 7 & 8 & 9 \end{vmatrix}$. (ii) $\begin{vmatrix} -2 & 3 & 4 \\ -1 & \enclose{circle}{-2} & 3 \\ 4 & 0 & 1 \end{vmatrix}$.

(iii) $\begin{vmatrix} x^2 & x & 1 \\ 2x & x^2 & -1 \\ \enclose{circle}{x^3} & 3x & x^2 \end{vmatrix}$.

4. Evaluate:

(i) $\begin{vmatrix} 7 & 9 & 11 \\ 2 & 4 & 6 \\ 5 & 7 & 9 \end{vmatrix}$. (ii) $\begin{vmatrix} 3 & 11 & 21 \\ 4 & -2 & 3 \\ -7 & 6 & 4 \end{vmatrix}$. (iii) $\begin{vmatrix} x^2 & x & 1 \\ y^2 & y & 1 \\ z^2 & z & 1 \end{vmatrix}$.

5. Solve the equation

$$\begin{vmatrix} x+3 & 1 & 5 \\ x-7 & 2 & 3 \\ 2x+1 & 5 & 6 \end{vmatrix} = 0.$$

6. Evaluate:

(i) $\begin{vmatrix} j & j2 & 3 \\ 4 & -j & 2 \\ 1 & -1 & j3 \end{vmatrix}$.

(ii) $\begin{vmatrix} 1-j & 2+j & 3+j \\ j & 1+j & 2-j \\ j2 & 4-j & 1+j2 \end{vmatrix}$, where $j = \sqrt{-1}$.

7. Show that $\begin{vmatrix} j & 5 & -1 \\ j^2 & 7 & -j \\ j^3 & 9 & 1 \end{vmatrix} = 0.$

8. If $x + jy = \begin{vmatrix} j & 1-j & 1+j \\ 2 & 2-j & 2+j \\ -j & 1+j2 & 1-j2 \end{vmatrix}$, find x and y.

Answers

1. (i) 33. (ii) 40. (iii) -157. (iv) -28.

2. (i) $\begin{vmatrix} 5 & 9 \\ 2 & 4 \end{vmatrix} = 2.$ (ii) $\begin{vmatrix} a & b \\ d & e \end{vmatrix} = ae - bd.$

(iii) $\begin{vmatrix} x & 1 \\ l & 1 \end{vmatrix} = x - l.$

3. (i) $-\begin{vmatrix} 1 & 2 \\ 7 & 8 \end{vmatrix} = 6.$ (ii) $+\begin{vmatrix} -2 & 4 \\ 4 & 1 \end{vmatrix} = -18.$

(iii) $+\begin{vmatrix} x & 1 \\ x^2 & -1 \end{vmatrix} = -x - x^2.$

4. (i) 0. (ii) -275. (iii) $x^2(y-z) + y^2(z-x) + z^2(x-y)$, which reduces to $(x-y)(y-z)(x-z)$.

5. $x = \dfrac{149}{2}$.

6. (i) $12(1 + j)$. (ii) $13 + j36$.

8. $x = 8$, $y = 12$.

13·4. Some properties of determinants

A knowledge of some of the simpler properties of determinants is often useful in simplifying a determinant before expansion. Some of these properties are also needed when dealing with matrices (see chapter 15).

Below are listed some of the properties. The first three are given without any verification; they may easily be verified by expansion of the given determinants. Although only listed for third-order determinants, these rules are valid for determinants of any order.

I. *Interchanging columns (or rows) an odd number of times changes the sign but not the numerical value; interchanging an even number of times leaves the sign and value unchanged.* Thus

$$\begin{vmatrix} b_1 & a_1 & c_1 \\ b_2 & a_2 & c_2 \\ b_3 & a_3 & c_3 \end{vmatrix} = -\begin{vmatrix} a_1 & b_1 & c_1 \\ a_2 & b_2 & c_2 \\ a_3 & b_3 & c_3 \end{vmatrix},$$

whilst

$$\begin{vmatrix} b_1 & c_1 & a_1 \\ b_2 & c_2 & a_2 \\ b_3 & c_3 & a_3 \end{vmatrix} = \begin{vmatrix} a_1 & b_1 & c_1 \\ a_2 & b_2 & c_2 \\ a_3 & b_3 & c_3 \end{vmatrix}.$$

II. *Multiplying each element of a column (or row) of a determinant by a constant, multiplies the value of the determinant by that constant.* Thus

$$\begin{vmatrix} ka_1 & b_1 & lc_1 \\ ka_2 & b_2 & lc_2 \\ ka_3 & b_3 & lc_3 \end{vmatrix} = kl\begin{vmatrix} a_1 & b_1 & c_1 \\ a_2 & b_2 & c_2 \\ a_3 & b_3 & c_3 \end{vmatrix}.$$

This rule sometimes helps to reduce the numbers before expansion. For example

$$\begin{vmatrix} 24 & 36 & 12 \\ -15 & -30 & 25 \\ 9 & 18 & 27 \end{vmatrix} = 12 \times 5 \times 9\begin{vmatrix} 2 & 3 & 1 \\ -3 & -6 & 5 \\ 1 & 2 & 3 \end{vmatrix}.$$

III. *A determinant is unaltered in value if its columns are written as rows, or vice-versa.* Thus

$$\begin{vmatrix} a_1 & b_1 & c_1 \\ a_2 & b_2 & c_2 \\ a_3 & b_3 & c_3 \end{vmatrix} = \begin{vmatrix} a_1 & a_2 & a_3 \\ b_1 & b_2 & b_3 \\ c_1 & c_2 & c_3 \end{vmatrix}.$$

IV. *If two columns (or rows) of a determinant are identical, or one is a multiple of the other, the value of the determinant is zero.* For example, if

$$\Delta = \begin{vmatrix} a_1 & ka_1 & c_1 \\ a_2 & ka_2 & c_2 \\ a_3 & ka_3 & c_3 \end{vmatrix},$$

then by II,

$$\Delta = k\begin{vmatrix} a_1 & a_1 & c_1 \\ a_2 & a_2 & c_2 \\ a_3 & a_3 & c_3 \end{vmatrix}.$$

Using rule I:

$$k\begin{vmatrix} a_1 & a_1 & c_1 \\ a_2 & a_2 & c_2 \\ a_3 & a_3 & c_3 \end{vmatrix} = -k\begin{vmatrix} a_1 & a_1 & c_1 \\ a_2 & a_2 & c_2 \\ a_3 & a_3 & c_3 \end{vmatrix},$$

on interchanging the first and second columns.

Thus $\Delta = -\Delta$, which can only be true if $\Delta = 0$. As a further example, consider Question 7 of Exercise 40.

$$\Delta = \begin{vmatrix} j & 5 & -1 \\ j^2 & 7 & -j \\ j^3 & 9 & 1 \end{vmatrix}.$$

Now using $j^2 = -1$, in column 3, $-1 = j.j$, $-j = j.j^2$, $1 = j.j^3$. Thus

$$\Delta = \begin{vmatrix} j & 5 & j.j \\ j^2 & 7 & j.j^2 \\ j^3 & 9 & j.j^3 \end{vmatrix}.$$

$\therefore$ $\Delta = 0$, as the third column is a multiple of the first.

V. *If a column (or row) of a determinant has each element the algebraic sum of n constituent elements, the determinant is equal to the algebraic sum of n determinants, in each of which the elements consist of single terms.* For example,

$$\begin{vmatrix} a_1 + d_1 - e_1 & b_1 & c_1 \\ a_2 + d_2 - e_2 & b_2 & c_2 \\ a_3 + d_3 - e_3 & b_3 & c_3 \end{vmatrix}$$

$$= \begin{vmatrix} a_1 & b_1 & c_1 \\ a_2 & b_2 & c_2 \\ a_3 & b_3 & c_3 \end{vmatrix} + \begin{vmatrix} d_1 & b_1 & c_1 \\ d_2 & b_2 & c_2 \\ d_3 & b_3 & c_3 \end{vmatrix} - \begin{vmatrix} e_1 & b_1 & c_1 \\ e_2 & b_2 & c_2 \\ e_3 & b_3 & c_3 \end{vmatrix}.$$

This is easily verified by actual expansion.

VI. *If the elements of any column (or row) are increased or diminished by equimultiples of corresponding elements of another column (or row), the value of the determinant is unaltered.* For example,

$$\begin{vmatrix} a_1 + kb_1 - lc_1 & b_1 & c_1 \\ a_2 + kb_2 - lc_2 & b_2 & c_2 \\ a_3 + kb_3 - lc_3 & b_3 & c_3 \end{vmatrix}$$

$$= \begin{vmatrix} a_1 & b_1 & c_1 \\ a_2 & b_2 & c_2 \\ a_3 & b_3 & c_3 \end{vmatrix} + k \begin{vmatrix} b_1 & b_1 & c_1 \\ b_2 & b_2 & c_2 \\ b_3 & b_3 & c_3 \end{vmatrix} - l \begin{vmatrix} c_1 & b_1 & c_1 \\ c_2 & b_2 & c_2 \\ c_3 & b_3 & c_3 \end{vmatrix}$$

(from rules II and V)

$$= \begin{vmatrix} a_1 & b_1 & c_1 \\ a_2 & b_2 & c_2 \\ a_3 & b_3 & c_3 \end{vmatrix} \quad \text{(from rule IV).}$$

This property is often very useful in simplifying a determinant before expansion, especially if a row or column can be reduced until its elements are unity. It is then very easy to make two of these elements zero. In the example given below c_1, c_2, c_3 and r_1, r_2, r_3 are used as abbreviations for column 1, 2, 3, ...; row 1, 2, 3, ..., etc.

EXAMPLE

Evaluate

$$\Delta = \begin{vmatrix} 31 & 91 & 47 \\ 14 & 29 & 30 \\ 21 & 36 & 37 \end{vmatrix}.$$

$$\Delta = \begin{vmatrix} 31 & 60 & 16 \\ 14 & 15 & 16 \\ 21 & 15 & 16 \end{vmatrix} \quad \begin{pmatrix} c_2 - c_1, \\ c_3 - c_1 \end{pmatrix}$$

$$= 16 \times 15 \begin{vmatrix} 31 & 4 & 1 \\ 14 & 1 & 1 \\ 21 & 1 & 1 \end{vmatrix}$$

$$= 16 \times 15 \begin{vmatrix} 17 & 3 & 0 \\ 14 & 1 & 1 \\ 7 & 0 & 0 \end{vmatrix} \quad \begin{pmatrix} r_1 - r_2, \\ r_3 - r_2 \end{pmatrix}$$

$$= 16 \times 15 \times 7 \begin{vmatrix} 3 & 0 \\ 1 & 1 \end{vmatrix}$$

$$= 16 \times 15 \times 7 \times 3$$

$$= \underline{5040.}$$

VII. *If the elements of any column (or row) are multiplied by the cofactors of the corresponding elements of another column (or row), the sum of the products is zero.* This property has been given mainly because it is required in the chapter on matrices (see § 15·82).

As an example consider the determinant

$$\Delta = \begin{vmatrix} a_1 & b_1 & c_1 \\ a_2 & b_2 & c_2 \\ a_3 & b_3 & c_3 \end{vmatrix}.$$

Now consider $a_1B_1 + a_2B_2 + a_3B_3$, where B_1, B_2, B_3 are the cofactors of b_1, b_2, b_3 (see § 13·31); then

$$B_1 = -\begin{vmatrix} a_2 & c_2 \\ a_3 & c_3 \end{vmatrix}, \quad B_2 = +\begin{vmatrix} a_1 & c_1 \\ a_3 & c_3 \end{vmatrix}, \quad B_3 = -\begin{vmatrix} a_1 & c_1 \\ a_2 & c_2 \end{vmatrix}.$$

$\therefore\ a_1B_1 + a_2B_2 + a_3B_3$

$$= -a_1\begin{vmatrix} a_2 & c_2 \\ a_3 & c_3 \end{vmatrix} + a_2\begin{vmatrix} a_1 & c_1 \\ a_3 & c_3 \end{vmatrix} - a_3\begin{vmatrix} a_1 & c_1 \\ a_2 & c_2 \end{vmatrix}$$

$$= -\begin{vmatrix} a_1 & a_1 & c_1 \\ a_2 & a_2 & c_2 \\ a_3 & a_3 & c_3 \end{vmatrix}$$

(this is easily verified on expansion by the 1st column)

$$= 0 \text{ (by rule IV).}$$

EXAMPLE

Simplify the determinant, and hence solve the equation

$$\begin{vmatrix} x-3 & x+5 & 1 \\ x+1 & 3 & x-1 \\ 4 & x+2 & x-3 \end{vmatrix} = 0.$$

$$\begin{vmatrix} x-3 & x+5 & 1 \\ x+1 & 3 & x-1 \\ 4 & x+2 & x-3 \end{vmatrix}$$

$$= \begin{vmatrix} x-3 & x+5 & 2x+3 \\ x+1 & 3 & 2x+3 \\ 4 & x+2 & 2x+3 \end{vmatrix} \quad (c_3 + (c_2 + c_1))$$

$$= (2x+3)\begin{vmatrix} x-3 & x+5 & 1 \\ x+1 & 3 & 1 \\ 4 & x+2 & 1 \end{vmatrix}$$

$$= (2x+3)\begin{vmatrix} x-3 & x+5 & 1 \\ 4 & -x-2 & 0 \\ -x+7 & -3 & 0 \end{vmatrix} \quad \begin{pmatrix} r_2 - r_1, \\ r_3 - r_1 \end{pmatrix}$$

$$= (2x+3)\begin{vmatrix} 4 & -(x+2) \\ 7-x & -3 \end{vmatrix},$$

expanding by the third column

$$= (2x + 3)\{-12 + (7 - x)(x + 2)\}$$

$$= (2x + 3)(2 + 5x - x^2).$$

$$\therefore (2x + 3)(2 + 5x - x^2) = 0.$$

Thus $x = -\frac{3}{2}$ or $\dfrac{5 \pm \sqrt{33}}{2}$,

giving $\underline{x = -1{\cdot}5, \quad 5{\cdot}3723 \quad \text{or} \quad -0{\cdot}3723.}$

Exercise 41

1. Show that $\begin{vmatrix} 2 & 5 & 3 \\ 4 & -9 & 6 \\ 6 & 15 & 9 \end{vmatrix} = 0.$

2. Evaluate $\begin{vmatrix} 2 & -1 & 7 \\ 4 & 1 & 5 \\ 6 & 1 & 11 \end{vmatrix}.$

3. Evaluate $\begin{vmatrix} 2 & -1 & 4 \\ 5 & 2 & 7 \\ 8 & 5 & 10 \end{vmatrix}.$

4. Evaluate $\begin{vmatrix} 39 & 41 & 42 \\ 37 & 39 & 40 \\ 16 & -3 & 19 \end{vmatrix}.$

5. Evaluate $\begin{vmatrix} 71 & 77 & 15 \\ 15 & 21 & -17 \\ 31 & 37 & -3 \end{vmatrix}.$

6. Evaluate $\begin{vmatrix} x + y & x & y \\ x & z + x & z \\ y & z & y + z \end{vmatrix}.$

7. Evaluate $\begin{vmatrix} 9 & 10 & 13 & 4 \\ 5 & 7 & -2 & 1 \\ 2 & 4 & 6 & 8 \\ 3 & 8 & -11 & 15 \end{vmatrix}$.

8. Evaluate $(3 + j2)\begin{vmatrix} 1+j & 0 & 0 \\ 0 & 3-j2 & 0 \\ 0 & 0 & 4+j2 \end{vmatrix}$.

9. Solve the equation

$$\begin{vmatrix} 15-2x & 11 & 10 \\ 11-3x & 17 & 16 \\ 7-x & 14 & 13 \end{vmatrix} = 0.$$

[L.U.]

10. Evaluate $\begin{vmatrix} 1-j & 2+j & 3-j \\ 1+j & 4-j & 2+j \\ 1-j2 & 1+j2 & 3-j2 \end{vmatrix} \div \begin{vmatrix} j & 1 & 2 \\ 1 & j & 2 \\ j & j & -3 \end{vmatrix}$

in the form $a + jb$.

Answers

1. Row 3 is three times row 1.

2. 12.

3. 0. [Subtract row 1 from rows 2 and 3; then new row 3 is a multiple of new row 2.]

4. 126. [Take $r_1 - r_2$ as a first step.]

5. 1632. [Start with $c_2 - c_1$.]

6. $4xyz$. [Start with $r_1 - r_2 - r_3$.]

7. 1692. [Try $c_2 - 2c_1$, $c_3 - 3c_1$, $c_4 - 4c_1$ and hence reduce to a third-order determinant.]

8. $26 + j78$.

9. $x = 4$.

10. $\frac{3}{58} + j\frac{11}{29}$.

13·5. Solution of a set of simultaneous linear equations

Let the equations be

$$a_1x + b_1y + c_1z = d_1, \tag{1}$$

$$a_2x + b_2y + c_2z = d_2, \tag{2}$$

$$a_3x + b_3y + c_3z = d_3. \tag{3}$$

Now
$$\begin{vmatrix} a_1x & b_1 & c_1 \\ a_2x & b_2 & c_2 \\ a_3x & b_3 & c_3 \end{vmatrix} \equiv \begin{vmatrix} a_1x + b_1y + c_1z & b_1 & c_1 \\ a_2x + b_2y + c_2z & b_2 & c_2 \\ a_3x + b_3y + c_3z & b_3 & c_3 \end{vmatrix}, \tag{4}$$

multiplying column 2 by y, column 3 by z and adding to column 1.

Now *for values of x, y and z which satisfy the equations,*

$$a_1x + b_1y + c_1z = d_1, \quad a_2x + b_2y + c_2z = d_2,$$

and
$$a_3x + b_3y + c_3z = d_3.$$

Putting these values in (4) gives:

$$\begin{vmatrix} a_1x & b_1 & c_1 \\ a_2x & b_2 & c_2 \\ a_3x & b_3 & c_3 \end{vmatrix} = \begin{vmatrix} d_1 & b_1 & c_1 \\ d_2 & b_2 & c_2 \\ d_3 & b_3 & c_3 \end{vmatrix},$$

i.e.
$$x\begin{vmatrix} a_1 & b_1 & c_1 \\ a_2 & b_2 & c_2 \\ a_3 & b_3 & c_3 \end{vmatrix} = \begin{vmatrix} d_1 & b_1 & c_1 \\ d_2 & b_2 & c_2 \\ d_3 & b_3 & c_3 \end{vmatrix}.$$

Let
$$\begin{vmatrix} a_1 & b_1 & c_1 \\ a_2 & b_2 & c_2 \\ a_3 & b_3 & c_3 \end{vmatrix} \equiv \Delta.$$

Then
$$x = \frac{\begin{vmatrix} d_1 & b_1 & c_1 \\ d_2 & b_2 & c_2 \\ d_3 & b_3 & c_3 \end{vmatrix}}{\Delta}. \tag{5}$$

In a similar manner it can easily be proved that

$$y = \frac{\begin{vmatrix} a_1 & d_1 & c_1 \\ a_2 & d_2 & c_2 \\ a_3 & d_3 & c_3 \end{vmatrix}}{\Delta} \quad \text{and} \quad z = \frac{\begin{vmatrix} a_1 & b_1 & d_1 \\ a_2 & b_2 & d_2 \\ a_3 & b_3 & d_3 \end{vmatrix}}{\Delta}. \qquad (6)$$

A rule can now be stated for writing down, on sight, the solutions to a given set of simultaneous linear equations:

Take all the number terms to the right-hand side of each equation and arrange the left-hand sides in order of columns x's, y's, z's. The denominator for each of the solutions, Δ, is the determinant formed by omitting x, y, z from the left-hand sides. The numerator for the solution for x is found by replacing the coefficients of x (the a's) in Δ by the constant terms (the d's); the numerator for the solution for y is found by replacing the coefficients of y in Δ (the b's) by the constant terms (the d's), and so on. This rule is quite general and can be used for solving simultaneous linear equations of any order. For example, if there were four unknowns and four equations, the determinants would be of the fourth order.

13·51. Worked examples

(1) Solve the equations:

$$3x - 4y + 2z - 1 = 0,$$
$$2x + 9z - 7y = 15,$$
$$4x - 2y - 3z + 9 = 0.$$

Rearranging these equations:

$$3x - 4y + 2z = 1,$$
$$2x - 7y + 9z = 15,$$
$$4x - 2y - 3z = -9.$$

Thus
$$\Delta = \begin{vmatrix} 3 & -4 & 2 \\ 2 & -7 & 9 \\ 4 & -2 & -3 \end{vmatrix}$$

$$= 3(21 + 18) - 2(12 + 4) + 4(-36 + 14)$$

$$= 117 - 32 - 88,$$

i.e. $\underline{\Delta = -3.}$

Using the rule given in (5) and (6) above:

$$x = \frac{\begin{vmatrix} 1 & -4 & 2 \\ 15 & -7 & 9 \\ -9 & -2 & -3 \end{vmatrix}}{-3}, \quad y = \frac{\begin{vmatrix} 3 & 1 & 2 \\ 2 & 15 & 9 \\ 4 & -9 & -3 \end{vmatrix}}{-3},$$

$$z = \frac{\begin{vmatrix} 3 & -4 & 1 \\ 2 & -7 & 15 \\ 4 & -2 & -9 \end{vmatrix}}{-3}.$$

Evaluating the determinants:

$$x = \frac{-3}{-3}, \quad y = \frac{-6}{-3}, \quad z = \frac{-9}{-3},$$

giving $\underline{x = 1, \quad y = 2, \quad z = 3.}$

(2) In determining an electrical resistance, the following equations occur:

$$Gg + Pp - Qq = e,$$

$$-Gg + (P - G)r - (Q + G)s = 0,$$

$$(P + Q)b + Qq + (Q + G)s = E.$$

Find, as the ratio of two determinants, the value of G independent of P and Q, and determine the relation between p, q, r and s if G be independent of E. [L.U.]

Although not essential to the question, the student will find it instructive to build up a circuit which will fit the given equations. It is assumed that capital letters refer to currents, small letters to resistances, and e and E are voltages. Fig. 68 below shows a possible circuit. As can be seen, it is a Wheatstone Bridge type of circuit with a battery (e) in one leg. G presumably stands for the galvanometer current.

The given equations are the back e.m.f. equations for the networks BCL, CDL and $ABCDF$ respectively.

Writing the given equations with G, P and Q considered as unknowns:

$$\begin{aligned} gG \quad &+ pP \quad &- qQ \quad &= e, \\ -(g+r+s)G &+ rP &- sQ \quad &= 0, \\ sG \quad &+ bP &+ (b+q+s)Q &= E. \end{aligned}$$

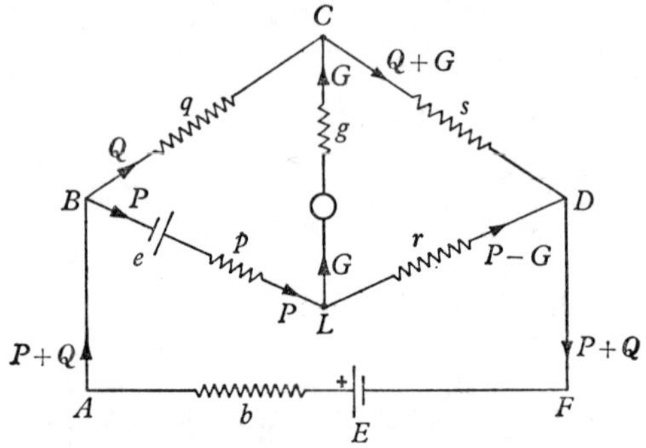

Fig. 68

Using the rule given in § 13·5 above:

$$G = \frac{\begin{vmatrix} e & p & -q \\ 0 & r & -s \\ E & b & (b+q+s) \end{vmatrix}}{\begin{vmatrix} g & p & -q \\ -(g+r+s) & r & -s \\ s & b & (b+q+s) \end{vmatrix}}. \qquad (1)$$

Now for G to be independent of E, the coefficient of E in the expansion of the determinant in the numerator of equation (1) must be zero (E does not occur in the denominator).

Thus $\begin{vmatrix} p & -q \\ r & -s \end{vmatrix} = 0$, giving $qr - ps = 0$,

or $\underline{p/q = r/s}$.

(This is the usual Wheatstone Bridge relationship for balance.)

13·52. Note on the linear dependence of a set of equations

Consider the equations

$$2x - 3y + 7 = 0, \quad x + 2y - 3 = 0 \quad \text{and} \quad x - 12y + 23 = 0.$$

It appears at first sight that there are three equations given from which to solve for only two unknowns. The solutions for x and y from the first two equations would not, in general, be likely to satisfy the third. In this case however, the third equation is *not* independent of the first two equations. Multiplying the first equation by 2 gives $4x - 6y + 14 = 0$; multiplying the second equation by 3 gives $3x + 6y - 9 = 0$; on subtraction these give $x - 12y + 23 = 0$, the third equation. Thus the values of x and y satisfying the first two equations will also satisfy the third.

In general, three equations such as

$$a_1x + b_1y + c_1 = 0, \quad a_2x + b_2y + c_2 = 0$$

and

$$a_3x + b_3y + c_3 = 0$$

are said to be *linearly dependent* equations if there exists a relation of the type

$$\lambda_1(a_1x + b_1y + c_1) + \lambda_2(a_2x + b_2y + c_2) + \lambda_3(a_3x + b_3y + c_3) = 0,$$

where λ_1, λ_2, λ_3 are constants, not all zero. The same pair of values of x and y then satisfies all three equations.

It can easily be shown that the condition for this is that

$$\Delta = \begin{vmatrix} a_1 & b_1 & c_1 \\ a_2 & b_2 & c_2 \\ a_3 & b_3 & c_3 \end{vmatrix} = 0. \tag{1}$$

EXAMPLE

Test the equations

$$3x - 7y + 5 = 0, \quad 4x + 2y - 7 = 0, \quad 18x - 8y - 11 = 0$$

for linear dependence.

From (1) above:
$$\Delta = \begin{vmatrix} 3 & -7 & 5 \\ 4 & 2 & -7 \\ 18 & -8 & -11 \end{vmatrix}$$
$$= 3(-78) - 4(117) + 18(39)$$
$$= -234 - 468 + 702$$
$$= 0.$$

The equations *are* therefore linearly dependent.

When writing down the equations for an electrical network, using Kirchhoff's mesh law, it is very easy to obtain linearly dependent equations (see also § 16·2(*a*)).

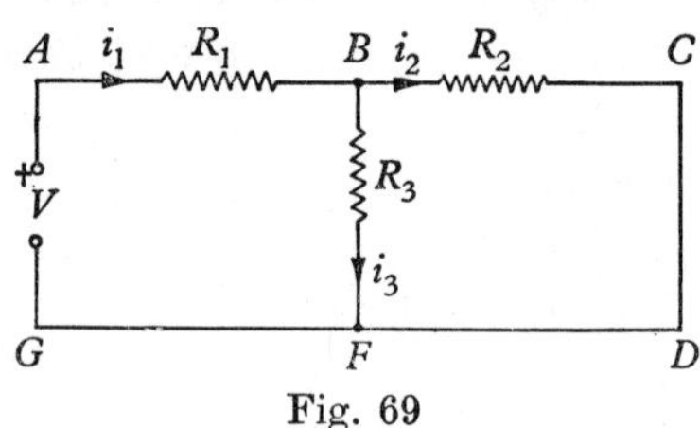

Fig. 69

For the purpose of solving the network, it is pointless to take any more equations than the maximum number which are independent.

For example, consider the circuit shown in fig. 69.

Taking the mesh $ABFG$:

$$V = R_1 i_1 + R_3 i_3. \tag{1}$$

Taking the mesh $BCDF$:

$$0 = R_2 i_2 - R_3 i_3. \tag{2}$$

Taking the mesh $ACDG$:

$$V = R_1 i_1 + R_2 i_2. \tag{3}$$

These three equations are linearly dependent; equation (3) is obviously the sum of equations (1) and (2).

Their interdependence arises from the fact that two of the meshes have a mutual impedance, R_3.

The third, independent, equation is found of course from Kirchhoff's current law; giving in this case, $i_1 - i_2 - i_3 = 0$.

EXERCISE 42

1. Solve by determinants the equations

$$7x + 5y - 13z + 4 = 0,$$
$$9x + 2y + 11z - 37 = 0,$$
$$3x - y + z - 2 = 0.$$

[L.U.]

2. Solve the equations

$$x + 2y + z = 4,$$
$$3x - 4y - 4z = 10,$$
$$5x + 3y + 7z = -9. \qquad \text{[L.U.]}$$

3. In solving for the currents in a simple resistance circuit, the following equations occurred:

$$3i_1 + 4i_3 = 7,$$
$$6i_2 + 5(i_1 + i_2 - i_3) = 7,$$
$$3i_1 - 6i_2 = 0.$$

Sketch a circuit that would satisfy these equations, and solve for i_1, i_2, i_3 by means of determinants.

4. Solve by determinants the equations

$$1{\cdot}5W_1 - 2{\cdot}5W_2 + 3W_3 = 21,$$
$$0{\cdot}5(W_1 - 2W_2) + 2(W_1 + W_3) = 24,$$
$$3W_1 + 0{\cdot}5(W_3 - W_1) = 15.$$

5. Given the equations

$$x + y + z = 1, \quad px + qy + rz = s, \quad p^2x + q^2y + r^2z = s^2,$$

find, as the ratio of two determinants, the value of x.

Show that if $s = q$ or $s = r$, the value of x is zero.

6. If

$$p(x - a) + 2x - y = 0,$$
$$p(y - a) - x + 2y - z = 0,$$
$$p(z - a) - y + 2z = 0,$$

express x as the ratio of two determinants and hence show that

$$x = \frac{ap(p + 3)}{(p^2 + 4p + 2)}. \qquad \text{[L.U.]}$$

7. Solve the equations

$$(1 + j2)i_1 - (2 + j3)i_2 = 100,$$
$$(2 + j3)i_2 - (3 + j4)i_3 = -50 + j50\sqrt{3},$$
$$i_1 + i_2 + i_3 = 0.$$

Find the modulus of i_1, i_2 and i_3.

(These are the 'complex' currents in a certain three-phase circuit.)

8. The following equations arise in solving for the (complex) currents in a certain 'parallel —T' circuit:

$$\frac{j}{\omega C}(i_1 + i_3) - \tfrac{1}{2}Ri_1 = E,$$

$$ri - \frac{j}{\omega C}i_3 - \tfrac{1}{2}Ri_1 = 0,$$

$$R(i_3 - i) - ri - \frac{j}{2\omega C}i_2 = 0,$$

$$R(i_3 - i_2 - i) + R(i_3 - i) - \frac{j}{\omega C}i_3 - \frac{j}{\omega C}(i_1 + i_3) = 0.$$

Writing $n = \omega RC$ and $k = r/R$, show that the solution for i is given by

$$i = -\frac{E}{R}\begin{vmatrix} -\frac{1}{2} & 0 & -j/n \\ 0 & -j/2n & 1 \\ -j/n & -1 & 2(1-j/n) \end{vmatrix}$$

$$\div \begin{vmatrix} 0 & (\frac{1}{2} - j/n) & 0 & -j/n \\ k & -\frac{1}{2} & 0 & -j/n \\ -(k+1) & 0 & -j/2n & 1 \\ -2 & -j/n & -1 & 2(1-j/n) \end{vmatrix},$$

and evaluate i in its $a + jb$ form.

9. The input terminals of a four-terminal network are A, B; the output terminals are C, D. The lines AC and BD each consist of two resistances, each of value r ohms, in series. The mid-points of the lines are connected by a resistance r' ohms. The output terminals are closed by a resistance of 600 ohms.

Calculate the values of r and r' which will reduce the voltage across CD to $1/3{\cdot}16$ of that applied at AB, while making the resistance, as measured between A and B, 600 ohms. [L.U.]

10. Fig. 70 shows three star-connected impedances with phase voltages e_1, e_2, e_3 as shown. Show that the currents i_1, i_2, i_3 are given by the equations

$$Z_1 i_1 - Z_2 i_2 = e_1 - e_2,$$

$$Z_2 i_2 - Z_3 i_3 = e_2 - e_3,$$

$$i_1 + i_2 + i_3 = 0.$$

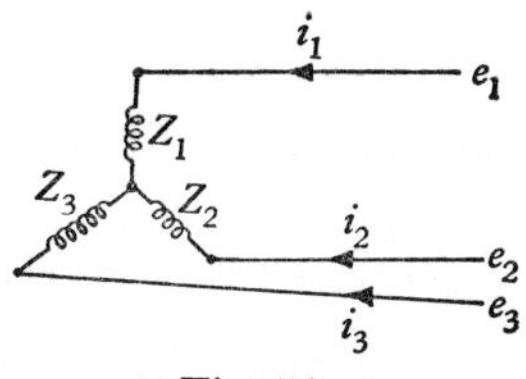

Fig. 70

Find the three currents.

11. Test for linear dependence the equations:

(i) $x + 2y - 3 = 0, \quad 3x - y + 1 = 0, \quad 4x - 2y + 3 = 0.$

(ii) $2x - y + 7 = 0, \quad 4x + 3y - 2 = 0,$
$2x + 9y - 25 = 0.$

12. Test for linear dependence the equations:

$$V_1 = R_1 i_1 + R_2 i_2,$$

$$V_1 - V_2 = R_1 i_1 + R_3 i_3,$$

$$V_2 = R_2 i_2 - R_3 i_3,$$

$$i_1 - i_2 - i_3 = 0.$$

Answers

1. $x = 1, y = 3, z = 2.$

2. $x = 2, y = 3, z = -4.$

3. $i_1 = \frac{21}{19}, i_2 = \frac{21}{38}, i_3 = \frac{35}{38}.$

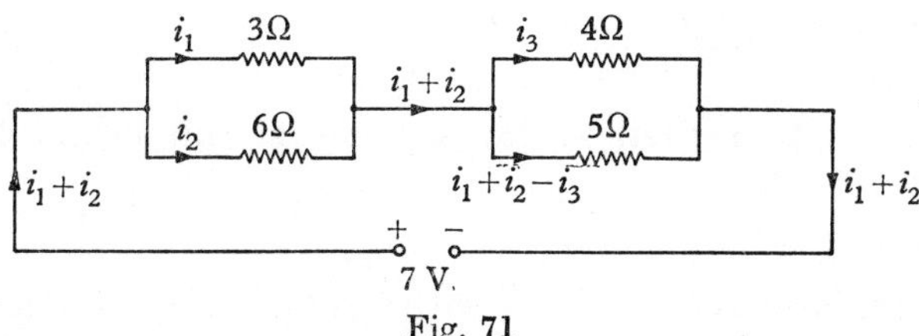

Fig. 71

4. $W_1 = 4, W_2 = 6, W_3 = 10.$

5. $$x = \begin{vmatrix} 1 & 1 & 1 \\ s & q & r \\ s^2 & q^2 & r^2 \end{vmatrix} \div \begin{vmatrix} 1 & 1 & 1 \\ p & q & r \\ p^2 & q^2 & r^2 \end{vmatrix}.$$

If $s = q$ or $s = r$, two columns of the determinant in the numerator are identical, therefore $x = 0$.

7. $i_1 \simeq 16{\cdot}3 - j11{\cdot}3, \quad |i_1| \simeq 19{\cdot}8.$

$i_2 \simeq -4{\cdot}5 + j17{\cdot}4, \quad |i_2| \simeq 18{\cdot}0.$

$i_3 \simeq -11{\cdot}8 - j6{\cdot}1, \quad |i_3| \simeq 13{\cdot}3.$

8. $$i = -\frac{E}{R} \cdot \frac{(n^2 - 1)\{(kn^2 - 2 - k) + j2n(1 + 2k)\}}{(kn^2 - 2 - k)^2 + 4n^2(1 + 2k)^2}$$

9. $r = 155{\cdot}8\ \Omega$, $r' = 422\ \Omega$.

[It is *not* claimed that the shortest method here is to use determinants; but any method used inherently involves the solution of simultaneous equations.]

10. $$i_1 = \frac{Z_2(e_1 - e_3) + Z_3(e_1 - e_2)}{Z_1Z_2 + Z_2Z_3 + Z_3Z_1},$$

$$i_2 = \frac{Z_1(e_2 - e_3) + Z_3(e_2 - e_1)}{Z_1Z_2 + Z_2Z_3 + Z_3Z_1},$$

$$i_3 = \frac{Z_1(e_3 - e_2) + Z_2(e_3 - e_1)}{Z_1Z_2 + Z_2Z_3 + Z_3Z_1}.$$

11. (i) Linearly independent.
(ii) Linearly dependent.

12. Linearly dependent.

13·6. Relaxation method for solving simultaneous linear equations

13·61. Foreword

The engineer is often faced with the task of solving a set of simultaneous linear equations in four or five unknowns, the solutions to which are not likely to be whole numbers. Such practical cases arise in solving for currents in electrical networks, in structure problems in mechanics, and in many other cases. The answers in such cases are usually only required to be accurate to a given number of decimal places. Because of this, several methods of approximation to a set of solutions have been evolved. The relaxation method is being given here because it will serve as an elementary introduction to methods which are being applied to many important topics in engineering.

13·62. General description of the method

Consider the equations

$$11x_1 - 2x_2 + 18x_3 - 69 = 0, \tag{1}$$

$$4x_1 + 9x_2 + 11x_3 - 19 = 0, \tag{2}$$

$$2x_1 - 3x_2 + 8x_3 - 32 = 0. \tag{3}$$

The fundamental idea of the method is to find values for x_1, x_2 and x_3, by successive approximations, which will make the left-hand sides of these equations closer and closer to zero. In more detail: for any arbitrarily chosen values of x_1, x_2, x_3, let R_1, R_2, R_3 denote the differences from zero of the left-hand sides of the equations, i.e.

$$11x_1 - 2x_2 + 18x_3 - 69 = R_1, \tag{4}$$

$$4x_1 + 9x_2 + 11x_3 - 19 = R_2, \tag{5}$$

$$2x_1 - 3x_2 + 8x_3 - 32 = R_3. \tag{6}$$

The values of x_1, x_2 and x_3 are systematically changed until R_1, R_2 and R_3 are as close to zero as any required degree of accuracy.

R_1, R_2, R_3 are called the *residuals*.

When employing the method, the first step is to construct a table showing the changes, ΔR, in the residuals caused by unit changes in x_1, x_2 and x_3. This is done merely as a convenient reckoning device. Table I shows the table constructed for equations (4), (5) and (6).

Table I

	ΔR_1	ΔR_2	ΔR_3
$\Delta x_1 = 1$	11	4	2
$\Delta x_2 = 1$	-2	9	-3
$\Delta x_3 = 1$	18	11	8

It will be seen that the changes in R_1, R_2, R_3 due to unit changes in the x values are merely the corresponding coefficients of the x terms in the three equations. For example, a change in x_2 of one means a change in R_2 of nine.

A second table is now constructed in which the residuals are listed for successively, deliberately chosen values of x_1, x_2 and x_3.

Table II

x_1	x_2	x_3	R_1	R_2	R_3
0	0	0	-69	-19	-32
7	0	0	8	9	-18
0	0	3	62	42	6
-6	0	0	-4	18	-6
0	-2	0	0	0	0
1	-2	3	0	0	0

The steps in the construction of Table II are as follows:

In the first row the first approximations to x_1, x_2 and x_3 are written. If three values can be guessed which satisfy the given equations fairly well, these are chosen; but if not, starting with all three as zero is as good a starting point as any other. When x_1, x_2 and x_3 are all taken as zero, the residuals are -69, -19, -32, from equations (4), (5) and (6).

The largest residual is $R_1 = -69$. The corresponding x value is x_1. Table I shows that a change in x_1 of $+7$ units will make R_1 into $-69 + 7.11 = -69 + 77 = 8$; and will also change R_2 into $-19 + 7.4 = 9$, and R_3 into $-32 + 7.2 = -18$. These results are entered in the second row of Table II. Two points should be noted at this stage:

(i) There is no particular point in reducing any residual exactly to zero, early in the working. A simple integer which does considerably reduce the largest remaining residual is chosen at each stage.

(ii) Usually it is better to over-correct than under-correct. For example, in the first stage above, R_1 has been over-corrected from -69 to $+8$.

Continuing with the table: the largest residual is now $R_3 = -18$. Considering the corresponding x-value, x_3, Table I shows that a change of $+3$ in x_3 will make R_3 into $-18 + 3.8 = 6$, R_2 into $9 + 3.11 = 42$ and R_1 into $8 + 3.18 = 62$. These values are entered in the third row. Note that the *changes* in the x-values are entered in the first three columns, but the *actual values* of the residuals are entered in the last three columns. Note also that only one x-value is changed at any one stage.

The figures in the fourth and fifth rows should now be easily followed.

The final residuals in this case are all zero. This means that the equations have been solved accurately. Adding each of the x-columns, it is seen that the values $x_1 = 1$, $x_2 = -2$ and $x_3 = 3$ give zero residuals. These are, therefore, the accurate solutions to the original equations.

The main features of the relaxation method are:

(i) Most of the arithmetic can be performed mentally, being multiplication by integers.

(ii) It is particularly effective when approximate solutions are known and it is required to find more accurate solutions.

(iii) The process can be halted quite easily when the solutions have reached any required degree of accuracy (see example below).

13·63. A longer worked example

Solve the following equations, correct to four decimal places, given that $x_1 = 0$, $x_2 = 0{\cdot}5$, $x_3 = 1{\cdot}5$ and $x_4 = 3$ are very approximate solutions:

$$\begin{aligned} 20x_1 + 3x_2 - 5x_3 + 2x_4 + 7 &= 0, \\ 3x_1 - 30x_2 + 4x_3 - 5x_4 + 20 &= 0, \\ 2x_2 + 21x_3 + 3x_4 - 42 &= 0, \\ 2x_1 - 4x_2 - 3x_3 + 25x_4 - 62 &= 0. \end{aligned}$$

Table I

	ΔR_1	ΔR_2	ΔR_3	ΔR_4
$\Delta x_1 = 1$	20	3	0	2
$\Delta x_2 = 1$	3	-30	2	-4
$\Delta x_3 = 1$	-5	4	21	-3
$\Delta x_4 = 1$	2	-5	3	25

As approximate solutions have been given, these are taken as the starting point in Table II, and the corresponding residuals calculated. For example, when $x_1 = 0$, $x_2 = 0{\cdot}5$, $x_3 = 1{\cdot}5$, $x_4 = 3$, the left-hand side of the first equation becomes $3.0{\cdot}5 - 5.1{\cdot}5 + 2.3 + 7 = 7$. The residual R_1 is thus taken as 7. In this way, the residuals for the first row in Table II are 7, -4, $-0{\cdot}5$ and 6·5. In this row it is seen that in order to make the largest residual smaller, fractional changes in x-values would have to be taken. To keep in terms of integers, therefore, each residual is multiplied by 100, as

Table II

x_1	x_2	x_3	x_4	R_1	R_2	R_3	R_4
0	0·5	1·5	3	7	− 4	− 0·5	6·5
0·00	0·50	1·50	3·00	700	− 400	− 50	650
− 36	0	0	0	− 20	− 508	− 50	578
0	0	0	− 24	− 68	− 388	− 122	− 22
0	− 13	0	0	− 107	2	− 148	30
0	0	8	0	− 147	34	20	6
8	0	0	0	13	58	20	22
0	2	0	0	19	− 2	24	14
0	0	− 1	0	24	− 6	3	17
− 1	0	0	0	4	− 9	3	15
− 0·2900	0·3900	1·5700	2·7600	400	− 900	300	1500
0	0	0	− 61	278	− 595	117	− 25
0	− 20	0	0	218	5	77	55
− 11	0	0	0	− 2	− 28	77	33
0	0	− 4	0	18	− 44	− 7	45
0	0	0	− 2	14	− 34	− 13	− 5
0	− 1	0	0	11	− 4	− 15	− 1
0	0	1	0	6	0	6	− 4
− 0·2911	0·3879	1·5697	2·7537				

shown in row 2. This will mean that the figures which follow refer to the first two decimal places. This process is quite general; whenever the numbers in the residual columns become so low that fractional changes in x-values would be required to lower them further, then the residuals are multiplied by 100 and the changes in x-values thenceforth refer to the next two decimal places. Thus in row 11 the residuals are multiplied again by 100. If a check is felt to be necessary it is most conveniently carried out at one of these stages. For example, it could be checked that values

$x_1 = -0{\cdot}29$, $x_2 = 0{\cdot}39$, $x_3 = 1{\cdot}57$ and $x_4 = 2{\cdot}76$ *do* give residuals 0·04, − 0·09, 0·03 and 0·15 in the four equations. (It would be instructive for the student actually to do this.)

In the last row of Table II above, it will be noticed that the residuals remaining are appreciably less than half the change in the residuals for the respective x-values (i.e. 20, − 30, 21 and 25; see Table I). It follows that the solutions $x_1 = -0{\cdot}2911$, $x_2 = 0{\cdot}3879$, $x_3 = 1{\cdot}5697$, $x_4 = 2{\cdot}7537$ are correct to 4 decimal places, as the number to be added or subtracted in the fifth decimal place would be less than 5.

13·64. Special cases

There are types of equations for which the residuals only diminish slowly or whose values oscillate. For these cases special methods have been developed which are outside the scope of this book.

The ideal state of affairs for rapid results is for the leading-diagonal coefficients to be large compared with the others (see the example in § 13·63 above).

In some cases this can be arranged by inter-combination of the given equations. For example, consider the equations

$$8x - 2y + z - 11 = 0, \tag{1}$$

$$x - 6y - z + 1 = 0, \tag{2}$$

$$9x - 50y + z + 35 = 0. \tag{3}$$

The first two equations do have their largest coefficients along the leading diagonal. In the third equation it would be desirable to have the coefficient of z larger than those of x and y. In this example this can easily be achieved. A new equation (3)(a) is made up as follows

(2) × 8 gives $\quad 8x - 48y - 8z + 8 = 0.$

(3) is $\quad 9x - 50y + z + 35 = 0.$

Subtracting these equations:

$$x - 2y + 9z + 27 = 0. \tag{3)(a}$$

The equations from which to obtain solutions would be taken as

$$8x - 2y + z - 11 = 0,$$
$$x - 6y - z + 1 = 0,$$
$$x - 2y + 9z + 27 = 0.$$

It will be seen that the largest coefficients now appear in the leading diagonal.

It is emphasized that this readjustment of the equations should be attempted only if the required result is fairly easy to obtain; otherwise it is a waste of time.

Exercise 43

1. Solve the equations given in § 13·64 above.

2. Find approximate values of x_1, x_2 and x_3 which satisfy

$$1{\cdot}93x_1 - 1{\cdot}07x_2 + 0{\cdot}625x_3 - 13{\cdot}4 = 0,$$
$$4x_1 + x_2 - 0{\cdot}046x_3 = 0,$$
$$x_1 + 3x_2 - 0{\cdot}346x_3 = 0.$$

3. Find approximate solutions for:

$$4x_1 + x_2 - 0{\cdot}6x_3 + 0{\cdot}15 = 0,$$
$$7x_1 + 28x_2 - 3x_3 + 0{\cdot}75 = 0,$$
$$0{\cdot}6x_1 + 0{\cdot}429x_2 - 0{\cdot}362x_3 + 0{\cdot}121 = 0.$$

4. Given that very approximate solutions are

$$x_1 = -0{\cdot}5, \quad x_2 = -0{\cdot}7, \quad x_3 = 0, \quad x_4 = -0{\cdot}2,$$

find solutions, correct to 5 decimal places, of the equations:

$$40x_1 - 20x_2 - 2x_3 + 2x_4 + 9 = 0,$$
$$-40x_1 + 44x_2 + 4x_3 - 4x_4 + 9 = 0,$$
$$-4x_1 + 4x_2 + 44x_3 - 40x_4 - 9 = 0,$$
$$2x_1 - 2x_2 - 20x_3 + 40x_4 + 9 = 0.$$

Answers

1. $x = 2$, $y = 1$, $z = -3$.
2. $x_1 = -0{\cdot}55$, $x_2 = 3{\cdot}55$, $x_3 = 29{\cdot}21$.
3. $x_1 = 0{\cdot}016$, $x_2 = 0{\cdot}009$, $x_3 = 0{\cdot}371$.
4. $x_1 = -0{\cdot}59799$, $x_2 = -0{\cdot}77006$, $x_3 = 0{\cdot}01434$, $x_4 = -0{\cdot}22643$.

CHAPTER 14

DOUBLE INTEGRALS AND SOME APPLICATIONS

14·1. Introduction

Unless proceeding to the advanced theory of subjects such as electrostatics, the practical electrical engineer is not often likely to meet multiple integrals. Only a very brief introduction to such integrals is therefore given here.

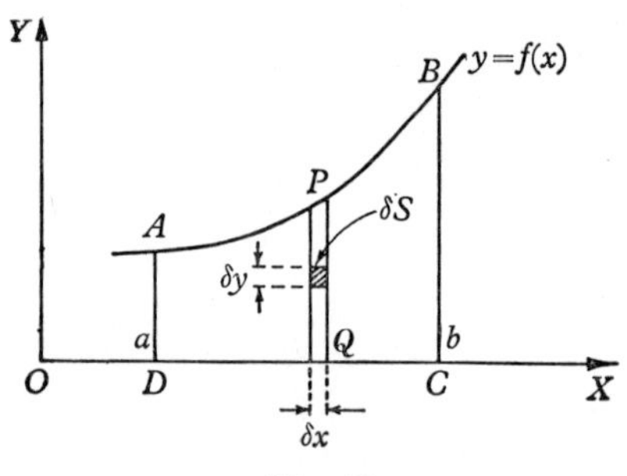

Fig. 72

As with ordinary integrals, double integrals are used as a means of performing a summation. Without doubt the student has performed double integration *mentally* every time an area, centre of gravity, etc., has been calculated by ordinary integration.

As an example, consider the area under the curve shown in fig. 72, from $x = a$ to $x = b$.

The normal procedure is to take the area of the strip PQ as $y\delta x$ and evaluate the total area as

$$\lim_{\delta x \to 0} \Sigma\, y\delta x = \int_a^b y\, dx.$$

Now in estimating the area of the strip, a summation has already been done mentally. The comparatively long strip PQ may be considered as being made up of very short elements of length δy and constant width δx. Thus the area of the strip is

$$\delta x \sum_{y=0}^{y} \delta y = \delta x \int_0^y dy = y\delta x.$$

The full process of finding the area under the curve is now written as

$$\lim_{\substack{\delta x \to 0 \\ \delta y \to 0}} \sum_{x=a}^{x=b} \delta x \left(\sum_0^y \delta y \right) = \int_a^b dx \left(\int_0^y dy \right),$$

where the integral in the curved bracket is worked first. Actually the brackets are usually omitted. Thus

$$\int_a^b dx \int_0^y dy = \int_a^b \Big[y \Big]_0^y dx = \int_a^b y\, dx.$$

The small element $\delta x \delta y$ is often written as δS, where S signifies area, and the phrase '$\int dS$ taken over the area $ABCD$' may be used.

Similarly, the first moment of δS about the y-axis may be written $x\delta S$. The total first moment of the area about the y-axis is then $\int x\, dS$ taken over the area. This is

$$\int_a^b \left(\int_0^y x\, dy \right) dx \quad \text{or} \quad \int_a^b dx \int_0^y x\, dy = \int_a^b xy\, dx.$$

In performing the integration in the bracket it must be remembered to treat x as constant when integrating with respect to y. (Referring to fig. 72, x is constant along the strip PQ.)

Obviously in the example considered no advantage is gained by using double integration, but in some problems there is a definite advantage.

Some books write $\int_a^b dx \int_0^y x\, dy$ as $\int_a^b \int_0^y x\, dx\, dy$, whilst others write it as $\int_a^b \int_0^y x\, dy\, dx$, thus following different conventions in designating that integration with respect to y must come first. This book will use the symbolism $\int_a^b dx \int_0^y x\, dy$, where the right-hand integration is done first.

Triple integrals arise in much the same manner, for example $\int_1^2 dx \int_x^{x^2} dy \int_y^{y^2} xyz\, dz$, but they will not be considered here.

EXAMPLES

(1) Evaluate $\int_0^2 dx \int_1^2 (x^2 + y^2)\, dy$.

$$\int_0^2 dx \int_1^2 (x^2 + y^2)\, dy = \int_0^2 \left[x^2 y + \tfrac{1}{3}y^3\right]_1^2 dx$$

$$= \int_0^2 \{(2x^2 + \tfrac{8}{3}) - (x^2 + \tfrac{1}{3})\}\, dx$$

$$= \int_0^2 (x^2 + \tfrac{7}{3})\, dx$$

$$= \left[\tfrac{1}{3}x^3 + \tfrac{7}{3}x\right]_0^2$$

$$= \tfrac{22}{3}.$$

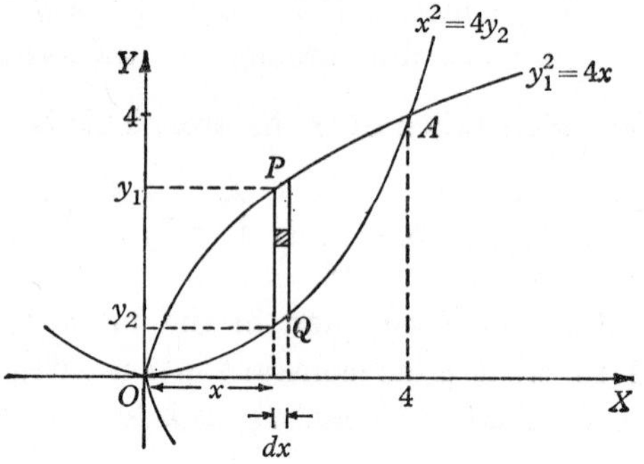

Fig. 73

(2) Find, by double integration, the area contained between the two curves $y^2 = 4x$ and $x^2 = 4y$.

Fig. 73 shows a sketch of the required area.

Where the curves cut, at O and A,

$$y^2 = 4x = \frac{x^4}{16},$$

giving $$64x - x^4 = x(64 - x^3) = 0.$$

Thus
$$\left.\begin{aligned} x &= 0 \quad \text{or} \quad 4, \\ y &= 0 \quad \text{or} \quad 4. \end{aligned}\right\} \tag{1}$$

For convenience one curve may be designated by $y_1^2 = 4x$ and the other by $x^2 = 4y_2$.

Again, for convenience, dx and dy will be used from the start, instead of δx and δy.

The area of the small element is $dx\,dy$.

Keeping x constant and letting y vary from y_2 to y_1, the area of the strip PQ is

$$dx \int_{y_2}^{y_1} dy = dx \int_{\frac{1}{4}x^2}^{2\sqrt{x}} dy.$$

Summing such strips from $x = 0$ to $x = 4$ gives the total area between the curves as

$$\begin{aligned} \int_0^4 dx \int_{\frac{1}{4}x^2}^{2\sqrt{x}} dy &= \int_0^4 \Big[y\Big]_{\frac{1}{4}x^2}^{2\sqrt{x}} dx \\ &= \int_0^4 (2\sqrt{x} - \tfrac{1}{4}x^2)\,dx \\ &= \Big[\tfrac{4}{3}x^{\frac{3}{2}} - \tfrac{1}{12}x^3\Big]_0^4 \\ &= \underline{\tfrac{16}{3}}. \end{aligned}$$

(3) Evaluate $\displaystyle\int_0^1 dr \int_0^{\frac{1}{2}\pi} r^2 \sin\theta\, d\theta.$

$$\begin{aligned} \int_0^1 dr \int_0^{\frac{1}{2}\pi} r^2 \sin\theta\, d\theta &= \int_0^1 \Big[-r^2\cos\theta\Big]_0^{\frac{1}{2}\pi} dr \\ &= \int_0^1 (-0 + r^2)\,dr = \Big[\tfrac{1}{3}r^3\Big]_0^1 = \underline{\tfrac{1}{3}}. \end{aligned}$$

Exercise 44

Evaluate the double integrals given below:

1. $\displaystyle\int_0^1 dy \int_0^2 xy\,dx.$ **2.** $\displaystyle\int_0^2 dy \int_0^1 (x^2 + y^2)\,dx.$

3. $\int_0^{\pi} d\theta \int_0^R r^3\,dr.$ 4. $\int_0^4 dx \int_0^{4x-x^2} dy.$

5. $\int_0^{\pi} d\phi \int_0^{\frac{1}{2}\pi} (2\cos\theta + 3\sin 2\theta)\,d\theta.$

6. $\int_0^2 dy \int_0^2 e^{x+y}\,dx.$

7. $\int_1^2 dx \int_0^{x^2} \frac{1}{x^3}\,dy.$ 8. $\int_{-1}^2 dx \int_{-2}^3 (y^2 - xy)\,dy.$

9. $\int_0^1 dr \int_0^{\frac{1}{4}\pi} r\cos^2\theta\,d\theta.$ 10. $\int_0^a dx \int_0^{\sqrt{(a^2-x^2)}} x\,dy.$

11. In Example 2 of § 14·1 show that the x coordinate of the centroid of the area is given by

$$\bar{x}\int_0^4 dx \int_{\frac{1}{4}x^2}^{2\sqrt{x}} dy = \int_0^4 dx \int_{\frac{1}{4}x^2}^{2\sqrt{x}} x\,dy.$$

Hence find $\bar{x}$.

ANSWERS

1. 1. **2.** $\frac{10}{3}$. **3.** $\frac{1}{4}\pi R^4$. **4.** $\frac{32}{3}$. **5.** 5π.

6. $(e^2 - 1)^2$. **7.** $\log_e 2$. **8.** $\frac{125}{4}$. **9.** $\frac{1}{8}(\frac{1}{2}\pi + 1)$.

10. $\frac{1}{3}a^3$. **11.** $\bar{x} = 1{\cdot}8$.

14·2. Change of order of integration

If *both* sets of limits for integration are *constants*, the order in which the integration is carried out is immaterial.

For example, reversing the order of integration in Question 1 of Exercise 44 above:

$$\int_0^2 dx \int_0^1 xy\,dy = \int_0^2 \left[\tfrac{1}{2}xy^2\right]_0^1 dx$$

$$= \int_0^2 \tfrac{1}{2}x\,dx = \left[\tfrac{1}{4}x^2\right]_0^2 = 1.$$

This is the same answer as before.

However, if any limit is *not* a constant, the order of integration *cannot* be reversed simply as above.

For example, reversing the order in Question 7 of Exercise 44 $\int_0^{x^2} dy \int_1^2 \frac{1}{x^3}\,dx$ is obtained; this is meaningless.

It cannot be over-emphasized that if a change in the order of integration is desired the area of integration should always be sketched. This is dealt with in the following sections (§§ 14·3 and 14·4).

14·3. A more detailed discussion of areas

Fig. 74 shows a closed, plane area S.

Let the equation of the lower part of the curve ACB be $y = y_1(x)$ and the equation of the upper part of the curve ADB be $y = y_2(x)$.

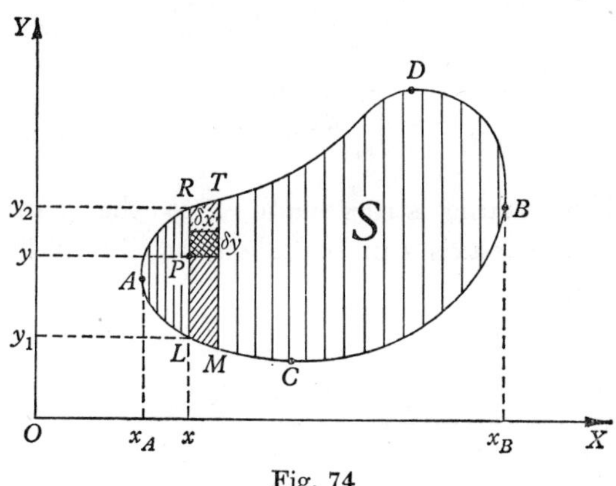

Fig. 74

Let P be any point (x, y) within the area.

Draw a small rectangle, with sides parallel to the axes, of which P is one corner. Let the lengths of the sides of the rectangle be δx and δy, as shown.

Now draw LR, MT parallel to the y-axis to form a strip $LMTR$. The width of the strip is constant, δx, and the length varies from y_1 to y_2. Thus the area of the strip is approximately

$$\sum_{y=y_1}^{y=y_2} \delta x\, \delta y.$$

To show that δx is constant in this summation, it may be written

$$\delta x \left(\sum_{y=y_1}^{y=y_2} \delta y \right).$$

Let A, B be the extremities of the width of the area S. The whole area can be covered by taking strips parallel to the strip $LMTR$, covering the width from $x = x_A$ to $x = x_B$. Thus

$$S \simeq \sum_{x_A}^{x_B} \delta x \left(\sum_{y_1}^{y_2} \delta y \right).$$

The smaller δx and δy, the closer the expression is to the accurate area.

In the limit, as $\delta x \to 0$ and $\delta y \to 0$,

$$S = \lim_{\substack{\delta x \to 0 \\ \delta y \to 0}} \sum_{x_A}^{x_B} \delta x \left(\sum_{y_1}^{y_2} \delta y \right)$$

$$= \int_{x_A}^{x_B} dx \int_{y_1}^{y_2} dy. \tag{1}$$

Instead of taking strips parallel to the y-axis, strips parallel to the x-axis can be taken, as shown in fig. 75.

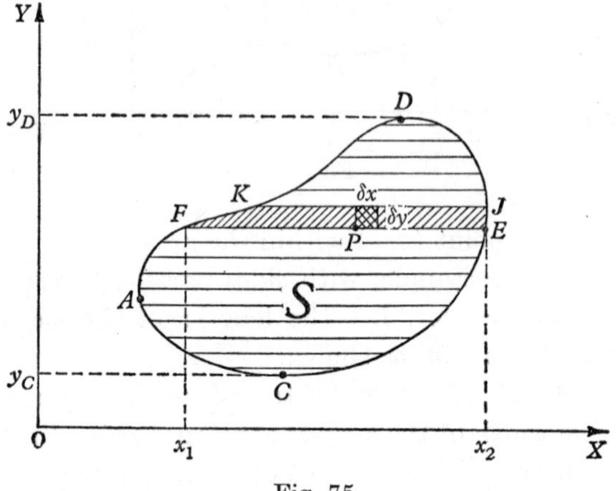

Fig. 75

Equation of l.h.s. of curve CAD: $x = x_1(y)$.

Equation of r.h.s. of curve CED: $x = x_2(y)$.

Area of strip $FKJE \simeq \delta y \left(\sum_{x_1}^{x_2} \delta x \right)$, δy being constant for the strip.

Adding parallel strips to cover the whole area:

$$S \simeq \sum_{y_C}^{y_D} \delta y \left(\sum_{x_1}^{x_2} \delta x \right),$$

where y_C, y_D are the extremities of the area in the direction of the y-axis.

As δx and δy both approach zero,

$$S = \int_{y_C}^{y_D} dy \int_{x_1}^{x_2} dx. \tag{2}$$

Equations (1) and (2) above are alternative methods for calculating the same area S. The limits of integration are not the same in both cases, but have to be determined separately. In practical cases this is best done by sketching a diagram of the area to be covered. x_A, x_B, and y_C, y_D will be numerical, but x_1, x_2 will in general be functions of y, and y_1, y_2 functions of x. This gives an insight into the remarks made in § 14·2 above.

14·4. Evaluation of double integrals over given areas

If $(\delta x \delta y)$ is first multiplied by a function of x and y, say $f(x, y)$, and the summation process is carried out over an area such as that in § 14·3 above, two equivalent double integrals are obtained:

$$\int_{y_C}^{y_D} dy \int_{x_1}^{x_2} f(x, y)\, dx = \int_{x_A}^{x_B} dx \int_{y_1}^{y_2} f(x, y)\, dy. \tag{1}$$

Either of these two integrals is regarded as evaluating $\iint f(x, y)\, dy\, dx$ over the given area S.

This is often written $\int f(x, y)\, dS$, taken over the area S.

The student should now carefully study the following examples.

EXAMPLE 1

Evaluate $\iint (x + y)\, dy\, dx$ over the rectangle bounded by the lines $x = 0$, $x = 2$; $y = 1$, $y = 2$:

(i) by taking strips parallel to the x-axis,

(ii) by taking strips parallel to the y-axis.

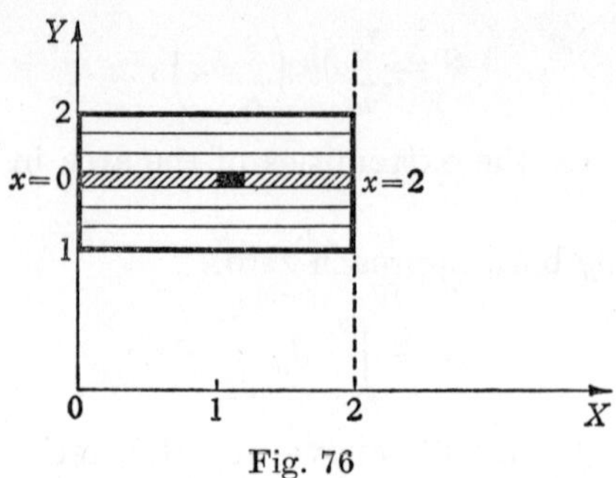

Fig. 76

(i) The stages may be summarized as follows:

(*a*) Draw a strip parallel to the x-axis.

(*b*) Mark in the x-values at the ends of the strip.

(*c*) Along this strip only x varies, from 0 to 2.

Summing along the strip gives

$$dy\left\{\int_0^2 (x+y)\,dx\right\}.$$

(*d*) Now move the strip parallel to its length to cover the given rectangle, y therefore varying from 1 to 2.

This summation gives

$$\int_1^2 dy\left\{\int_0^2 (x+y)\,dx\right\}.$$

Evaluating this integral:

$$\int_1^2 dy \int_0^2 (x+y)\,dx = \int_1^2 \left[\tfrac{1}{2}x^2 + yx\right]_0^2 dy$$

$$= \int_1^2 (2+2y)\,dy = \left[2y + y^2\right]_1^2 = \underline{5}.$$

(ii)

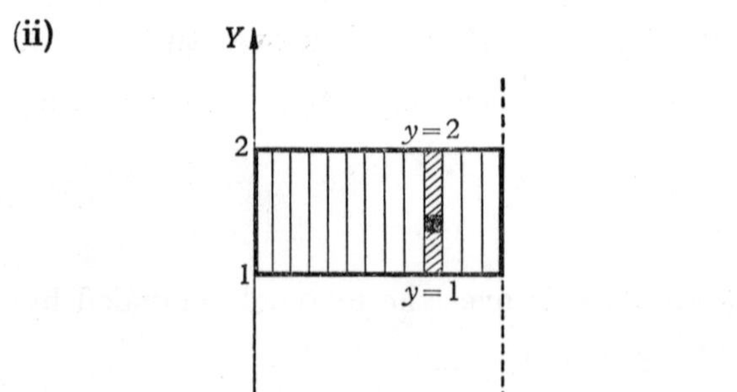

Fig. 77

(*a*) Draw a strip parallel to the y-axis.

(*b*) Mark in the y-values at the ends of the strip.

(*c*) Summing along the strip, where y varies from 1 to 2, gives

$$dx\left\{\int_1^2 (x+y)\,dy\right\}.$$

(*d*) Moving the strip parallel to its length, x varying from 0 to 2, gives

$$\int_0^2 dx\left\{\int_1^2 (x+y)\,dy\right\}.$$

Evaluating this integral:

$$\int_0^2 dx\int_1^2 (x+y)\,dy$$

$$=\int_0^2 \left[xy+\tfrac{1}{2}y^2\right]_1^2 dx$$

$$=\int_0^2 \{(2x+2)-(x+\tfrac{1}{2})\}\,dx=\int_0^2 (x+\tfrac{3}{2})\,dx$$

$$=\left[\tfrac{1}{2}x^2+\tfrac{3}{2}x\right]_0^2=\underline{5}.$$

With practice, all the steps in the argument need not be written down, the diagram being sufficient.

In many cases both orders of integration give rise to fairly easy integration, as in the case above; but sometimes drawing the strip in one direction gives an awkward integration. When this happens the alternative strip, parallel to the other axis, should be tried in case it gives an easier one.

EXAMPLE 2

Evaluate $\int y\,dS$ over the area of that part of the circle $x^2+y^2=a^2$ contained in the first quadrant.

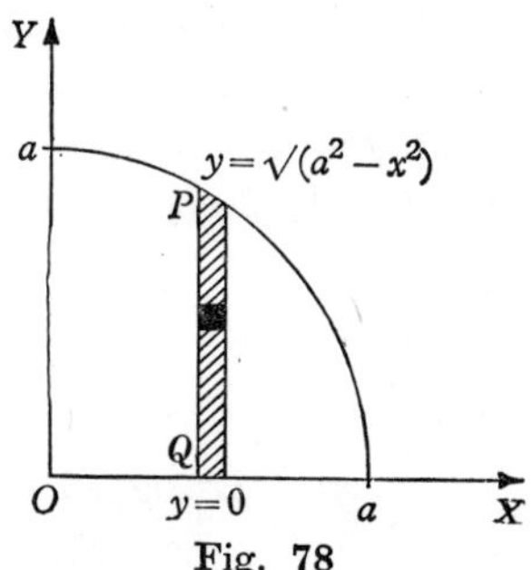

Fig. 78

The strip, PQ, is drawn parallel to OY. As $x^2+y^2=a^2$, then $y^2=a^2-x^2$.

In the first quadrant $y = +\sqrt{(a^2 - x^2)}$ at point P, $y = 0$ at point Q.

As PQ moves parallel to itself, x goes from 0 to a in covering the quarter-circle. Thus

$$\int y\,dS = \int_0^a dx \int_0^{\sqrt{(a^2-x^2)}} y\,dy$$

$$= \int_0^a dx \left[\tfrac{1}{2}y^2\right]_0^{\sqrt{(a^2-x^2)}}$$

$$= \int_0^a \tfrac{1}{2}(a^2 - x^2)\,dx = \tfrac{1}{2}\left[a^2x - \tfrac{1}{3}x^3\right]_0^a$$

$$= \underline{\tfrac{1}{3}a^3.}$$

EXAMPLE 3

Evaluate $\int \frac{x}{y}\,dS$ over the area of the triangle contained by the lines $x = 0$, $y = x$ and $y = 2 - x$.

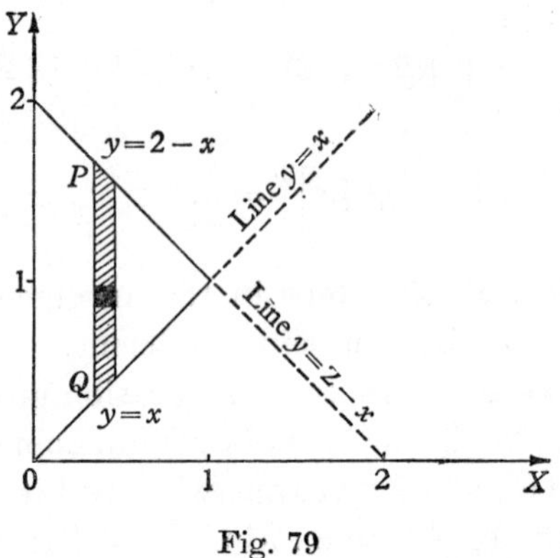

Fig. 79

Draw the strip, PQ, parallel to OY.

At P $y = 2 - x$; at Q $y = x$.

As PQ moves parallel to itself to cover the triangle, x varies from 0 to 1. Thus

$$\int \frac{x}{y}\,dS = \int_0^1 dx \int_x^{2-x} \frac{x}{y}\,dy = \int_0^1 dx \left[x \log_e y\right]_x^{2-x}$$

$$= \int_0^1 \{x \log_e (2 - x) - x \log_e x\}\,dx$$

$$= \int_0^1 x \log_e \left(\frac{2-x}{x}\right) dx = \int_0^1 x \log_e \left(\frac{2}{x} - 1\right) dx$$

$$= \left[\frac{x^2}{2} \log_e \left(\frac{2}{x} - 1\right)\right]_0^1 - \int_0^1 \frac{x^2}{2} \cdot \frac{1}{(2/x - 1)} \left(\frac{-2}{x^2}\right) dx$$

by parts

$$= 0 + \int_0^1 \frac{x}{(2-x)} dx$$

$$= \int_0^1 \left\{-1 + \frac{2}{(2-x)}\right\} dx = \Big[-x - 2\log_e(2-x)\Big]_0^1$$

$$= -1 + 2\log_e 2 = \underline{\log_e 4 - 1.}$$

Note. Consider $\lim_{x\to 0} x^n \log_e x$, $n > 0$. Put $x = e^{-t}$.

Then $x^n \log_e x = e^{-nt}(-t) = \dfrac{-t}{e^{nt}}$, and as $x \to 0$, $t \to \infty$.

Thus $\lim_{x\to 0} x^n \log_e x = \lim_{t\to\infty} \dfrac{-t}{e^{nt}} = \lim_{t\to\infty} \dfrac{-1}{ne^{nt}}$ (De L'Hôpital)

$$= \underline{0.}$$

Such a limit has been used in $\left[\frac{x^2}{2} \log_e \left(\frac{2}{x} - 1\right)\right]_0^1$ above.

Thus $$\left[\frac{x^2}{2} \log_e \left(\frac{2}{x} - 1\right)\right]_0^1 = \left[\frac{x^2}{2} \log_e \left(\frac{2-x}{x}\right)\right]_0^1$$

$$= \left[\frac{x^2}{2} \log_e (2-x)\right]_0^1 - \left[\frac{x^2}{2} \log_e x\right]_0^1$$

$$= 0 - \left[0 - \lim_{x\to 0} \frac{x^2}{2} \log_e x\right]$$

$$= 0.$$

The integration above was rather cumbersome, and the alternative method of taking strips parallel to OX will now be demonstrated. It will be found that no awkward limits arise in this case.

The strips now fall into two categories; those with an extremity on the line $y = x$, and those with an extremity on the line $y = 2 - x$ (i.e. $x = 2 - y$). The change takes place on the line $y = 1$ (AB).

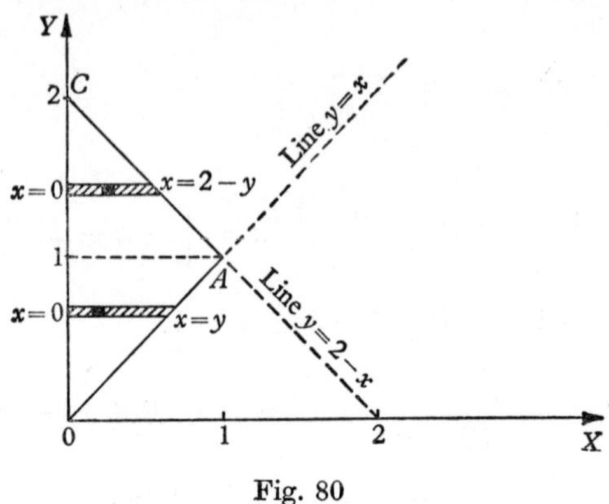

Fig. 80

The integration is thus performed in two stages:

Over area OAB. End-values for the strip are $x = 0$, $x = y$; y then varies from 0 to 1. Over this area then,

$$\int \frac{x}{y}\,dS = \int_0^1 dy \int_0^y \frac{x}{y}\,dx.$$

Over the area ABC. End-values for the strip are $x = 0$ and $x = 2 - y$; y then varies from 1 to 2. Over this area

$$\int \frac{x}{y}\,dS = \int_1^2 dy \int_0^{2-y} \frac{x}{y}\,dx.$$

Over the whole area.

$$\int \frac{x}{y}\,dS = \int_0^1 dy \int_0^y \frac{x}{y}\,dx + \int_1^2 dy \int_0^{2-y} \frac{x}{y}\,dx$$

$$= \int_0^1 \left[\frac{x^2}{2y}\right]_0^y dy + \int_1^2 \left[\frac{x^2}{2y}\right]_0^{2-y} dy$$

$$= \int_0^1 \frac{y}{2}\,dy + \int_1^2 \frac{(2-y)^2}{2y}\,dy$$

$$= \left[\frac{y^2}{4}\right]_0^1 + \int_1^2 \left(\frac{2}{y} - 2 + \frac{y}{2}\right) dy$$

$$= \tfrac{1}{4} + \left[2 \log_e y - 2y + \frac{y^2}{4}\right]_1^2$$

$$= \tfrac{1}{4} + 2 \log_e 2 - 2 + \tfrac{3}{4} = \underline{\log_e 4 - 1.}$$

Exercise 45

1. Evaluate $\iint \frac{1}{y}\,dx\,dy$ over the rectangle bounded by the lines $x = 0$, $x = 2$; $y = 1$, $y = 2$.

2. Evaluate $\iint x\,dx\,dy$ over that area of the circle $x^2 + y^2 = a^2$ contained in the first quadrant.

3. Find $\int y\,dS$ taken over the semicircular area bounded by the circle $x^2 + y^2 = a^2$ and the x-axis, in the first and second quadrants.

4. Evaluate $\iint x^2\,dx\,dy$ over the area bounded by the parabola $y^2 = 4x$ and the line $x = 1$.

5. Evaluate $\int (x^2 + y^2)\,dS$ over that half of the circle $x^2 + y^2 = a^2$ which lies above the x-axis.

6. Evaluate $\int (x^2 + y^2)\,dS$ over the area of the triangle whose vertices are the points (0, 1), (1, 1) and (1, 2).

Answers

1. $2 \log_e 2$. **2.** $\tfrac{1}{3}a^3$. **3.** $\tfrac{2}{3}a^3$. **4.** $\tfrac{8}{7}$. **5.** $\dfrac{\pi a^4}{4}$.

6. $1\tfrac{1}{6}$. [Note that the line joining the points (0, 1) and (1, 2) is $y = x + 1$.]

14·5. Double integrals in polar co-ordinates

Areas whose outlining curves are given by polar equations can be divided into elements by radial lines (see fig. 81).

If P is the point (r, θ) and R the point $(r + \delta r, \theta + \delta\theta)$, then $PQ = \delta r$ and $PT = r\delta\theta$.

The element $PQRT$ has an approximate area given by $\delta S = (r\delta\theta)\delta r$.

Let the two parts of the outlining curve, ACB and ADB, have equations $r = r_1(\theta)$, $r = r_2(\theta)$.

With θ kept constant, r varies from $r_1(\theta)$ to $r_2(\theta)$; afterwards, to cover the whole area, θ varies from θ_1 to θ_2.

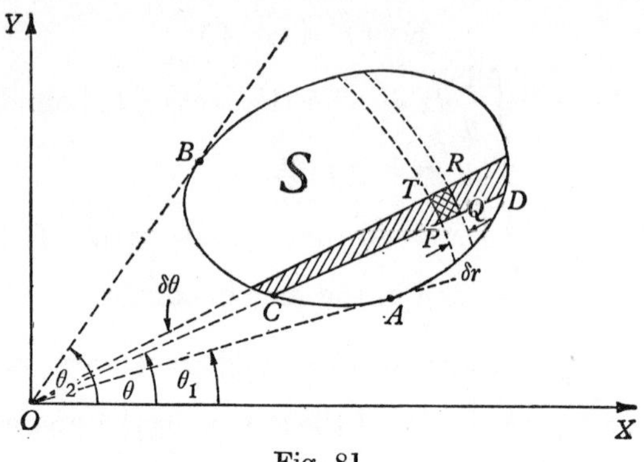

Fig. 81

The total area is therefore given by

$$S = \lim_{\substack{\delta\theta \to 0 \\ \delta r \to 0}} \sum_{\theta_1}^{\theta_2} \delta\theta \sum_{r_1}^{r_2} r\delta r = \int_{\theta_1}^{\theta_2} d\theta \int_{r_1}^{r_2} r\, dr.$$

$\int z\, dS$ taken over the area, where z is a function of r and θ, is given by $\int_{\theta_1}^{\theta_2} d\theta \int_{r_1}^{r_2} zr\, dr.$

No new ideas are involved; the main problem is to determine the correct limits for r and θ.

EXAMPLE 1

Interpret $\int_0^a dx \int_{\sqrt{(ax-x^2)}}^{\sqrt{(a^2-x^2)}} \frac{dy}{\sqrt{(a^2 - x^2 - y^2)}}$ as an integral taken over an area, showing the area in a diagram. Evaluate the integral by transforming to polar co-ordinates. [L.U.]

With x constant, y varies from $\sqrt{(ax - x^2)}$ to $\sqrt{(a^2 - x^2)}$. The outlining curves of the area are therefore $y = \sqrt{(a^2 - x^2)}$, giving $x^2 + y^2 = a^2$ and $y = \sqrt{(ax - x^2)}$, giving $x^2 - ax + y^2 = 0$, or $(x - \frac{1}{2}a)^2 + y^2 = \frac{1}{4}a^2$.

Both these curves are circles; one with centre (0, 0) and radius a, the other with centre $(\frac{1}{2}a, 0)$ and radius $\frac{1}{2}a$. As x varies from 0 to a and only the positive y-values are mentioned, the area must only be that in the first quadrant, see fig. 82.

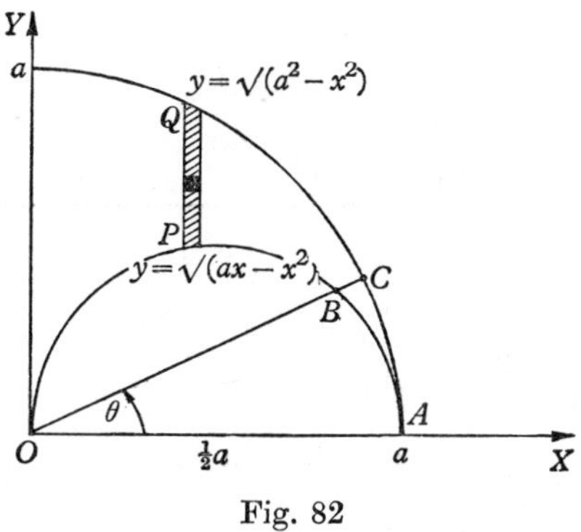

Fig. 82

The polar equation of the smaller circle is $r = a \cos \theta$; that of the larger circle is $r = a$.

With θ kept constant (for example, along line OBC), r varies from $a \cos \theta$ to a. To sweep out the given area, θ then has to vary from 0 to $\frac{1}{2}\pi$.

In polar co-ordinates $x = r \cos \theta, y = r \sin \theta$, thus $x^2 + y^2 = r^2$ and $\sqrt{(a^2 - x^2 - y^2)} = \sqrt{(a^2 - r^2)}$.

The given double integral therefore becomes

$$\int_0^{\frac{1}{2}\pi} d\theta \int_{a \cos \theta}^{a} \frac{r\, dr}{\sqrt{(a^2 - r^2)}}$$

$$= \int_0^{\frac{1}{2}\pi} \left[-\sqrt{(a^2 - r^2)} \right]_{a \cos \theta}^{a} d\theta$$

$$= \int_0^{\frac{1}{2}\pi} a \sin \theta\, d\theta = \left[-a \cos \theta \right]_0^{\frac{1}{2}\pi} = \underline{a}.$$

Example 2

Fig. 83 shows a long straight conductor carrying a current of 10 amperes and a coplanar circular area tangential to the conductor. The radius of the circle is 0·2 m. Find the total magnetic flux through the circular area in webers.

Take the line of the conductor as the initial line, and the point of contact as the pole (origin).

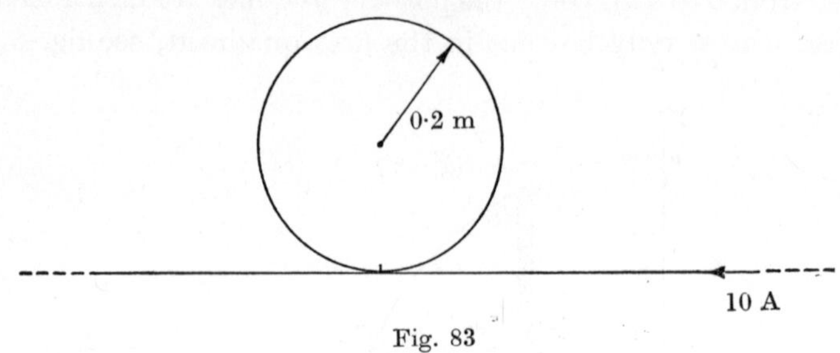

Fig. 83

The polar equation of the circle is then

$$r = 0{\cdot}4 \sin\theta \qquad (1)$$

(see fig. 84).

The magnetizing force at $P(r, \theta)$ is $\dfrac{i}{2\pi PM}$, where PM is measured in metres and i in amperes.

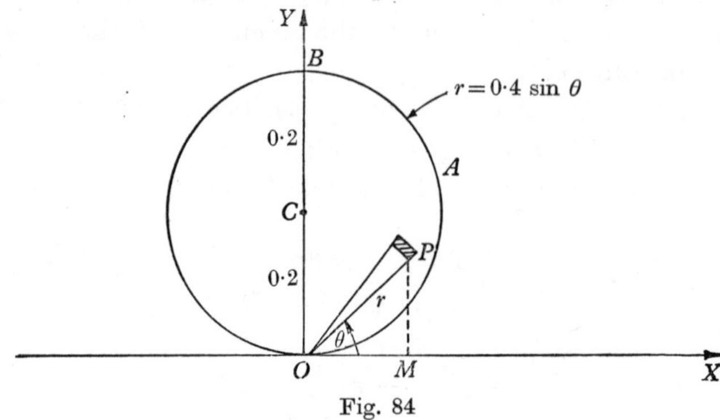

Fig. 84

Thus the magnetizing force at P is $\dfrac{10}{2\pi r \sin\theta}$, where r is measured in metres.

The density at P is therefore

$$\frac{10\mu_0}{2\pi r \sin\theta} \text{ tesla,}$$

where μ_0 is the permeability of free space.

The flux over a small area $r\,dr\,d\theta$ square metres round P is therefore

$$\frac{10\mu_0}{2\pi r\sin\theta}\times r\,dr\,d\theta \text{ webers,}$$

$$=\frac{5\mu_0}{\pi}\times\frac{dr\,d\theta}{\sin\theta}\text{ webers.}$$

By symmetry about OB, the total flux over the circle is $2\iint\frac{5\mu_0\,d\theta\,dr}{\pi\sin\theta}$, taken over the semi-circle OAB.

Using (1), this becomes

$$\frac{10\mu_0}{\pi}\int_0^{\frac{1}{2}\pi}d\theta\int_0^{0\cdot4\sin\theta}\frac{1}{\sin\theta}\,dr$$

$$=\frac{10\mu_0}{\pi}\int_0^{\frac{1}{2}\pi}\left[\frac{r}{\sin\theta}\right]_0^{0\cdot4\sin\theta}d\theta$$

$$=\frac{10\mu_0}{\pi}\int_0^{\frac{1}{2}\pi}0\cdot4\,d\theta$$

$$=2\mu_0$$

$$=8\pi\times10^{-7}\text{ webers (since }\mu_0=4\pi.10^{-7}).$$

Exercise 46

(These questions include some on Cartesian co-ordinates.)

1. Use double integration to find the area of the cardioid $r=a(1+\cos\theta)$.

2. Find the polar equations of the circles $x^2+y^2=a^2$ and $(x-a)^2+y^2=a^2$. Show that the area outside the first circle but inside the second is given by

$$2\int_0^{\frac{1}{3}\pi}d\theta\int_a^{2a\cos\theta}r\,dr.\quad\text{Evaluate this area.}$$

3. Evaluate $\displaystyle\int_0^a dx\int_x^a\frac{y\,dy}{\sqrt{(x^2+y^2)}}$ and state the region of the (x, y) plane to which it refers. Change the order of integration, evaluate and verify that the two values are the same. [L.U.]

4. Find the area inside the cardioid $r = 1 + \cos\theta$ and to the right of the line $4r\cos\theta = 3$.

5. Find the centroid of the area bounded by the cardioid $r = a(1 + \cos\theta)$.

6. Show by means of a diagram the area over which the double integral $\int_{\frac{1}{2}a}^{a} dx \int_{x}^{a} \frac{x\,dy}{(x^2 + y^2)^{\frac{3}{2}}}$ is taken.

By inverting the order of integration, or otherwise, show that the value of the integral is $\log_e\left(\frac{2 + \sqrt{5}}{1 + \sqrt{2}}\right) - \frac{1}{\sqrt{2}}\log_e 2$.

[L.U.]

7. Find the centroid of the right-hand loop of the lemniscate $r^2 = a^2\cos 2\theta$.

8. A long straight conductor carries a current of 10 amperes. Co-planar with it is a circular area, of radius 0·1 m, whose centre is at a distance of 0·2 m from the conductor. Find in webers, the total magnetic flux through the circle.

9. (i) Exhibit by a diagram the region in the x–y plane over which the repeated integral

$$\int_0^a dx \int_0^{\sqrt{(a^2 - x^2)}} \sin\left\{\frac{\pi}{a^2}(a^2 - x^2 - y^2)\right\} dy$$

extends. By conversion to polar co-ordinates, or otherwise, show that the value of this integral is $\frac{1}{2}a^2$.

(ii) Express as a repeated integral the volume contained between the cylinder $y^2 = 4ax$ and the planes $x + z = a$, $z = 0$. Show that this volume is $\frac{16}{15}a^2$. [L.U.]

ANSWERS

1. $\frac{3\pi a^2}{2}$. 2. $\frac{a^2}{6}(2\pi + 3\sqrt{3})$.

3. The region is the triangle bounded by the lines $x = 0$, $y = a$, $y = x$.

The new integral is $\int_0^a dy \int_0^y \frac{y\,dx}{\sqrt{(x^2 + y^2)}}$. The value is $\frac{1}{2}a^2\log(1 + \sqrt{2})$.

4. $\dfrac{\pi}{2}+\dfrac{9\sqrt{3}}{16}$. (The integral is $2\displaystyle\int_0^{\frac{1}{3}\pi} d\theta \int_{\frac{3}{4}\sec\theta}^{1+\cos\theta} r\,dr$.)

5. $\bar{x} = \frac{5}{6}a$.

6. The area is the triangle bounded by the lines $y = x$, $x = \frac{1}{2}a$, $y = a$.

The new integral is $\displaystyle\int_{\frac{1}{2}a}^{a} dy \int_{\frac{1}{2}a}^{y} \frac{x\,dx}{(x^2+y^2)^{\frac{3}{2}}}$.

7. $\bar{x} = \dfrac{\pi a\sqrt{2}}{8} \simeq 0{\cdot}56a$.

The area of the loop is $\frac{1}{2}a^2$ and its first moment round OY is $2\displaystyle\int_0^{\frac{1}{4}\pi} d\theta \int_0^{a\sqrt{(\cos 2\theta)}} r^2 \cos\theta\,dr$.

[*Note.* To evaluate $\displaystyle\int (\cos 2\theta)^{\frac{3}{2}} \cos\theta\,d\theta$, put $\cos 2\theta = 1 - 2\sin^2\theta$ and then $\sin\theta = \dfrac{1}{\sqrt{2}}\sin\phi$.]

8. $(2-\sqrt{3})4\pi \times 10^{-7}$ webers.
[The double integral to be evaluated is

$$\frac{10\mu_0}{\pi}\int_{\frac{1}{3}\pi}^{\frac{1}{2}\pi} d\theta \int_{0\cdot 2\sin\theta - 0\cdot 1\sqrt{(4\sin^2\theta-3)}}^{0\cdot 2\sin\theta + 0\cdot 1\sqrt{(4\sin^2\theta-3)}} \frac{dr}{\sin\theta}$$

which reduces to

$$\frac{2\mu_0}{\pi}\int_{\frac{1}{3}\pi}^{\frac{1}{2}\pi} \frac{\sqrt{(4\sin^2\theta-3)}}{\sin\theta}\,d\theta = \frac{2\mu_0}{\pi}\int_{\frac{1}{3}\pi}^{\frac{1}{2}\pi} \sqrt{(1-3\cot^2\theta)}\,d\theta.$$

This can be evaluated by the successive substitutions $\sin\phi = \sqrt{3}\cot\theta$ and $t = \tan\phi$.
Alternatively the question can be done using Cartesian co-ordinates.]

9. (i) $\displaystyle\int_0^a dr \int_0^{\frac{1}{2}\pi} r \sin\frac{\pi}{a^2}(a^2-r^2)\,.\,d\theta$.

(ii) $\displaystyle\int_0^a dx \int_{-2\sqrt{(ax)}}^{2\sqrt{(ax)}} (a-x)\,dy$.

CHAPTER 15

ELEMENTS OF MATRIX ALGEBRA

The employment of matrices in the analysis of electrical networks is playing an ever-increasing part in electrical engineering. This chapter and the following one form an introduction to matrix analysis of networks. Only simple cases will be considered and the student might think at first that the effort required to master matrix algebra outweighs its applications. However, it is stressed that in complicated networks the remarkable power of matrix analysis is almost indispensable. The present chapter will be devoted to elementary matrix algebra, in which it is necessary to acquire manipulative skill before applications to electrical networks can be easily carried out.

15·1. Definition and notation

A matrix is an array of numbers in rows and columns.

Common notations for a matrix are a pair of square, or a pair of curved brackets; inside these the elements are arrayed in rows and columns. Thus

$$\begin{bmatrix} a_1 & b_1 & c_1 \\ a_2 & b_2 & c_2 \end{bmatrix}$$

is a matrix of orders two and three (2 rows and 3 columns). More shortly it is called a 2×3 matrix (pronounced 'two by three'). It has $2 \times 3 = 6$ elements. Notice that the number of rows always precedes the number of columns.

In general, an $m \times n$ matrix is an array of mn elements consisting of m rows and n columns. If $m \neq n$, it is called a *rectangular* matrix: if $m = n$, it is called a *square* matrix of order m. Thus

$$\begin{bmatrix} a & b & c \\ l & m & n \\ p & q & r \end{bmatrix}$$

is a square matrix of order three.

The similarity in arrangement to a determinant is obvious, but confusion must not arise. A determinant *always* has the same number of rows as columns. More important still, a determinant has a definite numerical value which can be found by performing on its elements a definite set of operations. A matrix stands for nothing more than an ordered array of numbers. Operations with such arrays are obviously defined in such a manner that they are useful. For example, the matrix equation $[a][x] = [d]$ can serve to represent a complete set of linear equations; such brevity is an obvious convenience.

When no confusion can arise, a single letter can be used to denote a complete array, e.g. $[a_1 \quad a_2 \quad a_3]$ may be written $[a_r]$, or, as in some books, by a letter in Clarendon type $\mathbf{a}$.

15·2. Equality of two matrices

Two matrices are defined as being equal if, and only if, all corresponding elements are respectively equal to one another. This automatically dictates that they shall have the same number of rows and columns.

For example, if

$$\begin{bmatrix} e_1 \\ e_2 \end{bmatrix} = \begin{bmatrix} j\omega L_1 i_1 + j\omega M i_2 \\ j\omega L_2 i_2 + j\omega M i_2 \end{bmatrix},$$

then it follows that

$$e_1 = j\omega L_1 i_1 + j\omega M i_2 \quad \text{and} \quad e_2 = j\omega L_2 i_2 + j\omega M i_2.$$

Similarly, if

$$\begin{bmatrix} a & b & c \\ x & y & z \end{bmatrix} = \begin{bmatrix} 1 & 3 & -2 \\ -1 & 2 & 4 \end{bmatrix},$$

then $a = 1$, $b = 3$, $c = -2$, $x = -1$, $y = 2$ and $z = 4$.

It is impossible to have a matrix equation of the form

$$\begin{bmatrix} x & y \\ p & q \end{bmatrix} = \begin{bmatrix} 1 & 5 & 7 \\ 2 & 9 & 3 \\ 4 & 7 & 1 \end{bmatrix}$$

unless x, y, p, q are themselves considered as matrices.

15·3. Linear transformations

Consider the relationships expressed by

$$\left.\begin{aligned} x' &= a_1x + b_1y, \\ y' &= a_2x + b_2y, \end{aligned}\right\} \tag{1}$$

i.e. x' and y' are linear expressions in x and y.

Equations (1) are said to form a *linear transform* of (x, y) into (x', y').

The rules by which matrices are manipulated are designed especially for linear transformations and the general treatment of linear equations.

Consider the linear transform

$$\left.\begin{aligned} x'' &= c_1x' + d_1y', \\ y'' &= c_2x' + d_2y'. \end{aligned}\right\} \tag{2}$$

These equations transform (x', y') into (x'', y'').

Transforms (1) and (2), if performed in succession, will transform (x, y) into (x'', y'').

Substitution for x' and y' from (1) into (2) gives

$$x'' = c_1(a_1x + b_1y) + d_1(a_2x + b_2y),$$

$$y'' = c_2(a_1x + b_1y) + d_2(a_2x + b_2y),$$

i.e.

$$\left.\begin{aligned} x'' &= (a_1c_1 + a_2d_1)x + (b_1c_1 + b_2d_1)y, \\ y'' &= (a_1c_2 + a_2d_2)x + (b_1c_2 + b_2d_2)y. \end{aligned}\right\} \tag{3}$$

Now write the coefficients of x and y of equations (1) as rows of a 2×2 matrix, obtaining $\begin{bmatrix} a_1 & a_2 \\ b_1 & b_2 \end{bmatrix}$.

Do the same for equations (2) and (3), obtaining

$$\begin{bmatrix} c_1 & c_2 \\ d_1 & d_2 \end{bmatrix} \text{ and } \begin{bmatrix} a_1c_1 + a_2d_1 & a_1c_2 + a_2d_2 \\ b_1c_1 + b_2d_1 & b_1c_2 + b_2d_2 \end{bmatrix}.$$

Thus, successive transformations have implied that

$$\begin{bmatrix} a_1 & a_2 \\ b_1 & b_2 \end{bmatrix} \text{ operating on } \begin{bmatrix} c_1 & c_2 \\ d_1 & d_2 \end{bmatrix}$$

produces

$$\begin{bmatrix} a_1c_1 + a_2d_1 & a_1c_2 + a_2d_2 \\ b_1c_1 + b_2d_1 & b_1c_2 + b_2d_2 \end{bmatrix}.$$

This is the rule by which matrices are multiplied and the student will find it informative to read this paragraph again after reading § 15·6.

15·4. Double suffix notation

By the use of this notation the written work in matrix algebra can be much condensed. However, it will not be used often in this book as it takes some time to become really expert in its manipulation. An example of its use is given in § 15·6, showing how the rule for multiplication can be written very shortly if this notation is used.

Consider the matrix

$$\begin{bmatrix} a_{11} & a_{12} & a_{13} \\ a_{21} & a_{22} & a_{23} \\ a_{31} & a_{32} & a_{33} \end{bmatrix}.$$

The double suffixes show quite clearly the position of each element in the matrix. Thus a_{23} is the element in the second row and the third column. In general, a_{rs} would signify the element in the rth row and sth column. When it is clear that r and s vary from 1 to 3, the matrix given above could be referred to as $[a_{rs}]$.

15·5. Sum and difference of two matrices

For two matrices to be capable of addition or subtraction they must have the same number of rows and the same number of columns. They are then said to be *conformable* for addition or subtraction.

The sum (or difference) of two $m \times n$ matrices, say C and C', is another $m \times n$ matrix of which every element is the sum (or difference) of the two elements occupying corresponding places in C and C'.

Thus, if

$$C = \begin{bmatrix} a_{11} & a_{12} & a_{13} \\ a_{21} & a_{22} & a_{23} \end{bmatrix} \quad \text{and} \quad C' = \begin{bmatrix} a'_{11} & a'_{12} & a'_{13} \\ a'_{21} & a'_{22} & a'_{23} \end{bmatrix},$$

then $$C + C' = \begin{bmatrix} a_{11} + a'_{11} & a_{12} + a'_{12} & a_{13} + a'_{13} \\ a_{21} + a'_{21} & a_{22} + a'_{22} & a_{23} + a'_{23} \end{bmatrix}$$

and $$C - C' = \begin{bmatrix} a_{11} - a'_{11} & a_{12} - a'_{12} & a_{13} - a'_{13} \\ a_{21} - a'_{21} & a_{22} - a'_{22} & a_{23} - a'_{23} \end{bmatrix}.$$

Note. (i) The rule for addition gives $C + C + C = 3C$ as a matrix of which *every* element is three times the corresponding element in C. Contrast this with the rule for multiplying a determinant by three, where it is sufficient to multiply the elements of only a *single* row or column by three.

Thus in the example above

$$3C = \begin{bmatrix} 3a_{11} & 3a_{12} & 3a_{13} \\ 3a_{21} & 3a_{22} & 3a_{23} \end{bmatrix}.$$

(ii) If $A = B$ also signifies that $A - B = 0$; as the definition of equality gave A and B identical elements, then the definition of subtraction implies that a matrix is zero (called a *null* matrix) if, and only if, all its elements are zero.

For example, $\begin{bmatrix} 0 & 0 & 0 \\ 0 & 0 & 0 \end{bmatrix}$ is a null matrix.

Exercise 47

1. Write down the sums and differences of the following pairs of matrices:

(i) $[a] = \begin{bmatrix} 3 & 1 \\ 2 & -4 \end{bmatrix}, \quad [b] = \begin{bmatrix} 0 & 1 \\ 5 & 7 \end{bmatrix}.$

(ii) $[x] = \begin{bmatrix} a & 0 \\ 0 & b \end{bmatrix}, \quad [y] = \begin{bmatrix} 0 & a \\ b & 0 \end{bmatrix}.$

(iii) $[a] = \begin{bmatrix} 1 & 3 & 5 \\ 3 & 5 & 7 \end{bmatrix}, \quad [b] = \begin{bmatrix} 0 & -1 & -2 \\ 1 & 2 & 3 \end{bmatrix}.$

(iv) $[a] = [8 \quad 9 \quad 10], \quad [b] = [2 \quad 4 \quad 6].$

(v) $[a] = \begin{bmatrix} 1 & 4 \\ 2 & 5 \\ 3 & 6 \end{bmatrix}, \quad [b] = \begin{bmatrix} 4 & -4 \\ 5 & -5 \\ 6 & -6 \end{bmatrix}.$

2. Is it possible to add or subtract matrices $[a]$ and $[b]$ in the following cases?

(i) $[a]$ has 2 rows, $[b]$ has 3 rows,

(ii) $[a]$ has 3 columns, $[b]$ has 5 columns,

(iii) $[a]$ has 3 rows, $[b]$ has 3 columns.

3. If $\begin{bmatrix} x \\ y \\ z \end{bmatrix} = \begin{bmatrix} 5 \\ 2 \\ 3 \end{bmatrix} + \begin{bmatrix} 3 \\ 1 \\ 6 \end{bmatrix}$, find the values of x, y and z.

4. If $\begin{bmatrix} e_1 \\ e_2 \\ e_3 \end{bmatrix} = \begin{bmatrix} 5 + j2 \\ 4 + j \\ 1 + j2 \end{bmatrix} - \begin{bmatrix} 1 + j \\ 2 + j2 \\ 1 - j2 \end{bmatrix}$, evaluate e_1, e_2 and e_3.

5. If $\begin{bmatrix} i_1 \\ i_2 \\ i_3 \end{bmatrix} = 3[I_1] - 2[I_2]$,

where

$$[I_1] = \begin{bmatrix} 3x + 5y - z \\ 2x + y + z \\ 3x - 2y + 4z \end{bmatrix} \quad \text{and} \quad [I_2] = \begin{bmatrix} x - y - z \\ 2x + y + z \\ 3x - y - 2z \end{bmatrix},$$

find i_1, i_2 and i_3 in terms of x, y and z.

6. If $[a] = \begin{bmatrix} a_{11} & a_{12} & a_{13} \\ a_{21} & a_{22} & a_{23} \end{bmatrix}$, $[b] = \begin{bmatrix} b_{11} & b_{12} & b_{13} \\ b_{21} & b_{22} & b_{23} \end{bmatrix}$

and $[c] = \begin{bmatrix} c_{11} & c_{12} & c_{13} \\ c_{21} & c_{22} & c_{23} \end{bmatrix}$,

verify that $\{[a] + [b]\} - [c] = [a] + \{[b] - [c]\}$.

Note. This is an illustration of the fact that matrix summations follow the associative law of addition.

Answers

1. (i) $[a] + [b] = \begin{bmatrix} 3 & 2 \\ 7 & 3 \end{bmatrix}$, $[a] - [b] = \begin{bmatrix} 3 & 0 \\ -3 & -11 \end{bmatrix}$.

(ii) $[x] + [y] = \begin{bmatrix} a & a \\ b & b \end{bmatrix}$, $[x] - [y] = \begin{bmatrix} a & -a \\ -b & b \end{bmatrix}$.

(iii) $[a] + [b] = \begin{bmatrix} 1 & 2 & 3 \\ 4 & 7 & 10 \end{bmatrix}$, $[a] - [b] = \begin{bmatrix} 1 & 4 & 7 \\ 2 & 3 & 4 \end{bmatrix}$.

(iv) $[a] + [b] = [10 \quad 13 \quad 16]$, $[a] - [b] = [6 \quad 5 \quad 4]$.

(v) $[a] + [b] = \begin{bmatrix} 5 & 0 \\ 7 & 0 \\ 9 & 0 \end{bmatrix}$, $[a] - [b] = \begin{bmatrix} -3 & 8 \\ -3 & 10 \\ -3 & 12 \end{bmatrix}$.

2. (i) No. (ii) No. (iii) Only if $[a]$ has 3 columns and $[b]$ has 3 rows.

3. $x = 8$, $y = 3$, $z = 9$.

4. $e_1 = 4 + j$, $e_2 = 2 - j$, $e_3 = j4$.

5. $i_1 = 7x + 17y - z$, $i_2 = 2x + y + z$,
$i_3 = 3x - 4y + 16z$.

15·6. The matrix product

$$\overrightarrow{\begin{bmatrix} a & b & c \\ x & y & z \end{bmatrix}} \cdot \begin{bmatrix} l & p \\ m & q \\ n & r \end{bmatrix}\Bigg\downarrow = \begin{bmatrix} al + bm + cn & ap + bq + cr \\ xl + ym + zn & xp + yq + zr \end{bmatrix}. \quad (1)$$

$$[A] \quad \cdot \quad [B] \quad = \quad [C]$$

Consider the two matrices $[A]$ and $[B]$ shown above.

Take each element in the first *row* of $[A]$ and multiply it by the corresponding element in the first *column* of $[B]$; then add these products. The result is $al + bm + cn$.

It is called the *inner product* of row 1 of $[A]$ with column 1 of $[B]$. Now enter this inner product as the element in the first row, first column of a new matrix $[C]$.

Next form the inner product of row 1 of $[A]$ with column 2 of $[B]$. The result is $ap + bq + cr$. Enter this as the element in row 1, column 2 of $[C]$.

Next form the inner product of row 2 of $[A]$ with column 1 of $[B]$. The result is $xl + ym + zn$. Enter this as the element in row 2, column 1 of $[C]$.

Finally, form the inner product of row 2 of $[A]$ with column 2 of $[B]$. The result is $xp + yq + zr$. Enter this as the element in row 2, column 2 of $[C]$.

The new matrix [C] is defined to be the product [A] . [B], in that order.

Note the following points:

(i) the arrows drawn on the matrices remind one how the inner products are formed. If the student finds this helpful he is advised to draw the arrows until such time as the process of forming inner products becomes entirely automatic.

(ii) The number of *columns* in [A] was the same as the number of *rows* in [B]. If this had not been so, it would have been impossible to carry out in full the procedure described above. For example, consider

$$\underset{[A]}{\overrightarrow{\begin{bmatrix} a & b & c & d \\ x & y & z & w \end{bmatrix}}} \cdot \underset{[B]}{\begin{bmatrix} l & p \\ m & q \\ n & r \end{bmatrix}}\Big\downarrow .$$

Trying to form the inner product of row 1 of [A] with column 1 of [B], gives $al + bm + cn + d(?)$. There is no element to multiply d by unless [B] has a *fourth* row, which would be the same number as the number of *columns* of [A].

Before giving a more general definition of a matrix product it is informative to study the following example which uses double suffix notation:

$$\underset{[A]}{\overrightarrow{\begin{bmatrix} a_{11} & a_{12} & a_{13} \\ a_{21} & a_{22} & a_{23} \\ a_{31} & a_{32} & a_{33} \end{bmatrix}}} \cdot \underset{[B]}{\begin{bmatrix} b_{11} & b_{12} & b_{13} \\ b_{21} & b_{22} & b_{23} \\ b_{31} & b_{32} & b_{33} \end{bmatrix}}\Big\downarrow$$

$$= \underset{[C]}{\begin{bmatrix} a_{11}b_{11} + a_{12}b_{21} + a_{13}b_{31} & a_{11}b_{12} + a_{12}b_{22} + a_{13}b_{32} & a_{11}b_{13} + a_{12}b_{23} + a_{13}b_{33} \\ a_{21}b_{11} + a_{22}b_{21} + a_{23}b_{31} & a_{21}b_{12} + a_{22}b_{22} + a_{23}b_{32} & a_{21}b_{13} + a_{22}b_{23} + a_{23}b_{33} \\ a_{31}b_{11} + a_{32}b_{21} + a_{33}b_{31} & a_{31}b_{12} + a_{32}b_{22} + a_{33}b_{32} & a_{31}b_{13} + a_{32}b_{23} + a_{33}b_{33} \end{bmatrix}} . \qquad (2)$$

Notice that if a_{ij} represents the element in the ith row and jth column of $[A]$ and b_{ij} represents the element in the ith row and jth column of $[B]$, then the element in the ith row and jth column of $[C]$ is given by

$$c_{ij} = \sum_{r=1}^{r=3} a_{ir}b_{rj}.$$

The student should check this formula for every element of $[C]$, viz. $c_{11}, c_{12}, c_{13}, c_{21}, \ldots, c_{33}$. It is the basis of the definition of matrix multiplication to be found in more advanced and rigorous text-books.

Perhaps a more immediately helpful general definition is as follows:

(i) Two matrices, $[A]$ and $[B]$, can be multiplied to form the product $[A] . [B]$ if the number of columns in $[A]$ is the same as the number of rows in $[B]$. They are then said to be *conformable* for forming the product $[A] . [B]$.

(ii) If $[A]$ is a $(m \times n)$ matrix and $[B]$ is a $(n \times p)$ matrix, the product $[A] . [B]$ is a $(m \times p)$ matrix, $[C]$ say, in which the element in the rth row, sth column of $[C]$ is the inner product of the rth row of $[A]$ with the sth column of $[B]$.

It is essential that the student proceeds no further until he can form matrix products quickly and mechanically.

Worked examples

(1)
$$\begin{bmatrix} 3 & -1 \\ 2 & 3 \\ 1 & -2 \\ 0 & 4 \end{bmatrix} \cdot \begin{bmatrix} 4 & 1 & 3 \\ 2 & 0 & -1 \end{bmatrix}$$

$$= \begin{bmatrix} 3.4 + (-1).2 & 3.1 + (-1).0 & 3.3 + (-1)(-1) \\ 2.4 + 3.2 & 2.1 + 3.0 & 2.3 + 3.(-1) \\ 1.4 + (-2).2 & 1.1 + (-2).0 & 1.3 + (-2)(-1) \\ 0.4 + 4.2 & 0.1 + 4.0 & 0.3 + 4.(-1) \end{bmatrix}$$

$$= \begin{bmatrix} 10 & 3 & 10 \\ 14 & 2 & 3 \\ 0 & 1 & 5 \\ 8 & 0 & -4 \end{bmatrix}.$$

With practice the middle step is done mentally.

(2) If $[e] = \begin{bmatrix} e_1 \\ e_2 \end{bmatrix}$, $[z] = \begin{bmatrix} L_1D & MD \\ MD & L_2D \end{bmatrix}$, where $D \equiv d/dt$,

and $[i] = \begin{bmatrix} i_1 \\ i_2 \end{bmatrix}$, find $[z].[i]$. If $[e] = [z].[i]$, find e_1 and e_2.

$$[z].[i] = \overrightarrow{\begin{bmatrix} L_1D & MD \\ MD & L_2D \end{bmatrix}} . \begin{bmatrix} i_1 \\ i_2 \end{bmatrix}\Big\downarrow = \begin{bmatrix} L_1Di_1 + MDi_2 \\ MDi_1 + L_2Di_2 \end{bmatrix}$$

$$= \begin{bmatrix} L_1 \dfrac{di_1}{dt} + M \dfrac{di_2}{dt} \\[2ex] M \dfrac{di_1}{dt} + L_2 \dfrac{di_2}{dt} \end{bmatrix}.$$

If $[e] = [z].[i]$, then

$$\begin{bmatrix} e_1 \\ e_2 \end{bmatrix} = \begin{bmatrix} L_1 \dfrac{di_1}{dt} + M \dfrac{di_2}{dt} \\[2ex] M \dfrac{di_1}{dt} + L_2 \dfrac{di_2}{dt} \end{bmatrix}.$$

Thus
$$\underline{e_1 = L_1 \frac{di_1}{dt} + M \frac{di_2}{dt}}$$

and
$$\underline{e_2 = M \frac{di_1}{dt} + L_2 \frac{di_2}{dt}}.$$

The student will recognize these as the differential equations for a simple non-loss transformer.

(3) Find the product $[A] . [I]$ if

$$[A] = \begin{bmatrix} a_{11} & a_{12} & a_{13} \\ a_{21} & a_{22} & a_{23} \\ a_{31} & a_{32} & a_{33} \end{bmatrix} \quad \text{and} \quad I = \begin{bmatrix} 1 & 0 & 0 \\ 0 & 1 & 0 \\ 0 & 0 & 1 \end{bmatrix}.$$

$$[A].[I] = \begin{bmatrix} a_{11} & a_{12} & a_{13} \\ a_{21} & a_{22} & a_{23} \\ a_{31} & a_{32} & a_{33} \end{bmatrix} . \begin{bmatrix} 1 & 0 & 0 \\ 0 & 1 & 0 \\ 0 & 0 & 1 \end{bmatrix}$$

$$= \begin{bmatrix} a_{11}+0+0 & 0+a_{12}+0 & 0+0+a_{13} \\ a_{21}+0+0 & 0+a_{22}+0 & 0+0+a_{23} \\ a_{31}+0+0 & 0+a_{32}+0 & 0+0+a_{33} \end{bmatrix}$$

$$= \begin{bmatrix} a_{11} & a_{12} & a_{13} \\ a_{21} & a_{22} & a_{23} \\ a_{31} & a_{32} & a_{33} \end{bmatrix} = [A].$$

Such a matrix as I, in which all the elements are zero apart from the main diagonal consisting of unit elements, is called a *unit matrix* (see also § 15·62). It behaves in matrix multiplication in the same manner as unity in ordinary multiplication.

(4) Find the set of simultaneous linear equations which are given in matrix form by

$$\begin{bmatrix} 3 & -2 & 5 \\ 4 & 1 & -1 \\ 2 & -3 & -5 \end{bmatrix} . \begin{bmatrix} x \\ y \\ z \end{bmatrix} = \begin{bmatrix} 4 \\ 7 \\ -6 \end{bmatrix}.$$

Performing the multiplication gives

$$\begin{bmatrix} 3x - 2y + 5z \\ 4x + y - z \\ 2x - 3y - 5z \end{bmatrix} = \begin{bmatrix} 4 \\ 7 \\ -6 \end{bmatrix}.$$

Thus the equations are

$$3x - 2y + 5z = 4, \quad 4x + y - z = 7$$

and
$$2x - 3y - 5z = -6.$$

(5) Form the matrix equation which represents the simultaneous equations

$$a_{11}x + a_{12}y + a_{13}z = d_1,$$

$$a_{21}x + a_{22}y + a_{23}z = d_2,$$

$$a_{31}x + a_{32}y + a_{33}z = d_3.$$

On inspection of the result of Question 4 above, it is evident that the left-hand matrix is formed by the coefficients of x, y and z.

Thus $$\begin{bmatrix} a_{11} & a_{12} & a_{13} \\ a_{21} & a_{22} & a_{23} \\ a_{31} & a_{32} & a_{33} \end{bmatrix} . \begin{bmatrix} x \\ y \\ z \end{bmatrix} = \begin{bmatrix} d_1 \\ d_2 \\ d_3 \end{bmatrix}.$$

This can be checked by multiplying out the matrices.

This equation is often written briefly as

$$[a] . [x] = [d].$$

EXERCISE 48

1. Form the product $[a] . [b]$ when

(i) $[a] = \begin{bmatrix} 1 & 1 \\ 1 & 1 \end{bmatrix}$, $[b] = \begin{bmatrix} 1 & 3 \\ 2 & 4 \end{bmatrix}$;

(ii) $[a] = \begin{bmatrix} 1 & 3 \\ 3 & 2 \end{bmatrix}$, $[b] = \begin{bmatrix} 4 & 6 \\ 6 & 5 \end{bmatrix}$;

(iii) $[a] = \begin{bmatrix} 1 & 0 \\ 0 & 1 \end{bmatrix}$, $[b] = \begin{bmatrix} 0 & 2 \\ 2 & 0 \end{bmatrix}$.

2. When is it possible to form the product $[a] . [b]$ if

(i) $[a]$ has 2 rows, $[b]$ has 3 rows,

(ii) $[a]$ has 3 columns, $[b]$ has 5 columns,

(iii) $[a]$ has 3 rows, $[b]$ has 3 columns?

3. Form the product $[a] . [b]$ when

$$[a] = \begin{bmatrix} 1 & 2 & 3 \\ 4 & 5 & 6 \\ 2 & 1 & 3 \end{bmatrix}, \quad [b] = \begin{bmatrix} 2 & 1 \\ 4 & 3 \\ 6 & 5 \end{bmatrix}.$$

4. If $$\begin{bmatrix} e \\ 0 \end{bmatrix} = \begin{bmatrix} 1 & 0 & 1 \\ 0 & 1 & -1 \end{bmatrix} . \begin{bmatrix} e_a \\ e_b \\ e_c \end{bmatrix},$$

show that $$e = e_a + e_c \quad \text{and} \quad e_b = e_c.$$

5. Write down the matrix form of the equations

$$3x - 2y + z = 7, \quad x - 2y + 3z = 9, \quad 4x + 7y - 2z = -2.$$

6. Write down the equations which are represented by the matrix equation

$$\begin{bmatrix} 7 & 3 & -2 \\ 1 & 5 & 3 \\ -4 & 6 & 2 \end{bmatrix} \cdot \begin{bmatrix} x \\ y \\ z \end{bmatrix} = \begin{bmatrix} 2 \\ 19 \\ 7 \end{bmatrix}.$$

7. If
$$\begin{bmatrix} I_0 \\ I_\alpha \\ I_\beta \end{bmatrix} = \tfrac{1}{3}\begin{bmatrix} 1 & 1 & 1 \\ 2 & -1 & -1 \\ 0 & \tfrac{3}{2} & -\tfrac{3}{2} \end{bmatrix} \cdot \begin{bmatrix} I_a \\ I_b \\ I_c \end{bmatrix},$$
find I_0, I_α, I_β in terms of I_a, I_b, I_c.

8. Perform the matrix multiplication

$$\begin{bmatrix} 1+j & 2-j & 3+j2 \\ -2+j & 4+j3 & 5-j2 \end{bmatrix} \cdot \begin{bmatrix} 1-j & 2+j \\ 5-j3 & -2+j2 \\ 4+j3 & 1-j2 \end{bmatrix},$$

where $j = \sqrt{(-1)}$.

Answers

1. (i) $\begin{bmatrix} 3 & 7 \\ 3 & 7 \end{bmatrix}$. (ii) $\begin{bmatrix} 22 & 21 \\ 24 & 28 \end{bmatrix}$. (iii) $\begin{bmatrix} 0 & 2 \\ 2 & 0 \end{bmatrix}$.

2. (i) Only if $[a]$ has 3 columns.
(ii) Only if $[b]$ has 3 rows.
(iii) Only if $[b]$ has the same number of rows as $[a]$ has columns.

3.
$$\begin{bmatrix} 28 & 22 \\ 64 & 49 \\ 26 & 20 \end{bmatrix}.$$

5.
$$\begin{bmatrix} 3 & -2 & 1 \\ 1 & -2 & 3 \\ 4 & 7 & -2 \end{bmatrix} \cdot \begin{bmatrix} x \\ y \\ z \end{bmatrix} = \begin{bmatrix} 7 \\ 9 \\ -2 \end{bmatrix}.$$

6.
$$\begin{aligned} 7x + 3y - 2z &= 2, \\ x + 5y + 3z &= 19, \\ -4x + 6y + 2z &= 7. \end{aligned}$$

7. $I_0 = \frac{1}{3}(I_a + I_b + I_c), \quad I_\alpha = \frac{1}{3}(2I_a - I_b - I_c),$

$$I_\beta = \frac{1}{2}(I_b - I_c).$$

8. $\begin{bmatrix} 15 + j6 & 6 + j5 \\ 54 + j13 & -18 - j10 \end{bmatrix}.$

15·61. Post-multiplication and pre-multiplication

If $[A]$ is a (2×3) matrix and $[B]$ is a (3×4) matrix, then $[A]$ and $[B]$ are conformable for the product $[A].[B]$. However, according to the definition of the matrix product given in § 15·6, it is impossible to form the product $[B].[A]$, as the order of multiplying is then a (3×4) matrix with a (2×3) matrix. In this case $[B].[A]$ has no meaning. It is clear, therefore, that the commutative law, $ab = ba$, does *not* hold in general for matrix products. This is not as surprising as it may seem. The commutative law does not hold in vector products; that two successive operations obey the commutative law is in fact the exception rather than the rule. (For example, the son of Harry's mother is quite a different person from the mother of Harry's son! This could be written SM(H) $\neq$ MS(H).)

For the products $[A].[B]$ and $[B].[A]$ both to exist, if $[A]$ is a $(m \times n)$ matrix, $[B]$ must be a $(n \times m)$ matrix. Even then, in general, $[A].[B] \neq [B].[A]$.

For example, let

$$[A] = \begin{bmatrix} a_{11} & a_{12} \\ a_{21} & a_{22} \end{bmatrix} \quad \text{and} \quad [B] = \begin{bmatrix} b_{11} & b_{12} \\ b_{21} & b_{22} \end{bmatrix}.$$

Then $$[A].[B] = \begin{bmatrix} a_{11}b_{11} + a_{12}b_{21} & a_{11}b_{12} + a_{12}b_{22} \\ a_{21}b_{11} + a_{22}b_{21} & a_{21}b_{12} + a_{22}b_{22} \end{bmatrix},$$

but $$[B].[A] = \begin{bmatrix} b_{11} & b_{12} \\ b_{21} & b_{22} \end{bmatrix} . \begin{bmatrix} a_{11} & a_{12} \\ a_{21} & a_{22} \end{bmatrix}$$

$$= \begin{bmatrix} b_{11}a_{11} + b_{12}a_{21} & b_{11}a_{12} + b_{12}a_{22} \\ b_{21}a_{11} + b_{22}a_{21} & b_{21}a_{12} + b_{22}a_{22} \end{bmatrix}.$$

In general these two resulting matrices are quite different.

Great care must therefore be taken in the order in which any product is written. To emphasize this difference, in the product $[A].[B]$, $[A]$ is said to *pre-multiply* $[B]$, whilst in the product $[B].[A]$, $[A]$ is said to *post-multiply* $[B]$.

In special cases $[A].[B]$ does equal $[B].[A]$, but in this case $[A]$ and $[B]$ must satisfy special conditions. As these will not be needed in this book, they will not be investigated here.

EXAMPLES

(1) If $[A] = \begin{bmatrix} 2 & 1 \\ 4 & 3 \\ 6 & 5 \end{bmatrix}$ and $[B] = \begin{bmatrix} 1 & 2 & 3 \\ 4 & 5 & 6 \end{bmatrix}$, form the products $[A].[B]$ and $[B].[A]$, verifying that

$$[A].[B] \neq [B].[A].$$

$$[A].[B] = \begin{bmatrix} 2 & 1 \\ 4 & 3 \\ 6 & 5 \end{bmatrix} . \begin{bmatrix} 1 & 2 & 3 \\ 4 & 5 & 6 \end{bmatrix} = \begin{bmatrix} 6 & 9 & 12 \\ 16 & 23 & 30 \\ 26 & 37 & 48 \end{bmatrix}.$$

$$[B].[A] = \begin{bmatrix} 1 & 2 & 3 \\ 4 & 5 & 6 \end{bmatrix} . \begin{bmatrix} 2 & 1 \\ 4 & 3 \\ 6 & 5 \end{bmatrix} = \begin{bmatrix} 28 & 22 \\ 64 & 49 \end{bmatrix}.$$

Clearly $[A].[B] \neq [B].[A]$.

(2) If $[A] = \begin{bmatrix} 1 & 2 \\ 0 & 3 \end{bmatrix}$, $[B] = \begin{bmatrix} 1 & 0 & 5 \\ -2 & 3 & 0 \end{bmatrix}$,

$$[C] = \begin{bmatrix} 1 & 3 \\ -1 & -2 \\ 2 & 1 \end{bmatrix},$$

verify that $\{[A].[B]\}.[C] = [A].\{[B].[C]\}$.

This is an illustration of the fact that matrix products do obey the *associative* law of multiplication.

$$[A].[B] = \begin{bmatrix} -3 & 6 & 5 \\ -6 & 9 & 0 \end{bmatrix}.$$

Thus $$\{[A]\,.\,[B]\}\,.\,[C] = \begin{bmatrix} -3 & 6 & 5 \\ -6 & 9 & 0 \end{bmatrix} . \begin{bmatrix} 1 & 3 \\ -1 & -2 \\ 2 & 1 \end{bmatrix}$$

$$= \begin{bmatrix} 1 & -16 \\ -15 & -36 \end{bmatrix}.$$

$$[B]\,.\,[C] = \begin{bmatrix} 11 & 8 \\ -5 & -12 \end{bmatrix}.$$

Thus $$[A]\,.\,\{[B]\,.\,[C]\} = \begin{bmatrix} 1 & 2 \\ 0 & 3 \end{bmatrix} . \begin{bmatrix} 11 & 8 \\ -5 & -12 \end{bmatrix}$$

$$= \begin{bmatrix} 1 & -16 \\ -15 & -36 \end{bmatrix}.$$

Clearly $\{[A]\,.\,[B]\}\,.\,[C] = [A]\,.\,\{[B]\,.\,[C]\}.$

(3) If $$[A] = \begin{bmatrix} 1 & 0 \\ 0 & 2 \\ 1 & -3 \end{bmatrix}, \quad [B] = \begin{bmatrix} 2 & -1 \\ 3 & 0 \\ 1 & -2 \end{bmatrix}$$

$$\text{and } [C] = \begin{bmatrix} 1 & -1 & 2 \\ 2 & 1 & 3 \end{bmatrix},$$

verify that $[A].[C] - [B].[C] = \{[A] - [B]\}.[C]$.

This illustrates the fact that matrix products do obey the *distributive* law of multiplication.

$$[A].[C] = \begin{bmatrix} 1 & -1 & 2 \\ 4 & 2 & 6 \\ -5 & -4 & -7 \end{bmatrix},$$

$$[B].[C] = \begin{bmatrix} 0 & -3 & 1 \\ 3 & -3 & 6 \\ -3 & -3 & -4 \end{bmatrix}.$$

Therefore

$$[A].[C] - [B].[C] = \begin{bmatrix} 1 & 2 & 1 \\ 1 & 5 & 0 \\ -2 & -1 & -3 \end{bmatrix}. \qquad (1)$$

$$[A] - [B] = \begin{bmatrix} -1 & 1 \\ -3 & 2 \\ 0 & -1 \end{bmatrix},$$

therefore

$$\{[A] - [B]\}.[C] = \begin{bmatrix} -1 & 1 \\ -3 & 2 \\ 0 & -1 \end{bmatrix}.\begin{bmatrix} 1 & -1 & 2 \\ 2 & 1 & 3 \end{bmatrix}$$

$$= \begin{bmatrix} 1 & 2 & 1 \\ 1 & 5 & 0 \\ -2 & -1 & -3 \end{bmatrix}. \qquad (2)$$

Thus $\quad [A].[C] - [B].[C] = \{[A] - [B]\}.[C].$

15·62. Diagonal and unit matrices

A *diagonal* matrix is a square matrix, all the elements of which are zero except those in the leading diagonal.

For example $\quad \begin{bmatrix} 2 & 0 & 0 \\ 0 & 3 & 0 \\ 0 & 0 & -1 \end{bmatrix}.$

A *unit* matrix is a square matrix, of any order, which has unity as its leading diagonal and zero for all the other elements.

Thus $\quad [I] = \begin{bmatrix} 1 & 0 & 0 \\ 0 & 1 & 0 \\ 0 & 0 & 1 \end{bmatrix}$

is a unit matrix of order three.

If $[A]$ is any other square matrix of the same order as the unit matrix $[I]$, it is easily verified that

$$[A].[I] = [I].[A] = [A].$$

For example, take

$$[A] = \begin{bmatrix} a & b & c \\ d & e & f \\ g & h & i \end{bmatrix} \quad \text{and} \quad [I] = \begin{bmatrix} 1 & 0 & 0 \\ 0 & 1 & 0 \\ 0 & 0 & 1 \end{bmatrix},$$

then

$$[I].[A] = \begin{bmatrix} 1 & 0 & 0 \\ 0 & 1 & 0 \\ 0 & 0 & 1 \end{bmatrix} . \begin{bmatrix} a & b & c \\ d & e & f \\ g & h & i \end{bmatrix} = \begin{bmatrix} a & b & c \\ d & e & f \\ g & h & i \end{bmatrix} = [A]$$

and

$$[A].[I] = \begin{bmatrix} a & b & c \\ d & e & f \\ g & h & i \end{bmatrix} . \begin{bmatrix} 1 & 0 & 0 \\ 0 & 1 & 0 \\ 0 & 0 & 1 \end{bmatrix} = \begin{bmatrix} a & b & c \\ d & e & f \\ g & h & i \end{bmatrix} = [A].$$

It is also easily verified that

$$[I] = [I].[I] = [I].[I].[I] = \ldots.$$

Thus $[I]$ behaves in matrix multiplication the same way as unity does in arithmetical multiplication.

Exercise 49

1. Post-multiply $[B]$ by $[A]$ when

$$[A] = \begin{bmatrix} 1 & 2 \\ 3 & 4 \end{bmatrix}, \quad [B] = \begin{bmatrix} 4 & -1 \\ 2 & -3 \end{bmatrix}.$$

2. If possible, pre-multiply $[A]$ by $[B]$ in the following cases:

(i)

$$[A] = \begin{bmatrix} 1 & 2 & 3 \\ -1 & 2 & 0 \\ 4 & 0 & 1 \end{bmatrix}, \quad [B] = \begin{bmatrix} 1 & 2 & 3 \\ -2 & 4 & 1 \end{bmatrix}.$$

(ii)

$$[A] = \begin{bmatrix} 2 & 3 & 1 \\ 1 & 4 & 0 \\ -2 & 1 & -1 \end{bmatrix}, \quad [B] = \begin{bmatrix} 1 & 4 \\ 2 & 5 \\ 3 & 6 \end{bmatrix}.$$

3. If
$$[I] = \begin{bmatrix} 1 & 0 \\ 0 & 1 \end{bmatrix}, \quad [i] = \begin{bmatrix} \sqrt{-1} & 0 \\ 0 & -\sqrt{-1} \end{bmatrix},$$

$$[j] = \begin{bmatrix} 0 & 1 \\ -1 & 0 \end{bmatrix} \text{ and } [k] = \begin{bmatrix} 0 & \sqrt{-1} \\ \sqrt{-1} & 0 \end{bmatrix},$$

show that $[i].[j] = [k]$, $[j].[k] = [i]$, $[k].[i] = [j]$ and

$$[j].[i] = -[k], \quad [k].[j] = -[i], \quad [i].[k] = -[j].$$

4. With the same notation as in Question 3 above and taking $[A]^2$ as meaning $[A].[A]$, show that

$$[i]^2 = [j]^2 = [k]^2 = -[I].$$

5. If

$$[A] = \begin{bmatrix} a_1 & b_1 & c_1 \\ a_2 & b_2 & c_2 \\ a_3 & b_3 & c_3 \end{bmatrix} \text{ and } [B] = \begin{bmatrix} p & 0 & 0 \\ 0 & q & 0 \\ 0 & 0 & r \end{bmatrix},$$

evaluate the products $[A].[B]$ and $[B].[A]$.

6. Evaluate the continued product

$$\begin{bmatrix} 1 & 0 & 1 \\ 0 & 1 & -1 \end{bmatrix} \cdot \begin{bmatrix} Z_A & 0 & 0 \\ 0 & Z_B & 0 \\ 0 & 0 & Z_C \end{bmatrix} \cdot \begin{bmatrix} 1 & 0 \\ 0 & 1 \\ 1 & -1 \end{bmatrix}.$$

Answers

1. $\begin{bmatrix} 1 & 4 \\ -7 & -8 \end{bmatrix}.$

2. (i) $\begin{bmatrix} 11 & 6 & 6 \\ -2 & 4 & -5 \end{bmatrix}.$ (ii) Impossible.

5.
$$[A].[B] = \begin{bmatrix} pa_1 & qb_1 & rc_1 \\ pa_2 & qb_2 & rc_2 \\ pa_3 & qb_3 & rc_3 \end{bmatrix},$$

$$[B].[A] = \begin{bmatrix} pa_1 & pb_1 & pc_1 \\ qa_2 & qb_2 & qc_2 \\ ra_3 & rb_3 & rc_3 \end{bmatrix}.$$

(Note that post-multiplication by the diagonal matrix $[B]$ multiplies the *columns* of $[A]$ by p, q, r respectively, whilst pre-multiplication by $[B]$ multiplies the *rows* by p, q, r.)

6. $$\begin{bmatrix} Z_A + Z_C & -Z_C \\ -Z_C & Z_B + Z_C \end{bmatrix}.$$

15·7. The transpose of a matrix

If the rows and columns of a matrix are interchanged, the new matrix obtained is called the *transpose* of the original one.

Thus $\begin{bmatrix} 1 & 4 \\ 2 & 5 \\ 3 & 6 \end{bmatrix}$ has for its transpose $\begin{bmatrix} 1 & 2 & 3 \\ 4 & 5 & 6 \end{bmatrix}$.

If $[A]$ is any matrix, then the symbol for its transpose will be taken as $[A]_t$. Some books use the symbols $[\tilde{A}]$ or $[A]'$.

If $[A] = [A]_t$, the matrix is said to be *symmetrical*.

Obviously this can only occur if $[A]$ is a square matrix.

The symmetry necessary is shown by the following example:

$$[A] = \begin{bmatrix} a & h & g \\ h & b & f \\ g & f & c \end{bmatrix}, \quad [A]_t = \begin{bmatrix} a & h & g \\ h & b & f \\ g & f & c \end{bmatrix},$$

i.e. there must be symmetry about the principal diagonal. Clearly, the transpose of a transposed matrix is the original matrix, e.g.

$$[A] = \begin{bmatrix} a_1 & b_1 & c_1 \\ a_2 & b_2 & c_2 \end{bmatrix}, \quad [A]_t = \begin{bmatrix} a_1 & a_2 \\ b_1 & b_2 \\ c_1 & c_2 \end{bmatrix},$$

whilst $$\{[A]_t\}_t = \begin{bmatrix} a_1 & b_1 & c_1 \\ a_2 & b_2 & c_2 \end{bmatrix} = [A].$$

15·71. The determinant of a square matrix

The determinant of a square matrix is that determinant which has elements exactly corresponding to the elements of the matrix. Clearly, a rectangular matrix cannot have a corresponding determinant.

EXAMPLE

The determinant of $\begin{bmatrix} 1 & 2 & 3 \\ 0 & 2 & 1 \\ 3 & -1 & 0 \end{bmatrix}$ is $\begin{vmatrix} 1 & 2 & 3 \\ 0 & 2 & 1 \\ 3 & -1 & 0 \end{vmatrix} = -11$.

The determinant of the transpose of the matrix is

$\begin{vmatrix} 1 & 0 & 3 \\ 2 & 2 & -1 \\ 3 & 1 & 0 \end{vmatrix}$; that is, its rows and columns are interchanged.

Its value, of course, will still be -11.

EXERCISE 50

1. Write down the transposes of the following matrices:

(i) $\begin{bmatrix} 1 & 2 & 3 \\ -2 & 4 & 0 \\ 5 & 7 & -1 \end{bmatrix}$. (ii) $\begin{bmatrix} x & y & z \\ l & m & n \end{bmatrix}$. (iii) $\begin{bmatrix} 1 & 0 & 0 \\ 0 & 1 & 0 \\ 0 & 0 & 1 \end{bmatrix}$.

2. Write down a general diagonal matrix and show that it is equal to its own transpose.

3. If $[A] = \begin{bmatrix} 1 & 0 & 0 \\ 0 & 1 & -1 \end{bmatrix}$ and $[Z] = \begin{bmatrix} Z_1 & Z_2 \\ Z_3 & Z_4 \end{bmatrix}$,

evaluate the matrix product $[A]_t . [Z]$.

4. Write down the determinant of the matrix $\begin{bmatrix} 1 & -1 & 2 \\ 2 & -3 & 4 \\ 7 & 0 & 1 \end{bmatrix}$

and evaluate this determinant.

5. Write down the cofactor of the element in the 2nd row, 3rd column, of the determinant of the transpose of the matrix

$$\begin{bmatrix} a_1 & b_1 & c_1 \\ a_2 & b_2 & c_2 \\ a_3 & b_3 & c_3 \end{bmatrix}.$$

Evaluate this cofactor.

6. If

$$[C] = \begin{bmatrix} 1 & 0 & 0 \\ -1 & 1 & 0 \\ 0 & -1 & 1 \\ 0 & 0 & -1 \end{bmatrix} \quad \text{and} \quad [Z] = \begin{bmatrix} Z_A & 0 & 0 & 0 \\ 0 & Z_B & 0 & 0 \\ 0 & 0 & Z_C & 0 \\ 0 & 0 & 0 & Z_D \end{bmatrix},$$

find $[Z']$, where $[Z'] = [C]_t.[Z].[C]$.

Answers

1. (i) $\begin{bmatrix} 1 & -2 & 5 \\ 2 & 4 & 7 \\ 3 & 0 & -1 \end{bmatrix}$. (ii) $\begin{bmatrix} x & l \\ y & m \\ z & n \end{bmatrix}$. (iii) $\begin{bmatrix} 1 & 0 & 0 \\ 0 & 1 & 0 \\ 0 & 0 & 1 \end{bmatrix}$.

3. $\begin{bmatrix} Z_1 & Z_2 \\ Z_3 & Z_4 \\ -Z_3 & -Z_4 \end{bmatrix}$.

4. $\begin{vmatrix} 1 & -1 & 2 \\ 2 & -3 & 4 \\ 7 & 0 & 1 \end{vmatrix} = 13.$

5. $-\begin{vmatrix} a_1 & a_2 \\ c_1 & c_2 \end{vmatrix} = a_2c_1 - a_1c_2.$

6. $\begin{bmatrix} Z_A + Z_B & -Z_B & 0 \\ -Z_B & Z_B + Z_C & -Z_C \\ 0 & -Z_C & Z_C + Z_D \end{bmatrix}$.

15·8. The inverse of a square matrix and the solution of linear equations

Consider the set of simultaneous linear equations

$$\left.\begin{aligned} a_1x_1 + b_1x_2 + c_1x_3 &= d_1, \\ a_2x_1 + b_2x_2 + c_2x_3 &= d_2, \\ a_3x_1 + b_3x_2 + c_3x_3 &= d_3. \end{aligned}\right\} \tag{1}$$

These may be written in matrix form as

$$\begin{bmatrix} a_1 & b_1 & c_1 \\ a_2 & b_2 & c_2 \\ a_3 & b_3 & c_3 \end{bmatrix} \cdot \begin{bmatrix} x_1 \\ x_2 \\ x_3 \end{bmatrix} = \begin{bmatrix} d_1 \\ d_2 \\ d_3 \end{bmatrix}, \tag{2}$$

or more simply as

$$[a].[x] = [d]. \tag{3}$$

From § 13·5, the solutions to the equations in (1) are:

$$x_1 = \frac{\begin{vmatrix} d_1 & b_1 & c_1 \\ d_2 & b_2 & c_2 \\ d_3 & b_3 & c_3 \end{vmatrix}}{\Delta}, \quad x_2 = \frac{\begin{vmatrix} a_1 & d_1 & c_1 \\ a_2 & d_2 & c_2 \\ a_3 & d_3 & c_3 \end{vmatrix}}{\Delta}, \quad x_3 = \frac{\begin{vmatrix} a_1 & b_1 & d_1 \\ a_2 & b_2 & d_2 \\ a_3 & b_3 & d_3 \end{vmatrix}}{\Delta}, \tag{4}$$

where

$$\Delta = \begin{vmatrix} a_1 & b_1 & c_1 \\ a_2 & b_2 & c_2 \\ a_3 & b_3 & c_3 \end{vmatrix}. \tag{5}$$

Thus Δ is the determinant of the matrix $[a]$.

On partially expanding the determinants in (4), the solutions can be written

$$\left.\begin{aligned}
x_1 &= \frac{d_1\begin{vmatrix} b_2 & c_2 \\ b_3 & c_3 \end{vmatrix}}{\Delta} - \frac{d_2\begin{vmatrix} b_1 & c_1 \\ b_3 & c_3 \end{vmatrix}}{\Delta} + \frac{d_3\begin{vmatrix} b_1 & c_1 \\ b_2 & c_2 \end{vmatrix}}{\Delta}, \\
x_2 &= -\frac{d_1\begin{vmatrix} a_2 & c_2 \\ a_3 & c_3 \end{vmatrix}}{\Delta} + \frac{d_2\begin{vmatrix} a_1 & c_1 \\ a_3 & c_3 \end{vmatrix}}{\Delta} - \frac{d_3\begin{vmatrix} a_1 & c_1 \\ a_2 & c_2 \end{vmatrix}}{\Delta}, \\
x_3 &= \frac{d_1\begin{vmatrix} a_2 & b_2 \\ a_3 & b_3 \end{vmatrix}}{\Delta} - \frac{d_2\begin{vmatrix} a_1 & b_1 \\ a_3 & b_3 \end{vmatrix}}{\Delta} + \frac{d_3\begin{vmatrix} a_1 & b_1 \\ a_2 & b_2 \end{vmatrix}}{\Delta}.
\end{aligned}\right\} \tag{6}$$

Now consider the determinant of the *transpose* of $[a]$, that is,

$$\Delta_t = \begin{vmatrix} a_1 & a_2 & a_3 \\ b_1 & b_2 & b_3 \\ c_1 & c_2 & c_3 \end{vmatrix} \quad (= \Delta, \text{ numerically}).$$

On inspection of the solutions in (6), it will be seen that, if A_1 is the cofactor of a_1 in Δ_t, B_1 the cofactor of b_1, C_1 the cofactor of c_1, and so on, then the solutions in (6) can be written (see § 13·31),

$$x_1 = \frac{A_1\,d_1}{\Delta} + \frac{A_2\,d_2}{\Delta} + \frac{A_3\,d_3}{\Delta},$$

$$x_2 = \frac{B_1\,d_1}{\Delta} + \frac{B_2\,d_2}{\Delta} + \frac{B_3\,d_3}{\Delta},$$

$$x_3 = \frac{C_1\,d_1}{\Delta} + \frac{C_2\,d_2}{\Delta} + \frac{C_3\,d_3}{\Delta}.$$

In matrix form, therefore, the solutions are

$$\begin{bmatrix} x_1 \\ x_2 \\ x_3 \end{bmatrix} = \begin{bmatrix} \frac{A_1}{\Delta} & \frac{A_2}{\Delta} & \frac{A_3}{\Delta} \\ \frac{B_1}{\Delta} & \frac{B_2}{\Delta} & \frac{B_3}{\Delta} \\ \frac{C_1}{\Delta} & \frac{C_2}{\Delta} & \frac{C_3}{\Delta} \end{bmatrix} . \begin{bmatrix} d_1 \\ d_2 \\ d_3 \end{bmatrix}. \tag{7}$$

This may be written shortly

$$[x] = [a]^{-1}.[d], \tag{8}$$

where

$$[a]^{-1} = \begin{bmatrix} \frac{A_1}{\Delta} & \frac{A_2}{\Delta} & \frac{A_3}{\Delta} \\ \frac{B_1}{\Delta} & \frac{B_2}{\Delta} & \frac{B_3}{\Delta} \\ \frac{C_1}{\Delta} & \frac{C_2}{\Delta} & \frac{C_3}{\Delta} \end{bmatrix} = \frac{1}{\Delta} \begin{bmatrix} A_1 & A_2 & A_3 \\ B_1 & B_2 & B_3 \\ C_1 & C_2 & C_3 \end{bmatrix}. \tag{9}$$

The square matrix $[a]^{-1}$ is called the *inverse* of the square matrix $[a]$.

The inverse of any square matrix is built up in the same way, viz:

(i) Write down the transpose of the given square matrix.

(ii) Form a matrix whose elements are the corresponding cofactors of the elements in this transpose. This is called the *adjoint* of the matrix.

(iii) Divide each of these elements by the determinant of the original matrix.

The matrix thus obtained is the *inverse* of the original one. (Note that an inverse is impossible if $\Delta = 0$.)

On inspection of equations (3) and (8) above, the solution of a set of simultaneous linear equations can now be written very briefly in matrix form. Thus, if

$$\left.\begin{aligned}[a].[x] &= [d],\\ \text{then}\qquad [x] &= [a]^{-1}.[d].\end{aligned}\right\} \tag{10}$$

The solution is then completed apart from arithmetical manipulations.

The following examples should be studied carefully.

The written explanations can be cut down when the method has been grasped.

EXAMPLE 1

Using the matrix method, solve the equations

$$3i_1 - 4i_2 + i_3 = -2,$$

$$2i_1 + 3i_2 - 2i_3 = 2,$$

$$i_1 + 5i_2 - i_3 = 8.$$

In matrix form the equations are

$$\begin{bmatrix} 3 & -4 & 1 \\ 2 & 3 & -2 \\ 1 & 5 & -1 \end{bmatrix} . \begin{bmatrix} i_1 \\ i_2 \\ i_3 \end{bmatrix} = \begin{bmatrix} -2 \\ 2 \\ 8 \end{bmatrix}.$$

Writing this as $$[C].[i] = [d], \tag{1}$$

where

$$[C] = \begin{bmatrix} 3 & -4 & 1 \\ 2 & 3 & -2 \\ 1 & 5 & -1 \end{bmatrix}, \quad [i] = \begin{bmatrix} i_1 \\ i_2 \\ i_3 \end{bmatrix} \quad \text{and} \quad [d] = \begin{bmatrix} -2 \\ 2 \\ 8 \end{bmatrix}, \tag{2}$$

then the solution is given by

$$[i] = [C]^{-1}.[d]. \qquad (3)$$

The determinant of $[C]$ is given by

$$\Delta = \begin{vmatrix} 3 & -4 & 1 \\ 2 & 3 & -2 \\ 1 & 5 & -1 \end{vmatrix} = 28. \qquad (4)$$

The transpose of $[C]$ is given by

$$[C]_t = \begin{bmatrix} 3 & 2 & 1 \\ -4 & 3 & 5 \\ 1 & -2 & -1 \end{bmatrix}.$$

Thus the inverse of $[C]$ is given by (using (9) above)

$$[C]^{-1} = \frac{1}{\Delta}\begin{bmatrix} 7 & 1 & 5 \\ 0 & -4 & 8 \\ 7 & -19 & 17 \end{bmatrix} = \frac{1}{28}\begin{bmatrix} 7 & 1 & 5 \\ 0 & -4 & 8 \\ 7 & -19 & 17 \end{bmatrix}. \qquad (5)$$

From (2), (3) and (5):

$$\begin{bmatrix} i_1 \\ i_2 \\ i_3 \end{bmatrix} = \frac{1}{28}\begin{bmatrix} 7 & 1 & 5 \\ 0 & -4 & 8 \\ 7 & -19 & 17 \end{bmatrix} . \begin{bmatrix} -2 \\ 2 \\ 8 \end{bmatrix}.$$

Performing the matrix multiplication:

$$\begin{bmatrix} i_1 \\ i_2 \\ i_3 \end{bmatrix} = \frac{1}{28}\begin{bmatrix} 28 \\ 56 \\ 84 \end{bmatrix}.$$

Thus

$$\left.\begin{aligned} i_1 &= \tfrac{28}{28} = 1, \\ i_2 &= \tfrac{56}{28} = 2, \\ i_3 &= \tfrac{84}{28} = 3. \end{aligned}\right\}$$

It is *not* claimed that matrix methods shorten any of the arithmetic involved in solving simultaneous linear equations, but the symbolism is cut down.

Example 2

In a Wheatstone Bridge network the following equations arise, where i_g is the current through the galvanometer:

$$R_2 i_1 - R_1 i_2 + R_g i_g = 0,$$
$$R_4(i_1 - i_g) - R_3(i_2 + i_g) - R_g i_g = 0,$$
$$R_2 i_1 + R_4(i_1 - i_g) = e.$$

Find the value of i_g and deduce the conditions for balance.

On rearranging, the equations are:

$$R_2 i_1 - R_1 i_2 + R_g i_g = 0,$$
$$R_4 i_1 - R_3 i_2 - (R_3 + R_4 + R_g) i_g = 0,$$
$$(R_2 + R_4) i_1 - R_4 i_g = e.$$

In matrix form, these are:

$$\begin{bmatrix} R_2 & -R_1 & R_g \\ R_4 & -R_3 & -(R_3 + R_4 + R_g) \\ R_2 + R_4 & 0 & -R_4 \end{bmatrix} . \begin{bmatrix} i_1 \\ i_2 \\ i_g \end{bmatrix} = \begin{bmatrix} 0 \\ 0 \\ e \end{bmatrix}. \qquad (1)$$

This can be written briefly as $[Z].[i] = [e]$.

Let Δ be the determinant of $[Z]$.

Then

$$\Delta_t = \begin{vmatrix} R_2 & R_4 & R_2 + R_4 \\ -R_1 & -R_3 & 0 \\ R_g & -(R_3 + R_4 + R_g) & -R_4 \end{vmatrix}.$$

Thus

$$[Z]^{-1} = \frac{1}{\Delta}\begin{bmatrix} \cdots & \cdots & \cdots \\ \cdots & \cdots & \cdots \\ R_3(R_2 + R_4) & -R_1(R_2 + R_4) & R_1R_4 - R_2R_3 \end{bmatrix}.$$

Note that only the third row of this matrix has been evaluated, as i_g is the only current required.

Now $$[i] = [Z]^{-1}.[e],$$

i.e. $$\begin{bmatrix} i_1 \\ i_2 \\ i_g \end{bmatrix}$$

$$= \frac{1}{\Delta}\begin{bmatrix} \cdots & \cdots & \cdots \\ \cdots & \cdots & \cdots \\ R_3(R_2 + R_4) & -R_1(R_2 + R_4) & R_1R_4 - R_2R_3 \end{bmatrix} . \begin{bmatrix} 0 \\ 0 \\ e \end{bmatrix}.$$

Performing the multiplication to give i_g:

$$\underline{i_g = e(R_1R_4 - R_2R_3)/\Delta.}$$

For a balance, $i_g = 0$.
Thus, for a balance, $R_1R_4 - R_2R_3 = 0$, i.e.

$$\underline{\frac{R_1}{R_2} = \frac{R_3}{R_4}.}$$

Exercise 51

1. Form the inverses of the following matrices:

(i) $\begin{bmatrix} 1 & 3 \\ 2 & 4 \end{bmatrix}$. (ii) $\begin{bmatrix} a & b \\ c & d \end{bmatrix}$.

(iii) $\begin{bmatrix} 1 & 0 & 2 \\ 3 & 4 & 0 \\ 6 & -2 & 1 \end{bmatrix}$. (iv) $\begin{bmatrix} 5 & 2 & 4 \\ 3 & -1 & 2 \\ 1 & 4 & -3 \end{bmatrix}$.

2. An electrical network gives rise to the equations

$$e_1 = Z_1i_1 + Z_2i_2,$$
$$e_2 = Z_3i_1 + Z_4i_2.$$

Put these in matrix form and solve for i_1, i_2 using matrix methods.

3. Find the inverses of the following matrices:

(i) $\begin{bmatrix} 1 & 0 & 0 \\ 0 & 1 & 0 \\ 0 & 0 & 1 \end{bmatrix}$. (ii) $\begin{bmatrix} 0 & 0 & 1 \\ 1 & 0 & 0 \\ 0 & 1 & 0 \end{bmatrix}$. (iii) $\begin{bmatrix} 1 & 1 & 0 \\ 1 & -\frac{1}{2} & 1 \\ 1 & -\frac{1}{2} & -1 \end{bmatrix}$.

4. Solve the equations of Questions 1, 2 and 3 of Exercise 42, using matrix methods.

5. Write down, in matrix form, the equations

$$I_a = I_0 + I_1 + I_2,$$
$$I_b = I_0 + h^2I_1 + hI_2,$$
$$I_c = I_0 + hI_1 + h^2I_2,$$

where $h = e^{j(\frac{2}{3}\pi)}$ (i.e. $h^3 = 1$), and solve for I_0, I_1 and I_2.

Answers

1. (i) $$\frac{1}{-2}\begin{bmatrix} 4 & -3 \\ -2 & 1 \end{bmatrix} = \begin{bmatrix} -2 & \frac{3}{2} \\ 1 & -\frac{1}{2} \end{bmatrix}.$$

(ii) $$\frac{1}{(ad-bc)}\begin{bmatrix} d & -b \\ -c & a \end{bmatrix}.$$ This can be used as a formula for the inverse of a square matrix of 2nd order.

(iii) $$-\tfrac{1}{56}\begin{bmatrix} 4 & -4 & -8 \\ -3 & -11 & 6 \\ -30 & 2 & 4 \end{bmatrix}.$$

(iv) $$\tfrac{1}{49}\begin{bmatrix} -5 & 22 & 8 \\ 11 & -19 & 2 \\ 13 & -18 & -11 \end{bmatrix}.$$

2. $$i_1 = \frac{1}{(Z_1Z_4 - Z_2Z_3)}(Z_4e_1 - Z_2e_2),$$

$$i_2 = \frac{1}{(Z_1Z_4 - Z_2Z_3)}(Z_1e_2 - Z_3e_1).$$

3. (i) $\begin{bmatrix} 1 & 0 & 0 \\ 0 & 1 & 0 \\ 0 & 0 & 1 \end{bmatrix}$. (ii) $\begin{bmatrix} 0 & 1 & 0 \\ 0 & 0 & 1 \\ 1 & 0 & 0 \end{bmatrix}$. (iii) $\tfrac{1}{3}\begin{bmatrix} 1 & 1 & 1 \\ 2 & -1 & -1 \\ 0 & \frac{3}{2} & -\frac{3}{2} \end{bmatrix}$.

4. As for Exercise 42, Answers 1, 2 and 3.

5. $I_0 = \tfrac{1}{3}(I_a + I_b + I_c)$, $I_1 = \tfrac{1}{3}(I_a + hI_b + h^2I_c)$,
$I_2 = \tfrac{1}{3}(I_a + h^2I_b + hI_c)$.

(*Note.* These are the symmetrical, 3-phase components in terms of the line currents.)

15·81. Singular matrices

If the determinant, Δ, of a square matrix is zero, the matrix is called *singular*. It is obvious that in this case the matrix has no inverse (see § 15·8 (9)).

Referring to the method of solution of simultaneous equations given in § 15·8, it is seen that if the determinant Δ is zero, then the process of solution breaks down. In practice this means that the equations are *not* independent. One, or more, can be obtained from the others. For example, in the case of the three equations

$$2x - 4y + 7z = 5, \quad x + 2y - 3z = 2$$

and

$$x - 6y + 10z = 3,$$

the last equation is merely the difference of the first two equations. Thus in reality there are only two equations given, from which it is impossible to solve for specific values of x, y and z.

Any rectangular matrix is also called *singular*.

The reason for this can be demonstrated by once more referring to a set of linear relationships.

Consider the relationships

$$\left.\begin{aligned} x_1' &= a_1x_1 + b_1x_2, \\ x_2' &= a_2x_1 + b_2x_2, \\ x_3' &= a_3x_1 + b_3x_2. \end{aligned}\right\} \tag{1}$$

There are *three* variables, x_1', x_2', x_3', given as linear expressions in *two* variables x_1, x_2.

(1) may be written in matrix form:

$$[x'] = [C].[x], \tag{2}$$

where

$$[x'] = \begin{bmatrix} x_1' \\ x_2' \\ x_3' \end{bmatrix}, \quad [C] = \begin{bmatrix} a_1 & b_1 \\ a_2 & b_2 \\ a_3 & b_3 \end{bmatrix} \quad \text{and} \quad [x] = \begin{bmatrix} x_1 \\ x_2 \end{bmatrix}. \tag{3}$$

Now considered as equations in *three* variables, equations (1) may be written

$$\begin{aligned} x_1' &= a_1x_1 + b_1x_2 + 0.x_3, \\ x_2' &= a_2x_1 + b_2x_2 + 0.x_3, \\ x_3' &= a_3x_1 + b_3x_2 + 0.x_3. \end{aligned}$$

In theory, these can be solved for x_1, x_2, x_3 by the method of § 15·8, and may be written

$$\begin{bmatrix} x_1' \\ x_2' \\ x_3' \end{bmatrix} = [M] \, . \begin{bmatrix} x_1 \\ x_2 \\ x_3 \end{bmatrix},$$

where

$$[M] = \begin{bmatrix} a_1 & b_1 & 0 \\ a_2 & b_2 & 0 \\ a_3 & b_3 & 0 \end{bmatrix}. \qquad (4)$$

Thus, effectively, $\begin{bmatrix} a_1 & b_1 \\ a_2 & b_2 \\ a_3 & b_3 \end{bmatrix}$ is equivalent to $\begin{bmatrix} a_1 & b_1 & 0 \\ a_2 & b_2 & 0 \\ a_3 & b_3 & 0 \end{bmatrix}$.

The latter matrix has zero as its determinant and is therefore singular; thus, also, the rectangular matrix is called singular. In the theory of linear equations rectangular matrices may be taken as equivalent to square matrices, formed by the inclusion of rows (or columns) of zeros.

Summary. (i) A non-singular square matrix has one inverse which is given by equation (9) of § 15·8.

(ii) A singular matrix, square or rectangular, has no inverse.

15·82. The product of a matrix and its inverse

Let $[a] = \begin{bmatrix} a_1 & b_1 & c_1 \\ a_2 & b_2 & c_2 \\ a_3 & b_3 & c_3 \end{bmatrix}$ be a non-singular matrix.

Then $[a]_t = \begin{bmatrix} a_1 & a_2 & a_3 \\ b_1 & b_2 & b_3 \\ c_1 & c_2 & c_3 \end{bmatrix}$ and $[a]^{-1} = \dfrac{1}{\Delta}\begin{bmatrix} A_1 & A_2 & A_3 \\ B_1 & B_2 & B_3 \\ C_1 & C_2 & C_3 \end{bmatrix}$, where A_1, B_1, C_1, ..., are the cofactors of a_1, b_1, c_1, ... in $[a]_t$ and Δ is the determinant of $[a]$. Thus

$$[a].[a]^{-1} = \frac{1}{\Delta}\begin{bmatrix} a_1 & b_1 & c_1 \\ a_2 & b_2 & c_2 \\ a_3 & b_3 & c_3 \end{bmatrix} . \begin{bmatrix} A_1 & A_2 & A_3 \\ B_1 & B_2 & B_3 \\ C_1 & C_2 & C_3 \end{bmatrix}$$

$$= \frac{1}{\Delta}\begin{bmatrix} a_1A_1+b_1B_1+c_1C_1 & a_1A_2+b_1B_2+c_1C_2 & a_1A_3+b_1B_3+c_1C_3 \\ a_2A_1+b_2B_1+c_2C_1 & a_2A_2+b_2B_2+c_2C_2 & a_2A_3+b_2B_3+c_2C_3 \\ a_3A_1+b_3B_1+c_3C_1 & a_3A_2+b_3B_2+c_3C_2 & a_3A_3+b_3B_3+c_3C_3 \end{bmatrix}.$$

Using the properties of cofactors of a determinant given in § 13·4, VII,

$$[a].[a]^{-1} = \frac{1}{\Delta}\begin{bmatrix} \Delta & 0 & 0 \\ 0 & \Delta & 0 \\ 0 & 0 & \Delta \end{bmatrix}$$

$$= \begin{bmatrix} 1 & 0 & 0 \\ 0 & 1 & 0 \\ 0 & 0 & 1 \end{bmatrix} = [I],$$

a unit matrix of the same order as $[a]$.

Similarly $[a]^{-1}.[a] = [I]$.

This result can easily be extended to square matrices of any order.

It follows that: *The product of a non-singular matrix with its own inverse is a unit matrix of the same order.*

This is often used to define the inverse of a square matrix.

15·9. Cancellation of common factors in products of matrices with constant elements

Cancellation of a common factor in a matrix equation has to be done very carefully. This is because of the non-commutative property of the general matrix product and also because of the fact that a singular matrix has no inverse. The general idea behind cancellation of matrices is to multiply both sides of the equation by the inverse of the matrix to be cancelled, if this procedure is possible.

Only matrices whose elements are constants will be considered in this paragraph. The following examples bring out the main points:

(*a*) Let $[A].[C] = [B].[C]$, where $[C]$ is a *non-singular* matrix (i.e. $[C]$ is a square matrix which has an inverse).

Post-multiply both sides of the equation by $[C]^{-1}$.

Then $$[A].[C].[C]^{-1} = [B].[C].[C]^{-1}.$$

Thus $$[A].[I] = [B].[I] \quad \text{(see § 15·82)},$$

giving $$[A] = [B].$$

(*b*) Let $[A].[C] = [C].[B]$, where $[C]$ is a non-singular matrix.

Post-multiplying by $[C]^{-1}$:

$$[A].[C].[C]^{-1} = [C].[B].[C]^{-1}.$$

Thus $$[A].[I] = [C].[B].[C]^{-1}.$$

Now in general $[C].[B].[C]^{-1} \neq [C].[C]^{-1}.[B]$

(non-commutative property), therefore in general $[A] \neq [B]$.

(*c*) If the common factor $[C]$ is a singular matrix, $[C]^{-1}$ does not exist, thus if $[A].[C] = [B].[C]$, where $[C]$ is a singular matrix, $[C]$ can *never* be cancelled.

The following rule can now be stated:

A common factor $[C]$ *can be cancelled as in ordinary algebra, provided* $[C]$ *is a non-singular matrix and occupies the same relative position, at one of the ends, in both products.*

15·91. Cancellation of a matrix which has variable elements

Let $[A].[C] \equiv [B].[C]$, where $[A]$, $[B]$ have constant elements and $[C]$ has elements $x_1, x_2, \ldots$ which can assume any values. Taking the equation as an *identity* (i.e. it is true for *all* values of $x_1, x_2, \ldots$) and writing the equation as $([A] - [B]).[C] \equiv [0]$; then the product on the left-hand side must give a null matrix. As this must be true for all values of $x_1, x_2, \ldots$ in $[C]$, it follows that $[A] - [B] = [0]$, that is, $[A] = [B]$.

Similarly, if $[C].[A] = [C].[B]$, where $[A]$, $[B]$ have constant elements and $[C]$ has variable elements, it follows that $[A] = [B]$.

Thus a *common factor* $[C]$, *all the elements of which are variables, can always be cancelled* (*whether* $[C]$ *is singular or not*) *if* $[C]$ *has the same relative position, at one of the ends, in the products.*

15·92. The reversal rule for transposed matrices

The following rule, which is often useful, can be proved rigorously, but here is illustrated by an actual example.

If a matrix $[C]$ *is a product of two matrices* $[A]$ *and* $[B]$, *then the transpose of* $[C]$ *is equal to the product of the transposes of* $[A]$ *and* $[B]$ *in reverse order*, i.e. if

$$[C] = [A].[B],$$

then
$$[C]_t = [B]_t.[A]_t.$$

Let $$[A] = \begin{bmatrix} a_1 & b_1 \\ a_2 & b_2 \end{bmatrix} \quad \text{and} \quad [B] = \begin{bmatrix} l_1 \\ m_1 \end{bmatrix}.$$

If $[C] = [A].[B]$, then $$[C] = \begin{bmatrix} a_1 l_1 + b_1 m_1 \\ a_2 l_1 + b_2 m_1 \end{bmatrix}.$$

Now $$[A]_t = \begin{bmatrix} a_1 & a_2 \\ b_1 & b_2 \end{bmatrix}, \quad [B]_t = [l_1 \quad m_1]$$

and $$[C]_t = [a_1 l_1 + b_1 m_1 \quad a_2 l_1 + b_2 m_1].$$

Thus $$[B]_t.[A]_t = [l_1 \quad m_1].\begin{bmatrix} a_1 & a_2 \\ b_1 & b_2 \end{bmatrix}$$
$$= [a_1 l_1 + b_1 m_1 \quad a_2 l_1 + b_2 m_1] = [C]_t.$$

Exercise 52

1. Which of the following matrices are singular?

(i) $\begin{bmatrix} 3 & 2 \\ 1 & 4 \\ 5 & 3 \end{bmatrix}$. (ii) $\begin{bmatrix} 5 & 3 & 1 \\ 10 & 6 & 2 \\ 4 & 0 & 1 \end{bmatrix}$. (iii) $\begin{bmatrix} 5 & 1 & 4 \\ 2 & 0 & 3 \\ 1 & -1 & 2 \end{bmatrix}$.

(iv) $\begin{bmatrix} 6 & 7 & 2 \\ 3 & -1 & 4 \\ 0 & 2 & 3 \end{bmatrix}$. (v) $\begin{bmatrix} 4 & 1 & 0 \\ 2 & 3 & 7 \end{bmatrix}$. (vi) $\begin{bmatrix} 1 & 2 & 3 \\ 4 & 3 & 1 \\ 6 & 7 & 7 \end{bmatrix}$.

2. If $[C] = [A].[B]$, form the matrix $[C]_t$ in the following cases, using the reversal rule:

(i) $[A] = \begin{bmatrix} 1 & 2 \\ 3 & 4 \end{bmatrix}, \quad [B] = \begin{bmatrix} 3 & 2 & 1 \\ -1 & -2 & 3 \end{bmatrix}.$

(ii) $[A] = \begin{bmatrix} 1 & -1 & 2 \\ 2 & 3 & -4 \end{bmatrix}, \quad [B] = \begin{bmatrix} 5 & 7 \\ 2 & -1 \\ 3 & -4 \end{bmatrix}.$

(iii) $[A] = \begin{bmatrix} 1 & 0 & -1 \\ 0 & 1 & -1 \\ 0 & 0 & 1 \\ 1 & 1 & 0 \end{bmatrix}, \quad [B] = \begin{bmatrix} 1 & 0 & -1 & 1 \\ 0 & 0 & 1 & 1 \\ -1 & 1 & 0 & 0 \end{bmatrix}.$

3. In the following equations the matrices have constant elements and $[C]$ is a non-singular matrix. In which cases can $[C]$ be cancelled?

(i) $[A].[B].[C] = [D].[Z].[C].$

(ii) $[A].[C].[B] = [M].[C].[E].$

(iii) $[C].[A].[M] = [L].[C].[Z].$

4. In the following equations $[C]$ has all its elements variable and the equations are satisfied for all values of these elements. What relationships can be deduced, and does $[C]$ have to be a non-singular matrix?

(i) $[M].[C] = ([A] - [B]).[C].$

(ii) $([L] + [M]).[C] = ([L] - [M]).[C].$

5. Write down any three matrices for which the product $[A].[B].[C]$ is possible, and verify that $\{[A].[B].[C]\}_t = [C]_t.[B]_t.[A]_t$.

6. Write down any diagonal matrix and show that its inverse is another diagonal matrix with its elements the reciprocals of the elements of the original matrix.

7. (i) If $[a \quad b].\begin{bmatrix} 0 & 1 \\ 2 & 2 \end{bmatrix}.\begin{bmatrix} 2 & 1 \\ 1 & 1 \end{bmatrix} = [1 \quad 2].\begin{bmatrix} 1 & 1 \\ 1 & 0 \end{bmatrix}.\begin{bmatrix} 2 & 1 \\ 1 & 1 \end{bmatrix},$

find a and b, and verify that

$$[a \quad b].\begin{bmatrix} 0 & 1 \\ 2 & 2 \end{bmatrix} = [1 \quad 2].\begin{bmatrix} 1 & 1 \\ 1 & 0 \end{bmatrix}.$$

(ii) If
$$\underset{[A]}{[1 \quad 2]}.\underset{[C]}{\begin{bmatrix} 1 & 1 \\ 1 & 0 \end{bmatrix}}.\underset{[B]}{\begin{bmatrix} 2 & 1 \\ 1 & 1 \end{bmatrix}} = \underset{[M]}{[a \quad b]}.\underset{[C]}{\begin{bmatrix} 1 & 1 \\ 1 & 0 \end{bmatrix}}.\underset{[E]}{\begin{bmatrix} 1 & 2 \\ 1 & 1 \end{bmatrix}},$$

find a and b, and show that $[A].[B] \neq [M].[E]$.

Answers

1. (i), (ii), (v), (vi).

2. (i) $\begin{bmatrix} 1 & 5 \\ -2 & -2 \\ 7 & 15 \end{bmatrix}$. (ii) $\begin{bmatrix} 9 & 4 \\ 0 & 27 \end{bmatrix}$.

(iii) $\begin{bmatrix} 2 & 1 & -1 & 1 \\ -1 & -1 & 1 & 0 \\ -1 & 1 & 0 & 0 \\ 1 & 1 & 0 & 2 \end{bmatrix}$.

3. (i) only.

4. $[C]$ need not be a non-singular matrix.

(i) $[M] = [A] - [B]$.

(ii) $[M]$ is a null matrix.

7. (i) $a = -2$, $b = 3/2$.

(ii) $a = 10$, $b = -13$.

15·10. Partitioning of matrices and complex numbers as matrices

So far the elements of a matrix have been numbers, either real or complex. It is often convenient to use matrices in which the elements themselves are matrices. In this way the amount of written work may often be shortened.

For example if

$$[Z] = \left[\begin{array}{c:ccc} Z_{11} & Z_{12} & Z_{13} & Z_{14} \\ \hdashline Z_{21} & Z_{22} & Z_{23} & Z_{24} \\ Z_{31} & Z_{32} & Z_{33} & Z_{34} \\ Z_{41} & Z_{42} & Z_{43} & Z_{44} \end{array}\right], \tag{1}$$

groups of elements may be partitioned off in any convenient manner. The dotted lines in (1) above show one such chosen manner.

Thus $[Z]$ could be written

$$[Z] = \begin{bmatrix} Z_{aa} & Z_{ab} \\ Z_{ba} & Z_{bb} \end{bmatrix}, \tag{2}$$

where

$$[Z_{aa}] = [Z_{11}], \quad [Z_{ab}] = [Z_{12} \quad Z_{13} \quad Z_{14}],$$

$$[Z_{ba}] = \begin{bmatrix} Z_{21} \\ Z_{31} \\ Z_{41} \end{bmatrix} \quad \text{and} \quad [Z_{bb}] = \begin{bmatrix} Z_{22} & Z_{23} & Z_{24} \\ Z_{32} & Z_{33} & Z_{34} \\ Z_{42} & Z_{43} & Z_{44} \end{bmatrix}.$$

In calculations involving matrices of complex numbers the elements themselves can be treated as complex numbers or, if more convenient, the matrix may be 'expanded' into one containing real numbers only. The matrix form of a complex number is easily understood when it is remembered that a single complex number equation really involves *two* equations in real numbers.

Thus $a + jb = 3 + j4$ implies the two equations $a = 3$ and $b = 4$.

Again, if $\quad V_p + jV_q = (R + jX)(I_p + jI_q), \qquad (3)$

then
$$\left.\begin{aligned} V_p &= RI_p - XI_q \\ V_q &= XI_p + RI_q \end{aligned}\right\}. \tag{4}$$
and

Equations (4) can be represented by the matrix equation

$$\begin{bmatrix} V_p \\ V_q \end{bmatrix} = \begin{bmatrix} R & -X \\ X & R \end{bmatrix} \cdot \begin{bmatrix} I_p \\ I_q \end{bmatrix}. \tag{5}$$

Equation (3) is the vector equation

$$V = ZI \tag{6}$$

expressed in rectangular co-ordinates.

It appears then that the complex operator $Z = R + jX$ has for its matrix form $\begin{bmatrix} R & -X \\ X & R \end{bmatrix}$.

In general $a + jb$ may be represented in matrix form by

$$\begin{bmatrix} a & -b \\ b & a \end{bmatrix}. \tag{7}$$

Further justification for the matrix form of a complex number is given in the examples below.

EXAMPLES

(1) $(a + jb) + (c + jd) = (a + c) + j(b + d) = e$ say.

In matrix form this becomes

$$[e] = \begin{bmatrix} a & -b \\ b & a \end{bmatrix} + \begin{bmatrix} c & -d \\ d & c \end{bmatrix} = \begin{bmatrix} (a+c) & -(b+d) \\ (b+d) & (a+c) \end{bmatrix}.$$

(2) $(a + jb)(c + jd) = (ac - bd) + j(bc + da) = e$ say.

In matrix form this becomes

$$[e] = \begin{bmatrix} a & -b \\ b & a \end{bmatrix} \cdot \begin{bmatrix} c & -d \\ d & c \end{bmatrix} = \begin{bmatrix} (ac - bd) & -(bc + da) \\ (bc + da) & (ac - bd) \end{bmatrix}.$$

(3) The magnitude of a complex number represented by a matrix is the square root of the determinant of the matrix.

Thus, if $\quad Z = a + jb, \quad |Z| = \sqrt{(a^2 + b^2)}.$

In matrix form $\quad [Z] = \begin{bmatrix} a & -b \\ b & a \end{bmatrix}.$

The square root of the determinant of the matrix is

$$\begin{vmatrix} a & -b \\ b & a \end{vmatrix}^{\frac{1}{2}} = (a^2 + b^2)^{\frac{1}{2}}.$$

(4) If the elements of a matrix product are complex numbers, the matrices may be dealt with by replacing the complex numbers by their matrix form, applying the usual laws of matrix algebra, and then inserting the complex values in the final matrix. If this is done care must be taken to retain the character of each complex number.

Taking Question 8 of Exercise 48, let

$$[M] = \begin{bmatrix} 1+j & 2-j & 3+j2 \\ -2+j & 4+j3 & 5-j2 \end{bmatrix} \cdot \begin{bmatrix} 1-j & 2+j \\ 5-j3 & -2+j2 \\ 4+j3 & 1-j2 \end{bmatrix}$$

$$= \begin{bmatrix} a & b & c \\ d & e & f \end{bmatrix} \cdot \begin{bmatrix} g & k \\ h & l \\ i & m \end{bmatrix} \text{ say}$$

$$= \begin{bmatrix} ag+bh+ci & ak+bl+cm \\ dg+eh+fi & dk+el+fm \end{bmatrix}. \qquad (1)$$

Replace each element by the matrix form of the complex number it represents.

Thus $a = \begin{bmatrix} 1 & -1 \\ 1 & 1 \end{bmatrix}$, $b = \begin{bmatrix} 2 & 1 \\ -1 & 2 \end{bmatrix}$ etc.

Then $ag = \begin{bmatrix} 1 & -1 \\ 1 & 1 \end{bmatrix} \cdot \begin{bmatrix} 1 & 1 \\ -1 & 1 \end{bmatrix} = \begin{bmatrix} 2 & 0 \\ 0 & 2 \end{bmatrix}$,

$$bh = \begin{bmatrix} 2 & 1 \\ -1 & 2 \end{bmatrix} \cdot \begin{bmatrix} 5 & 3 \\ -3 & 5 \end{bmatrix} = \begin{bmatrix} 7 & 11 \\ -11 & 7 \end{bmatrix},$$

$$ci = \begin{bmatrix} 3 & -2 \\ 2 & 3 \end{bmatrix} \cdot \begin{bmatrix} 4 & -3 \\ 3 & 4 \end{bmatrix} = \begin{bmatrix} 6 & -17 \\ 17 & 6 \end{bmatrix}.$$

Thus $ag + bh + ci = \begin{bmatrix} 15 & -6 \\ 6 & 15 \end{bmatrix}$.

Proceeding in this way

$$[M] = \left[\begin{array}{cc:cc} 15 & -6 & 6 & -5 \\ 6 & 15 & 5 & 6 \\ \hdashline 54 & -13 & -18 & 10 \\ 13 & 54 & -10 & -18 \end{array}\right]$$

$$= \underline{\begin{bmatrix} 15+j6 & 6+j5 \\ 54+j13 & -18-j10 \end{bmatrix}}.$$

The working could be written:

$$[M] = \left[\begin{array}{cc:cc:cc} 1 & -1 & 2 & 1 & 3 & -2 \\ 1 & 1 & -1 & 2 & 2 & 3 \\ \hdashline -2 & -1 & 4 & -3 & 5 & 2 \\ 1 & -2 & 3 & 4 & -2 & 5 \end{array}\right]$$

$$\cdot \left[\begin{array}{cc:cc} 1 & 1 & 2 & -1 \\ -1 & 1 & 1 & 2 \\ \hdashline 5 & 3 & -2 & -2 \\ -3 & 5 & 2 & -2 \\ \hdashline 4 & -3 & 1 & 2 \\ 3 & 4 & -2 & 1 \end{array}\right].$$

The multiplication is then carried out as in the form of (1) above, retaining the character of each complex number, giving

$$[M] = \left[\begin{array}{cc:cc} 15 & -6 & 6 & -5 \\ 6 & 15 & 5 & 6 \\ \hdashline 54 & -13 & -18 & 10 \\ 13 & 54 & -10 & -18 \end{array}\right].$$

Note. Subdivision of matrices must NOT be used in an attempt to make two non-conformable matrices conformable.

For example:

$$\left[\begin{array}{cc:c} 5 & 6 & 7 \\ 2 & 3 & 1 \\ \hdashline 4 & 1 & 2 \end{array}\right] \cdot \begin{bmatrix} 4 & 3 \\ 2 & 5 \end{bmatrix} = \begin{bmatrix} a & b \\ c & d \end{bmatrix} \cdot \begin{bmatrix} 4 & 3 \\ 2 & 5 \end{bmatrix}$$

$$= \begin{bmatrix} 4a + 2b & 3a + 5b \\ 4c + 2d & 3c + 5d \end{bmatrix}.$$

This mistake may easily be made if it is forgotten that a, b, c, d are matrices themselves. The original two matrices are non-conformable for multiplication, and in this sense it is impossible to form the product $4a$ because this is really $\begin{bmatrix} 5 & 6 \\ 2 & 3 \end{bmatrix}$. [4], which is a 'product' of matrices non-conformable for multiplication.

If partitioning has been used in a matrix problem to make the written work shorter, in any numerical computation the sub-divided matrices must be expanded completely.

Exercise 53

1. Show that equations (4) of § 15·10 can be represented by the matrix equation

$$[V_p \quad V_q] = [I_p \quad I_q] . \begin{bmatrix} R & X \\ -X & R \end{bmatrix} .$$

Deduce that $a + jb$ may be represented in matrix form by

$$\begin{bmatrix} a & b \\ -b & a \end{bmatrix} \text{ and its conjugate by } \begin{bmatrix} a & -b \\ b & a \end{bmatrix} .$$

(*Note.* With this notation the vector equation would be $V = IZ$.)

2. If $[Z] = \begin{bmatrix} a & -b \\ b & a \end{bmatrix}$ and $[\bar{Z}]$ is its conjugate, show that

$$[Z].[\bar{Z}] = (a^2 + b^2)\,[I].$$

3. Write down the matrix form for $1 + 0j$, $0 + j$, $-1 + 0j$ and $0 - j$.

Using matrix notation verify that $j(a + jb) = -b + ja$.

4. The symmetrical component operator h is $-0{\cdot}5 + j\dfrac{\sqrt{3}}{2}$.

Put this in matrix form and verify that h^3 gives a unit matrix.

5. Using the matrix form of a complex number perform the product

$$\begin{bmatrix} a & b \\ c & d \end{bmatrix} . \begin{bmatrix} e & f \\ g & h \end{bmatrix} ,$$

where $a = 1 + j$, $b = 2 - j$, $c = 3 - j2$, $d = -1 + j$. $e = -1 - j$, $f = 3 - 0j$, $g = -2 + j2$ and $h = 3 + j3$.

6. If a non-singular matrix $[M]$ is partitioned in any manner to give a matrix $\begin{bmatrix} A & B \\ C & D \end{bmatrix}$, where $[A]$, $[B]$, $[C]$, $[D]$ are the 'sub-matrices', the inverse of $[M]$ is given by

$$[M]^{-1} = \begin{bmatrix} F & -F.B.E \\ -E.C.F & E+E.C.F.B.E \end{bmatrix},$$

where $[E] = [D]^{-1}$ and $[F] = \{[A]-[B].[E].[C]\}^{-1}$.

Verify this in the case of

$$[M] = \left[\begin{array}{cc:cc} 1 & 2 & -2 & 3 \\ -1 & 1 & 4 & 3 \\ \hdashline 1 & -1 & 1 & 0 \\ 0 & 1 & 2 & 3 \end{array}\right] = \begin{bmatrix} A & B \\ C & D \end{bmatrix}.$$

Answers

3. $\begin{bmatrix} 1 & 0 \\ 0 & 1 \end{bmatrix}$, $\begin{bmatrix} 0 & -1 \\ 1 & 0 \end{bmatrix}$, $\begin{bmatrix} -1 & 0 \\ 0 & -1 \end{bmatrix}$, $\begin{bmatrix} 0 & 1 \\ -1 & 0 \end{bmatrix}$.

4. $h = \begin{bmatrix} -0{\cdot}5 & -\dfrac{\sqrt{3}}{2} \\ \dfrac{\sqrt{3}}{2} & -0{\cdot}5 \end{bmatrix}$, $h^3 = \begin{bmatrix} 1 & 0 \\ 0 & 1 \end{bmatrix}$.

5. $\left[\begin{array}{cc:cc} -2 & -4 & 12 & -6 \\ 4 & -2 & 6 & 12 \\ \hdashline -5 & 5 & 3 & 6 \\ -5 & -5 & -6 & 3 \end{array}\right] = \begin{bmatrix} -2+j4 & 12+j6 \\ -5-j5 & 3-j6 \end{bmatrix}$.

6.

$$[M]^{-1} = \tfrac{1}{3}\begin{bmatrix} 6 & 9 & 6 & -15 \\ 9 & 15 & 6 & -24 \\ 3 & 6 & 3 & -9 \\ -5 & -9 & -4 & 15 \end{bmatrix}.$$

CHAPTER 16

AN INTRODUCTION TO THE APPLICATION OF MATRICES TO NETWORK ANALYSIS

16·1. Introduction

Now that computers are becoming more readily available the method of solution of comparatively large networks by means of fundamental circuit equations will probably come to be used quite frequently, as much of the tedious arithmetical calculations can be done by the computer. With this in mind matrix methods are important. It is not so much that they cut down the arithmetical calculations involved (in most cases they do not) but that with large and complicated networks they offer a purely mechanical method of setting up the equations for solution. Only a brief introduction is given here. It should be sufficient to enable the student to read text-books which specialize in network analysis.

It may be appropriate at this stage to give a brief résumé of the methods by which circuit equations are built up.

Note that in this chapter, for a.c. networks, when E, I and Z are used for voltage, current and impedance respectively, their complex values will be intended. If two or more driving voltages occur it will be assumed that they are of the same frequency, their complex values being capable of use in the same equation. The resulting solution will represent, by complex values, the steady-state conditions.

The solution to a network problem can be divided roughly into three parts:

(i) drawing of a network diagram,

(ii) formation of the network equations,

(iii) the solution of the network equations.

The number of equations necessary to solve a linear network can be obtained by the application of Ohm's and Kirchhoff's laws. They are applicable to the steady-state condition for a single frequency.

For revision purposes the laws are stated below.

(1) *Ohm's law (branch law).* The (vector) voltage drop V corresponding to a (vector) current I in an impedance Z is given by the (vector) equation

$$V = ZI.$$

(2) *Kirchhoff's first law (junction or node law).* The (vector) sum of all currents entering any junction in a network is zero, i.e. $\Sigma I = 0$.

Another way of expressing this is that the (vector) sum of the currents entering a junction (node) is equal to the (vector) sum of those leaving it.

(3) *Kirchhoff's second law (mesh law).* The (vector) sum of all the driving voltages (e.m.f.'s) in any closed path (mesh) in a network is equal to the (vector) sum of the products of the impedances and the currents in the constituent branches, i.e. $\Sigma E = \Sigma ZI$.

16·2. Formation of network equations

There are three main methods of setting up the network equations. They are (*a*) the branch-current method, (*b*) the

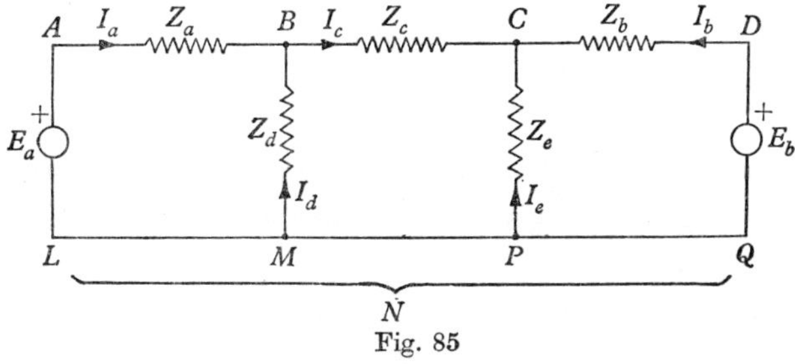

Fig. 85

mesh-current method and (*c*) the nodal-voltage method. The first two will be illustrated here; the third method should then be easily followed in any standard text-book.

(*a*) *Branch-current method.* This is the simplest method, but is the longest when applied to the larger networks.

Consider the simple network shown in fig. 85.

The driving voltages, impedances and branch currents have been marked.

The junction law for currents can be applied at all junctions (nodes). The equation obtained at the last junction considered can be derived from the other equations, so that the number of independent equations is one less than the number of junctions. The mesh law is then applied to as many closed paths as is necessary to obtain the required number of equations to solve for the unknowns (usually the currents). The closed paths can be chosen in any manner provided that *all* branches have been used in obtaining the required number of equations.

If the currents I_a, I_b, I_c, I_d, I_e are the unknowns required in fig. 85, then five independent equations are needed.

The junction law at B and C gives:

$$I_c = I_a + I_d, \tag{1}$$

$$I_c + I_b + I_e = 0. \tag{2}$$

Note that the points L, M, P, Q form only one node N. The junction law here would give $I_a + I_d + I_e + I_b = 0$, which is merely equation (1) subtracted from equation (2).

Thus three more independent equations are needed.

Using meshes $ABML$, $BCPM$, and $CDQP$, the mesh law gives:

$$E_a = Z_a I_a - Z_d I_d, \tag{3}$$

$$0 = Z_d I_d + Z_c I_c - Z_e I_e, \tag{4}$$

$$E_b = Z_b I_b - Z_e I_e. \tag{5}$$

In these three equations all branches have been used.

The five equations given above are sufficient to solve for the five currents.

If meshes $ACPL$ and $ADQL$ had been used, they would merely have given combinations of equations (3), (4) and (5), as can easily be verified.

(*b*) *Maxwell's mesh-current* (*or loop-current*) *method*. This method reduces the computation involved, especially in larger networks. The same network as in fig. 85 will be used as an example.

A general description of the method follows.

The whole network is divided into loops (or meshes). These are usually numbered for easy reference. A 'fictitious' current is then assumed to be flowing round each mesh. By convention this is usually taken positive if flowing clockwise. The 'passive' voltages around each loop, due to its own current and to those in meshes which have a mutual impedance with it are equated to the driving voltages. These equations are solved for the mesh currents and the actual branch currents are then easily found. The mesh-current method really involves defining mesh currents which automatically satisfy the junction law. It therefore eliminates all the corresponding junction law equations from the main calculations.

A mesh-current diagram for the network of fig. 85 is shown in fig. 86.

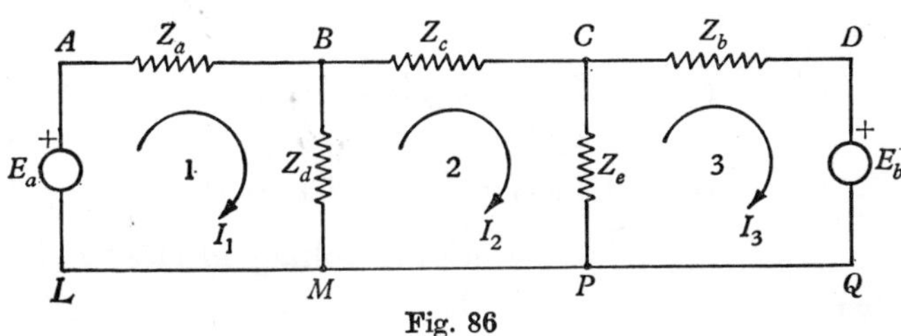

Fig. 86

There are three meshes for the three unknown mesh currents. Meshes $ACPL$ and $ADQL$ could have been chosen as two of the three meshes required, but the mesh currents would have been more difficult to mark clearly on the diagram.

Note that all mesh currents entering a junction also leave it, so that the junction law is automatically satisfied. For example, at B mesh currents I_1 and I_2 both enter and leave.

Applying the mesh law to the three meshes:

$$E_a = (Z_a + Z_d)I_1 - Z_dI_2, \tag{1}$$

$$0 = -Z_dI_1 + (Z_d + Z_c + Z_e)I_2 - Z_eI_3, \tag{2}$$

$$-E_b = -Z_eI_2 + (Z_b + Z_e)I_3. \tag{3}$$

As a more detailed description, equation (1) is built up as follows. Considering the mesh current I_1; it flows through all the impedances in mesh 1. The total 'passive' voltage due to this is $(Z_a + Z_d)I_1$. Mesh 1 also has a mutual impedance Z_d with mesh 2, and the mesh current I_2 flows anti-clockwise through Z_d referred to mesh 1. Thus the 'passive' voltage due to this is $-Z_dI_2$.

The driving voltage in mesh 1 is E_a, clockwise. Equation (1) above therefore follows.

When equations (1), (2) and (3) above have been solved for the mesh currents I_1, I_2, I_3 the actual branch currents are easily obtained as:

$$\left.\begin{aligned} I_a = I_1,\quad I_c = I_2,\quad I_b &= -I_3,\\ I_d = I_2 - I_1 \quad \text{and} \quad I_e &= I_3 - I_2.\end{aligned}\right\} \qquad (4)$$

(refer to figs. 85 and 86).

In a more general form equations (1)–(3) above may be written

$$E_1 = \quad Z_{11}I_1 - Z_{12}I_2 - Z_{13}I_3, \qquad (5)$$
$$E_2 = -Z_{21}I_1 + Z_{22}I_2 - Z_{23}I_3, \qquad (6)$$
$$E_3 = -Z_{31}I_1 - Z_{32}I_2 + Z_{33}I_3, \qquad (7)$$

where, in general, E_p is the driving voltage in mesh p, in the clockwise direction; Z_{pp} is the total impedance round the mesh and is called the *mesh self-impedance*; Z_{pq} is the total impedance in mesh p that is common also to mesh q and is called the *mesh mutual-impedance* between meshes p and q. For a bilateral network (i.e. one composed of impedances which have the same value either way the current flows through them) $Z_{pq} = Z_{qp}$, and the impedance coefficients in equations (5) to (7) will be symmetrical about the diagonal formed by the self-impedance terms. If the 'clockwise positive' current convention is used the non-diagonal elements will always be negative.

The student should check equations (1)–(3) against equations (5)–(7).

Note that equations (5)–(7) can be written in matrix form as $[E] = [Z].[I]$, where

$$[E] = \begin{bmatrix} E_1 \\ E_2 \\ E_3 \end{bmatrix}, \quad [Z] = \begin{bmatrix} Z_{11} & -Z_{12} & -Z_{13} \\ -Z_{21} & Z_{22} & -Z_{23} \\ -Z_{31} & -Z_{32} & Z_{33} \end{bmatrix}, \quad [I] = \begin{bmatrix} I_1 \\ I_2 \\ I_3 \end{bmatrix}.$$

The methods used for solving the linear equations occurring in network problems are usually based on one of three principles:

(i) *Elimination.* This is the usual method taught in elementary mathematics.

(ii) *Permutation.* This includes the methods of determinants or matrices.

(iii) *Iteration.* The Relaxation method in § 13·6 is an example of this principle.

16·21. Numerical example

Calculate the value of each branch current in the network shown in fig. 87.

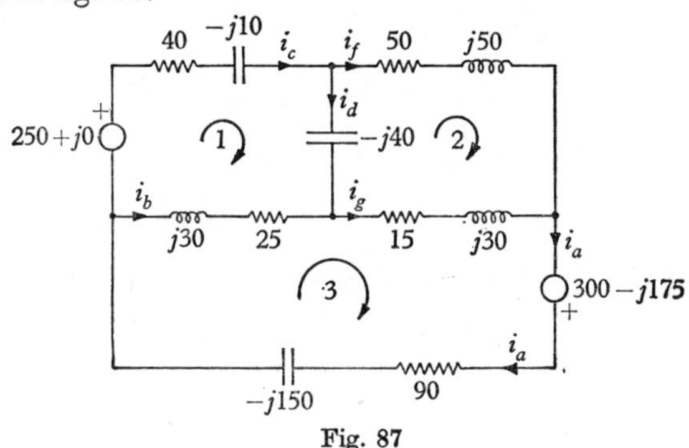

Fig. 87

Three meshes are selected as shown in fig. 87 and the mesh-current diagram is shown in detail in fig. 88 below.

The equations giving the branch currents in terms of the mesh currents are:

$$\left.\begin{aligned} i_a &= I_3, & i_b &= I_3 - I_1, \\ i_c &= I_1, & i_d &= I_1 - I_2, \\ i_f &= I_2, & i_g &= I_3 - I_2. \end{aligned}\right\} \quad (1)$$

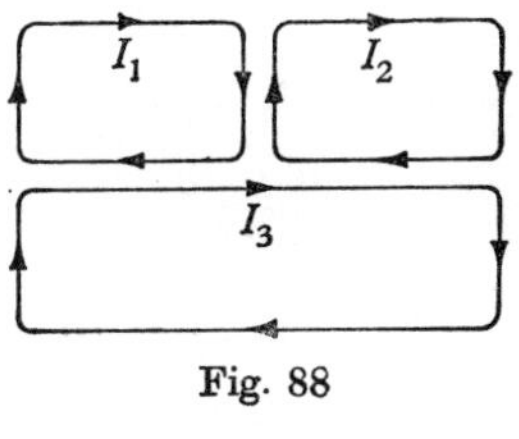

Fig. 88

Applying the mesh law to the three meshes:

$$250 + j0 = (40 - j10 - j40 + 25 + j30)I_1 - (-j40)I_2 - (25 + j30)I_3,$$

$$0 = (50 + j50 + 15 + j30 - j40)I_2 - (-j40)I_1 - (15 + j30)I_3,$$

$$300 - j175 = (90 - j150 + 25 + j30 + 15 + j30)I_3 - (25 + j30)I_1 - (15 + j30)I_2.$$

These give:

$$250 = (65 - j20)I_1 + j40I_2 - (25 + j30)I_3,$$

$$0 = j40I_1 + (65 + j40)I_2 - (15 + j30)I_3,$$

$$300 - j175 = -(25 + j30)I_1 - (15 + j30)I_2 + (130 - j90)I_3.$$

That is:

$$50 = (13 - j4)I_1 + j8I_2 - (5 + j6)I_3, \quad (2)$$

$$0 = j8I_1 + (13 + j8)I_2 - (3 + j6)I_3, \quad (3)$$

$$60 - j35 = -(5 + j6)I_1 - (3 + j6)I_2 + (26 - j18)I_3. \quad (4)$$

Solving these equations for I_1, I_2, I_3:

$$I_1 = \frac{\begin{vmatrix} 50 & j8 & -(5 + j6) \\ 0 & 13 + j8 & -(3 + j6) \\ 60 - j35 & -(3 + j6) & (26 - j18) \end{vmatrix}}{\Delta},$$

$$I_2 = \frac{\begin{vmatrix} 13 - j4 & 50 & -(5 + j6) \\ j8 & 0 & -(3 + j6) \\ -(5 + j6) & 60 - j35 & (26 - j18) \end{vmatrix}}{\Delta},$$

$$I_3 = \frac{\begin{vmatrix} 13 - j4 & j8 & 50 \\ j8 & 13 + j8 & 0 \\ -(5 + j6) & -(3 + j6) & 60 - j35 \end{vmatrix}}{\Delta},$$

where $\Delta = \begin{vmatrix} 13 - j4 & j8 & -(5 + j6) \\ j8 & 13 + j8 & -(3 + j6) \\ -(5 + j6) & -(3 + j6) & (26 - j18) \end{vmatrix}.$

These give:

$$\left.\begin{aligned} I_1 &\simeq 2{\cdot}93 + j1{\cdot}90, \\ I_2 &\simeq 0{\cdot}53 - j0{\cdot}97, \\ I_3 &\simeq 1{\cdot}97 + j1{\cdot}07. \end{aligned}\right\} \quad (5)$$

From (1) and (5):

$i_a = 1{\cdot}97 + j1{\cdot}07,$ $i_b = -0{\cdot}96 - j0{\cdot}83,$

$i_c = 2{\cdot}93 + j1{\cdot}90,$ $i_d = 2{\cdot}40 + j2{\cdot}87,$

$i_f = 0{\cdot}53 - j0{\cdot}97,$ $i_g = 1{\cdot}44 + j2{\cdot}04.$

EXERCISE 54

Write equations for the following networks (figs. 89, 90, 91) using mesh currents. Solve the equations and deduce the values of the branch currents.

1.

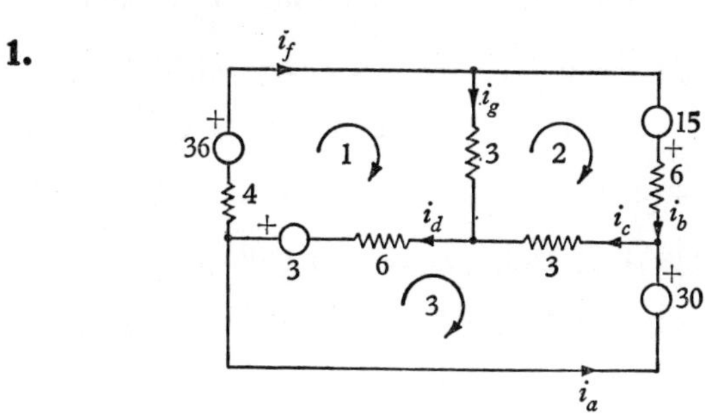

Fig. 89

2.

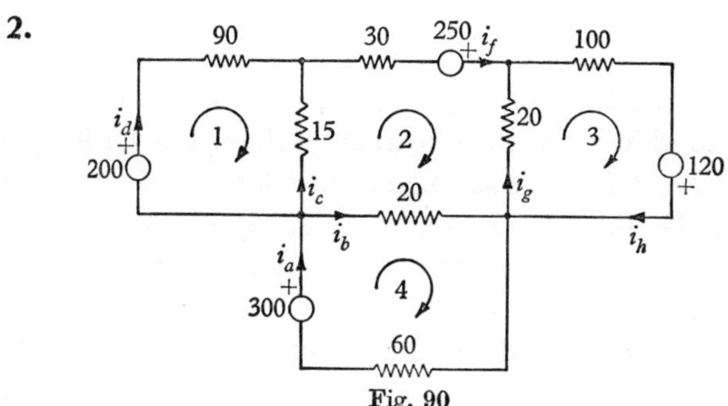

Fig. 90

3.

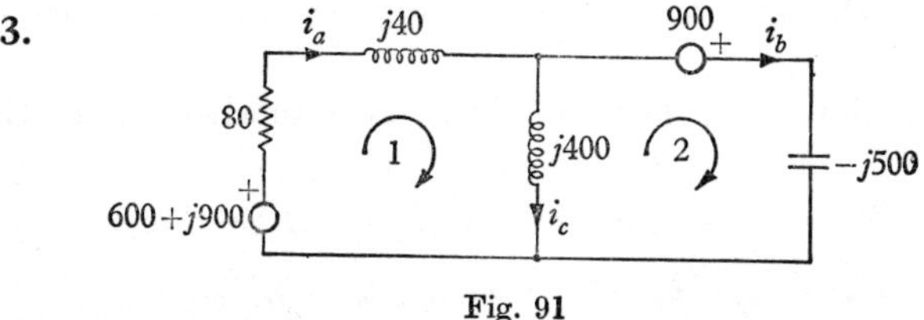

Fig. 91

ANSWERS

1. $13I_1 - 3I_2 - 6I_3 = 39, \quad - I_1 + 4I_2 - I_3 = 5,$
$- 2I_1 - I_2 + 3I_3 = - 11.$

$I_1 = 2{\cdot}797, \quad I_2 = 1{\cdot}635, \quad I_3 = - 1{\cdot}257.$

$i_a = 1{\cdot}257, \quad i_b = 1{\cdot}635, \quad i_c = 2{\cdot}892, \quad i_d = 4{\cdot}054,$

$i_f = 2{\cdot}797, \quad i_g = 1{\cdot}162.$

2. $- I_2 + 4I_4 = 15, \quad 21I_1 - 3I_2 = 40, \quad - I_2 + 6I_3 = 6,$
$- 3I_1 + 17I_2 - 4I_3 - 4I_4 = 50.$

$I_1 = 2{\cdot}621, \quad I_2 = 5{\cdot}013, \quad I_3 = 1{\cdot}836, \quad I_4 = 5{\cdot}003.$

$i_a = 5{\cdot}003, \quad i_b = - 0{\cdot}01, \quad i_c = 2{\cdot}392, \quad i_d = 2{\cdot}621,$

$i_f = 5{\cdot}013, \quad i_g = - 3{\cdot}177, \quad i_h = 1{\cdot}836.$

3. $30 + j45 = (4 + j22)I_1 - j20I_2, \quad 9 = - j4I_1 - jI_2.$

$I_1 = 0{\cdot}383 + j1{\cdot}486, \quad I_2 = - 1{\cdot}532 + j3{\cdot}056.$

$i_a = 0{\cdot}383 + j1{\cdot}486, \quad i_b = - 1{\cdot}532 + j3{\cdot}056,$

$i_c = 1{\cdot}915 - j1{\cdot}57.$

16·3. The Kron mesh method and the use of matrices

16·31. Introduction

A very simple network will be used to bring out the main points of this method. It may seem to the student that there is nothing to gain by the use of this method, but he is again reminded that it is in really complicated networks that the principles to be discussed here are extremely useful in providing a purely mechanical routine for setting up the equations to be solved.

Consider the network shown in fig. 92 (i). Given the driving voltages e_a, e_b, e_c and the impedances z_a, z_b, z_c, it is required to find the steady-state currents i_a, i_b, i_c. Using the mesh-current method, meshes $ABFH$ and $BCDF$ are chosen and shown in fig. 92 (ii).

Applying the mesh law to these two meshes gives the equations:

$$E_1 = e_a + e_c = (z_a + z_c)I_1 - z_cI_2, \tag{1}$$

$$E_2 = e_b - e_c = - z_cI_1 + (z_b + z_c)I_2. \tag{2}$$

The branch currents are then given in terms of the mesh currents by the equations:

$$\left.\begin{aligned} i_a &= I_1, \\ i_b &= \quad\quad I_2, \\ i_c &= I_1 - I_2. \end{aligned}\right\} \tag{3}$$

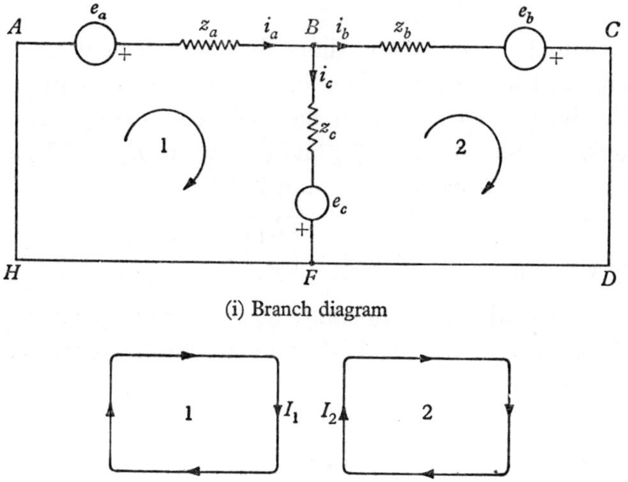

(i) Branch diagram

(ii) Mesh diagram

Fig. 92

Now equations (3) can be written in matrix form as

$$[i] = [C].[I], \tag{4}$$

where

$$[i] = \begin{bmatrix} i_a \\ i_b \\ i_c \end{bmatrix}, \tag{5}$$

$$[I] = \begin{bmatrix} I_1 \\ I_2 \end{bmatrix}, \tag{6}$$

and

$$[C] = \begin{bmatrix} 1 & 0 \\ 0 & 1 \\ 1 & -1 \end{bmatrix}. \tag{7}$$

(The student should verify this by performing the matrix multiplication.)

The way in which $[C]$ is built up from the coefficients of I_1, I_2 in equations (3) has been made obvious by the lay-out of these equations.

$[i]$ is called the *branch-current matrix*, $[I]$ the *mesh-current matrix* and $[C]$ the *transformation matrix*. $[C]$ is alternatively called the *connexion matrix*.

It will be seen later that this transformation matrix dominates the whole problem; once it is known, the calculation of the branch currents can be made completely automatic.

Referring to equations (1) and (2), where E_1 and E_2 are the driving voltages in meshes 1 and 2, let

$$[E] = \begin{bmatrix} E_1 \\ E_2 \end{bmatrix}. \tag{8}$$

$[E]$ is called the *mesh-e.m.f. matrix*; each element is the algebraic sum of the e.m.f.'s round the meshes 1 and 2.

Also take

$$[Z] = \begin{bmatrix} z_a + z_c & -z_c \\ -z_c & z_b + z_c \end{bmatrix}. \tag{9}$$

$[Z]$ is called the *mesh-impedance* matrix. The elements in its main diagonal are the total series impedances of meshes 1 and 2, whilst the other elements are mutual impedances of the meshes. (z_c is common to the meshes, and the negative sign occurs because the positive direction of the currents in the meshes are opposed.) Equations (1) and (2) can now be written in matrix form as

$$[E] = [Z].[I]. \tag{10}$$

(This should be verified by the student.)

From this equation

$$[I] = [Z]^{-1}.[E], \tag{11}$$

where $[Z]^{-1}$ is the inverse of $[Z]$.

Equation (11) is sometimes written as

$$[I] = [Y].[E],$$

where $$[Y] = [Z]^{-1}$$

is called the mesh-admittance matrix.

When equation (11) has been solved for $[I]$, $[i]$ is easily found from (4) above.

The network of fig. 92 is a very simple one and the matrices $[E]$ and $[Z]$ could easily be written down on sight from the figure, or from equations (1) and (2).

For more complicated networks there are two formulae which can be developed by which $[E]$ and $[Z]$ can be derived automatically from simple matrices obtained from individual branches and the transformation matrix $[C]$.

First the *branch-e.m.f. matrix* $[e]$ is formed; the branches a, b, c are treated separately and the total applied driving voltages (e.m.f.'s) in each branch are listed vertically beneath each other.

Thus
$$[e] = \begin{matrix} a \\ b \\ c \end{matrix}\begin{bmatrix} e_a \\ e_b \\ e_c \end{bmatrix}. \tag{12}$$

A square matrix is now made up by listing, *in the same order*, the self and mutual impedances for each branch. This is called the *branch-impedance matrix*. In this case

$$[z] = \begin{matrix} \\ a \\ b \\ c \end{matrix}\begin{matrix} \begin{matrix} a & b & c \end{matrix} \\ \begin{bmatrix} z_a & 0 & 0 \\ 0 & z_b & 0 \\ 0 & 0 & z_c \end{bmatrix} \end{matrix}. \tag{13}$$

The letters outside the matrix brackets serve as convenient 'sign-posts' but may be omitted if desired.

The full branch-impedance matrix for a 3-branch network would be

$$[z] = \begin{matrix} \\ a \\ b \\ c \end{matrix}\begin{matrix} \begin{matrix} a & b & c \end{matrix} \\ \begin{bmatrix} z_{aa} & z_{ab} & z_{ac} \\ z_{ba} & z_{bb} & z_{bc} \\ z_{ca} & z_{cb} & z_{cc} \end{bmatrix} \end{matrix},$$

where z_{aa} is the self-impedance of branch a (in this case z_a), and z_{ba} is the mutual impedance of branches b and a (in this case zero), and so on. The zeros in equation (13) signify that in this case all the branch mutual-impedances are zero.

From (1) and (2) above the relations between the mesh driving voltages and branch driving voltages are:

$$E_1 = e_a + e_c \quad \text{and} \quad E_2 = e_b - e_c. \tag{14}$$

In terms of the matrices $[E]$ and $[e]$, these relations can be written in the form

$$[E] = \begin{bmatrix} 1 & 0 & 1 \\ 0 & 1 & -1 \end{bmatrix} \cdot [e]. \tag{15}$$

The student should verify this, using equations (8) and (12).

The significant thing is that the 2×3 matrix in (15) is the *transpose* of the transformation matrix $[C]$ given in equation (7), i.e.

$$\begin{bmatrix} 1 & 0 & 1 \\ 0 & 1 & -1 \end{bmatrix} = [C]_t,$$

and equation (15) becomes

$$\underline{[E] = [C]_t . [e].} \tag{16}$$

Next consider the matrix denoted by

$$[C]_t . [z] . [C].$$

$$[C]_t . [z] . [C] = \begin{bmatrix} 1 & 0 & 1 \\ 0 & 1 & -1 \end{bmatrix} . \begin{bmatrix} z_a & 0 & 0 \\ 0 & z_b & 0 \\ 0 & 0 & z_c \end{bmatrix} . \begin{bmatrix} 1 & 0 \\ 0 & 1 \\ 1 & -1 \end{bmatrix}$$

$$= \begin{bmatrix} 1 & 0 & 1 \\ 0 & 1 & -1 \end{bmatrix} . \begin{bmatrix} z_a & 0 \\ 0 & z_b \\ z_c & -z_c \end{bmatrix}$$

$$= \begin{bmatrix} z_a + z_c & -z_c \\ -z_c & z_b + z_c \end{bmatrix}$$

$$= [Z] \quad \text{(see equation (9))}.$$

Thus

$$\underline{[C]_t . [z] . [C] = [Z].} \tag{17}$$

Equations (16) and (17) are really the crux of the matter. From them $[E]$ and $[Z]$ can be calculated in terms of the easily formulated branch-impedance and branch-e.m.f. matrices.

Then the formula of equation (11) can be used to find the mesh-currents and hence the branch currents.

The formulae given in (16) and (17) hold for all cases of bilateral networks, though a formal proof of this is beyond the scope of this book.

Before applying it to another network, the Kron mesh method will be briefly recapitulated:

(i) Given the network diagram, the branch-e.m.f. and branch-impedance matrices, $[e]$ and $[z]$, are written down.

(ii) A suitable mesh diagram is drawn, and from it the transformation matrix, $[C]$, is obtained by writing down the equations connecting the branch and mesh currents.

(iii) From the formulae in (16) and (17), the mesh-e.m.f. and mesh-impedance matrices, $[E]$ and $[Z]$ are obtained.

(iv) The formula in (11) is then used to find $[I]$.

(v) The branch currents are then found from the relation $[i] = [C].[I]$.

The whole procedure can be combined into one formula, namely,

$$[i] = [C].\{[C]_t.[z].[C]\}^{-1}.[C]_t.[e],$$

as the student can verify, but it is best to use the separate steps.

16·32. A simple bridge network

In fig. 93 (i) below the driving voltages e_a, e_b, e_c, e_d, e_f, e_g are taken as given and also the impedances z_a, z_b, z_c, z_d, z_f, z_g.

By convention the positive directions of the branch currents are usually taken as issuing out of the 'positive' side of the voltages.

In fig. 93 (ii) the three chosen meshes are shown.

Taking the branch matrices in order a, b, c, d, f, g:

$$[e] = \begin{bmatrix} e_a \\ e_b \\ e_c \\ e_d \\ e_f \\ e_g \end{bmatrix}, \tag{1}$$

$$[z] = \begin{array}{c} \\ a \\ b \\ c \\ d \\ f \\ g \end{array}\begin{array}{c} \begin{array}{cccccc} a & b & c & d & f & g \end{array} \\ \left[\begin{array}{cccccc} z_a & 0 & 0 & 0 & 0 & 0 \\ 0 & z_b & 0 & 0 & 0 & 0 \\ 0 & 0 & z_c & 0 & 0 & 0 \\ 0 & 0 & 0 & z_d & 0 & 0 \\ 0 & 0 & 0 & 0 & z_f & 0 \\ 0 & 0 & 0 & 0 & 0 & z_g \end{array}\right] \end{array}. \tag{2}$$

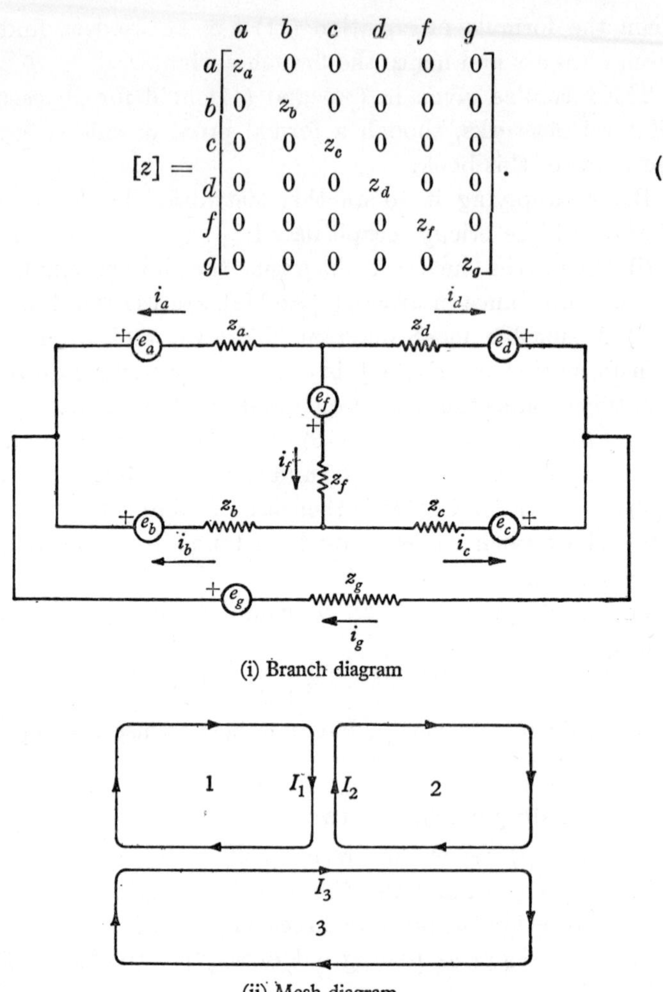

(i) Branch diagram

(ii) Mesh diagram

Fig. 93

(It is assumed that there is no coupling between the branches and therefore no mutual impedances.)

The relations between the branch and mesh currents are:

$$\left.\begin{aligned} i_a &= -I_1, \\ i_b &= I_1 - I_3, \\ i_c &= - I_2 + I_3, \\ i_d &= I_2, \\ i_f &= I_1 - I_2, \\ i_g &= I_3. \end{aligned}\right\} \tag{3}$$

Thus, taking

$$[i] = \begin{bmatrix} i_a \\ i_b \\ i_c \\ i_d \\ i_f \\ i_g \end{bmatrix} \quad \text{and} \quad [I] = \begin{bmatrix} I_1 \\ I_2 \\ I_3 \end{bmatrix} \tag{4}$$

and

$$[C] = \begin{bmatrix} -1 & 0 & 0 \\ 1 & 0 & -1 \\ 0 & -1 & 1 \\ 0 & 1 & 0 \\ 1 & -1 & 0 \\ 0 & 0 & 1 \end{bmatrix}, \tag{5}$$

then
$$[i] = [C].[I]. \tag{6}$$

(Note how $[C]$ is formed from the layout of equations (3).)

From (5),
$$[C]_t = \begin{bmatrix} -1 & 1 & 0 & 0 & 1 & 0 \\ 0 & 0 & -1 & 1 & -1 & 0 \\ 0 & -1 & 1 & 0 & 0 & 1 \end{bmatrix}. \tag{7}$$

$[E]$ and $[Z]$ can now be found, using

$$[E] = [C]_t.[e] \quad \text{and} \quad [Z] = [C]_t.[z].[C].$$

Finally, $[I] = [Z]^{-1}.[E]$ gives the mesh currents.

16·33. A simple 3-phase, 4-wire static network

Fig. 94 shows a 3-phase, 4-wire star load supplied from the star-connected secondary windings of a 3-phase transformer. Given e_a, e_b, e_c and z_{a1}, z_{b1}, z_{c1}, z_{a2}, etc., it is required to find the currents i_a, i_b, i_c and i_n. It is assumed, for this example, that all mutual impedances are negligible.

It is not assumed that the voltages, impedances or currents are necessarily balanced.

For the purposes of the mesh diagram and to obey the convention that clockwise flow is positive the network of fig. 94 is best drawn as shown in fig. 95 (i).

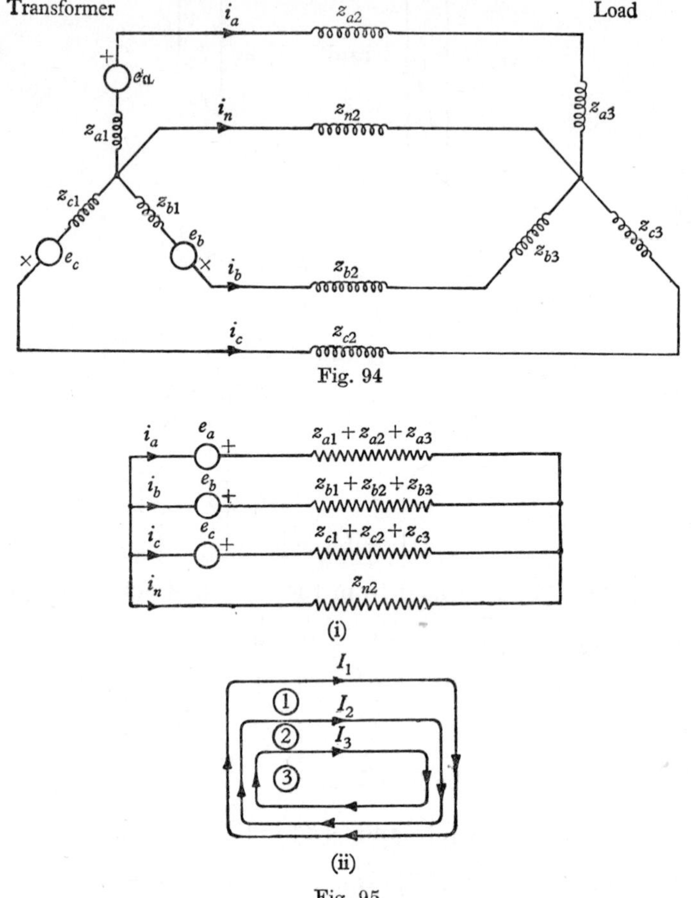

Fig. 94

(i)

(ii)

Fig. 95

The mesh-current diagram is shown in fig. 95 (ii).

The relations between the branch and mesh currents are:

$$\left.\begin{aligned} i_a &= I_1, \\ i_b &= \qquad\quad I_2, \\ i_c &= \qquad\qquad\quad I_3, \\ i_n &= -I_1 - I_2 - I_3. \end{aligned}\right\} \qquad (1)$$

The transformation matrix $[C]$, given by the coefficients of the mesh currents, is

$$[C] = \begin{bmatrix} 1 & 0 & 0 \\ 0 & 1 & 0 \\ 0 & 0 & 1 \\ -1 & -1 & -1 \end{bmatrix} \tag{2}$$

and
$$[i] = [C].[I]. \tag{3}$$

The branch-e.m.f. matrix is

$$[e] = \begin{bmatrix} e_a \\ e_b \\ e_c \\ 0 \end{bmatrix}, \tag{4}$$

whilst the branch-impedance matrix is

$$[z] = \begin{bmatrix} z_{a1}+z_{a2}+z_{a3} & 0 & 0 & 0 \\ 0 & z_{b1}+z_{b2}+z_{b3} & 0 & 0 \\ 0 & 0 & z_{c1}+z_{c2}+z_{c3} & 0 \\ 0 & 0 & 0 & z_{n2} \end{bmatrix} \tag{5}$$

The mesh-impedance matrix is $[Z] = [C]_t.[z].[C]$, and the mesh-e.m.f. matrix is

$$[E] = [C]_t.[e].$$

Finally,
$$[I] = [Z]^{-1}.[E].$$

It is interesting to work out the matrix form of $[Z]$.

$$[Z] = \begin{bmatrix} 1 & 0 & 0 & -1 \\ 0 & 1 & 0 & -1 \\ 0 & 0 & 1 & -1 \end{bmatrix}.[z].\begin{bmatrix} 1 & 0 & 0 \\ 0 & 1 & 0 \\ 0 & 0 & 1 \\ -1 & -1 & -1 \end{bmatrix}.$$

Using (5), this becomes

$$[Z] = \begin{bmatrix} 1 & 0 & 0 & -1 \\ 0 & 1 & 0 & -1 \\ 0 & 0 & 1 & -1 \end{bmatrix}$$

$$\cdot \begin{bmatrix} z_{a1} + z_{a2} + z_{a3} & 0 & 0 \\ 0 & z_{b1} + z_{b2} + z_{b3} & 0 \\ 0 & 0 & z_{c1} + z_{c2} + z_{c3} \\ -z_{n2} & -z_{n2} & -z_{n2} \end{bmatrix},$$

$$\text{i.e.} \quad [Z] = \begin{bmatrix} z_{a1} + z_{a2} + z_{a3} + z_{n2} & z_{n2} & z_{n2} \\ z_{n2} & z_{b1} + z_{b2} + z_{b3} + z_{n2} & z_{n2} \\ z_{n2} & z_{n2} & z_{c1} + z_{c2} + z_{c3} + z_{n2} \end{bmatrix}.$$

Note that the separate elements z_{n2} occur as positive because the directions of the marked currents in the meshes are the same. (Contrast § 16·31 (9).)

Now the separate elements z_{n2} in this matrix are in the positions that would be occupied by mutual impedances in a branch-impedance matrix (see §16·31).

This shows that an unbalanced network free from mutual coupling has the property that the neutral impedance can be replaced by a mutual impedance (z_n) between the phases and a corresponding increase in the self-impedance of each phase. (Shown by the terms such as $z_{a1} + z_{a2} + z_{a3} + z_n$.)

16·34. A simple network containing coupling between branches

Fig. 96 (i) shows a network containing two couplings. The driving voltage e_a is given as are also the self-impedances z_{aa}, z_{bb}, z_{cc} and the mutual impedances z_{ab}, z_{ba}, z_{bc}, z_{cb}.

Fig. 96 (ii) shows the mesh-current diagram.

The relations between branch and mesh currents are:

$$\left.\begin{aligned} i_a &= I_1, \\ i_b &= \quad -I_2, \\ i_c &= \quad\quad I_3. \end{aligned}\right\} \tag{1}$$

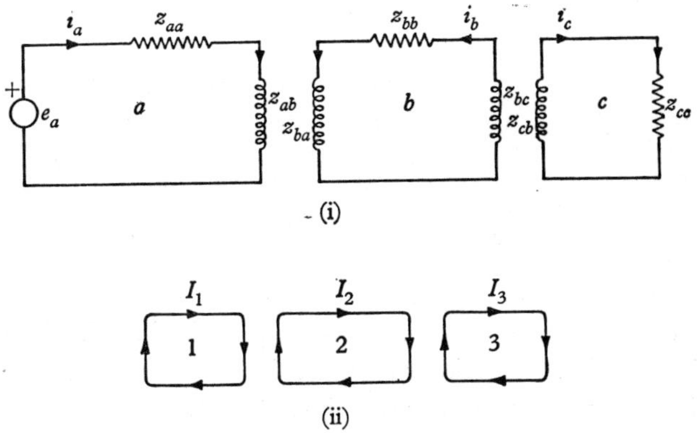

(i)

(ii)

Fig. 96

The transformation matrix is

$$[C] = \begin{bmatrix} 1 & 0 & 0 \\ 0 & -1 & 0 \\ 0 & 0 & 1 \end{bmatrix} \text{ and } [C]_t = \begin{bmatrix} 1 & 0 & 0 \\ 0 & -1 & 0 \\ 0 & 0 & 1 \end{bmatrix}. \tag{2}$$

The branch-e.m.f.-matrix is $[e] = \begin{bmatrix} e_a \\ 0 \\ 0 \end{bmatrix}$ and the mesh-e.m.f. matrix is

$$[E] = [C]_t \cdot [e] = \begin{bmatrix} e_a \\ 0 \\ 0 \end{bmatrix}. \tag{3}$$

The branch-impedance matrix is

$$[z] = \begin{bmatrix} z_{aa} & z_{ab} & 0 \\ z_{ba} & z_{bb} & z_{bc} \\ 0 & z_{cb} & z_{cc} \end{bmatrix}. \tag{4}$$

29

The mesh-impedance matrix is given by $[Z] = [C]_t \cdot [z] \cdot [C]$, i.e.

$$[Z] = \begin{bmatrix} 1 & 0 & 0 \\ 0 & -1 & 0 \\ 0 & 0 & 1 \end{bmatrix} \cdot \begin{bmatrix} z_{aa} & z_{ab} & 0 \\ z_{ba} & z_{bb} & z_{bc} \\ 0 & z_{cb} & z_{cc} \end{bmatrix} \cdot \begin{bmatrix} 1 & 0 & 0 \\ 0 & -1 & 0 \\ 0 & 0 & 1 \end{bmatrix}$$

$$= \begin{bmatrix} 1 & 0 & 0 \\ 0 & -1 & 0 \\ 0 & 0 & 1 \end{bmatrix} \cdot \begin{bmatrix} z_{aa} & -z_{ab} & 0 \\ z_{ba} & -z_{bb} & z_{bc} \\ 0 & -z_{cb} & z_{cc} \end{bmatrix},$$

giving

$$[Z] = \begin{bmatrix} z_{aa} & -z_{ab} & 0 \\ -z_{ba} & z_{bb} & -z_{bc} \\ 0 & -z_{cb} & z_{cc} \end{bmatrix}. \tag{5}$$

$[I]$ can now be calculated as $[I] = [Z]^{-1} . [E]$. Actually in this case there is little or no point in using a transformation matrix as the three meshes are the actual three sub-networks. In this case the equation $[i] = [z]^{-1} . [e]$ could be used on sight. However it is interesting to note that equation (5) for $[Z]$ is an illustration of the fact, previously mentioned, that if the 'clockwise positive' convention is observed for mesh currents, the non-diagonal elements of $[Z]$ will be negative, if the mesh currents are opposed in neighbouring meshes.

Taking the network as symmetric (i.e. $z_{ab} = z_{ba}$ etc.) and

$$z_{aa} = R_1 + j\omega L_1, \quad z_{bb} = R_2 + j\omega(L_2 + L_3),$$
$$z_{cc} = R_4 + j\omega L_4$$

and

$$z_{ab} = z_{ba} = j\omega M_1, \quad z_{bc} = z_{cb} = j\omega M_2,$$

then

$$[Z] = \begin{bmatrix} R_1 + j\omega L_1 & -j\omega M_1 & 0 \\ -j\omega M_1 & R_2 + j\omega(L_2 + L_3) & -j\omega M_2 \\ 0 & -j\omega M_2 & R_4 + j\omega L_4 \end{bmatrix}.$$

As the currents i_a, i_c are usually the only ones required, in more advanced work the network b is eliminated by taking the networks a and c to be coupled by a generalized mutual inductance. This cuts down the calculations involved.

Note. In connexion with inductively coupled circuits, the coupling reactance is taken as positive in the case when the coupled currents are directed alike, negative in the case when the currents are oppositely directed.

16·35. Calculation of a numerical case

Fig. 97 (i) shows a 3-phase, 3-wire star load supplied by voltages e_a, e_b, e_c. The complex values of the impedances are as shown, and there is no mutual coupling between them. It is desired to calculate the currents i_a, i_b, i_c given that $e_a = 50 + j28{\cdot}87$, $e_b = -50 + j28{\cdot}87$, $e_c = 0 - j57{\cdot}74$.

Note that in 3-phase work such as this, e_a, e_b, e_c are the *phase* voltages; the line voltages are actually given by the mesh-e.m.f. matrix $[E]$.

Fig. 97 (ii) shows the mesh-current diagram.

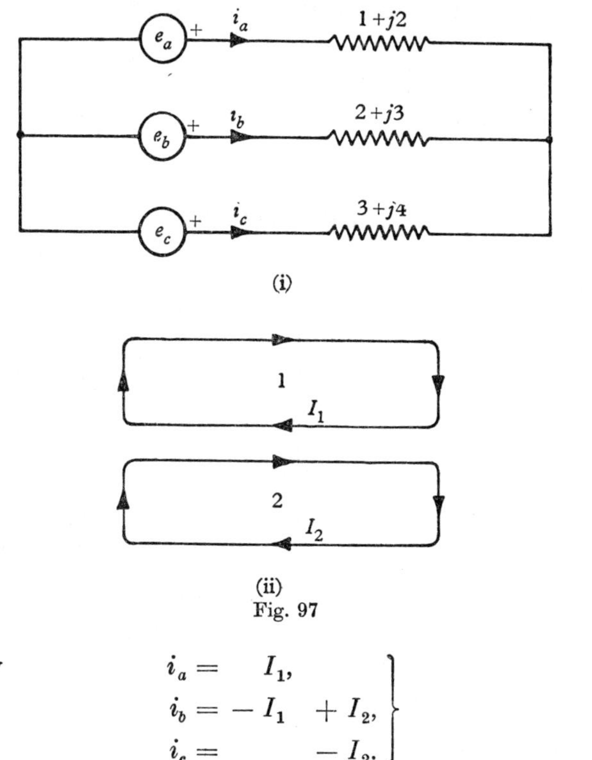

Fig. 97

Now
$$\left.\begin{aligned} i_a &= I_1, \\ i_b &= -I_1 + I_2, \\ i_c &= -I_2. \end{aligned}\right\} \qquad (1)$$

Thus
$$[i] = [C].[I], \qquad (2)$$

where the transformation matrix is

$$[C] = \begin{bmatrix} 1 & 0 \\ -1 & 1 \\ 0 & -1 \end{bmatrix} \text{ and } [C]_t = \begin{bmatrix} 1 & -1 & 0 \\ 0 & 1 & -1 \end{bmatrix}. \quad (3)$$

The mesh-e.m.f. matrix is given by

$$[E] = [C]_t \cdot [e] = \begin{bmatrix} 1 & -1 & 0 \\ 0 & 1 & -1 \end{bmatrix} \cdot \begin{bmatrix} 50 + j28{\cdot}87 \\ -50 + j28{\cdot}87 \\ 0 - j57{\cdot}74 \end{bmatrix},$$

i.e.

$$[E] = \begin{bmatrix} 100 + j0 \\ -50 + j86{\cdot}61 \end{bmatrix}. \quad (4)$$

The branch-impedance matrix is given by

$$[z] = \begin{bmatrix} 1 + j2 & 0 & 0 \\ 0 & 2 + j3 & 0 \\ 0 & 0 & 3 + j4 \end{bmatrix}.$$

The mesh-impedance matrix is therefore

$$[Z] = \begin{bmatrix} 1 & -1 & 0 \\ 0 & 1 & -1 \end{bmatrix} \cdot \begin{bmatrix} 1 + j2 & 0 & 0 \\ 0 & 2 + j3 & 0 \\ 0 & 0 & 3 + j4 \end{bmatrix} \cdot \begin{bmatrix} 1 & 0 \\ -1 & 1 \\ 0 & -1 \end{bmatrix}$$

$$= \begin{bmatrix} 1 + j2 & -(2 + j3) & 0 \\ 0 & 2 + j3 & -(3 + j4) \end{bmatrix} \cdot \begin{bmatrix} 1 & 0 \\ -1 & 1 \\ 0 & -1 \end{bmatrix},$$

giving

$$[Z] = \begin{bmatrix} 3 + j5 & -(2 + j3) \\ -(2 + j3) & 5 + j7 \end{bmatrix}. \quad (5)$$

Now

$$\Delta [Z] = \begin{vmatrix} 3 + j5 & -(2 + j3) \\ -(2 + j3) & 5 + j7 \end{vmatrix} = -15 + j34,$$

and $$\lfloor Z \rfloor_t = \begin{bmatrix} 3 + j5 & -(2 + j3) \\ -(2 + j3) & 5 + j7 \end{bmatrix}.$$

Thus $$[Z]^{-1} = \frac{1}{-15 + j34} \begin{bmatrix} 5 + j7 & 2 + j3 \\ 2 + j3 & 3 + j5 \end{bmatrix}$$

$$= \frac{-(15 + j34)}{1381} \begin{bmatrix} 5 + j7 & 2 + j3 \\ 2 + j3 & 3 + j5 \end{bmatrix}.$$

Now $[I] = [Z]^{-1}.[E]$, therefore

$$[I] = \frac{-(15 + j34)}{1381} \begin{bmatrix} 5 + j7 & 2 + j3 \\ 2 + j3 & 3 + j5 \end{bmatrix} . \begin{bmatrix} 100 + j0 \\ -50 + j86{\cdot}61 \end{bmatrix},$$

i.e.

$$\begin{bmatrix} I_1 \\ I_2 \end{bmatrix} = \frac{1}{1381} \begin{bmatrix} 163 - j275 & 72 - j113 \\ 72 - j113 & 125 - j177 \end{bmatrix} . \begin{bmatrix} 100 + j0 \\ -50 + j86{\cdot}61 \end{bmatrix}$$

$$= \frac{1}{1381} \begin{bmatrix} 22487 - j15614 \\ 16280 + j8376 \end{bmatrix}.$$

This gives $$\left.\begin{aligned} I_1 &\simeq 16{\cdot}3 - j11{\cdot}3 \\ I_2 &\simeq 11{\cdot}8 + j6{\cdot}1. \end{aligned}\right\} \quad (6)$$
and

From (2) and (6):

$$\begin{bmatrix} i_a \\ i_b \\ i_c \end{bmatrix} = \begin{bmatrix} 1 & 0 \\ -1 & 1 \\ 0 & -1 \end{bmatrix} . \begin{bmatrix} 16{\cdot}3 - j11{\cdot}3 \\ 11{\cdot}8 + j6{\cdot}1 \end{bmatrix},$$

giving

$$\underline{i_a = 16{\cdot}3 - j11{\cdot}3, \quad i_b = -4{\cdot}5 + j17{\cdot}4.}$$

$$\underline{i_c = -11{\cdot}8 - j6{\cdot}1.}$$

In conclusion, it is stressed again that no claim is made that matrix analysis provides any shorter method for the elementary circuits dealt with here. In practice such circuits would be done by ordinary methods.

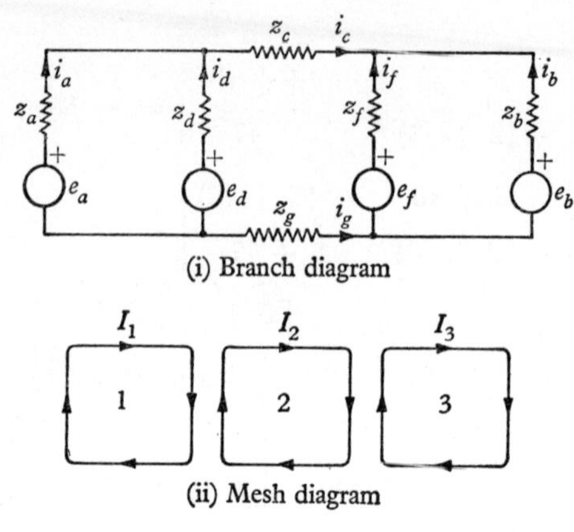

(i) Branch diagram

(ii) Mesh diagram

Fig. 98

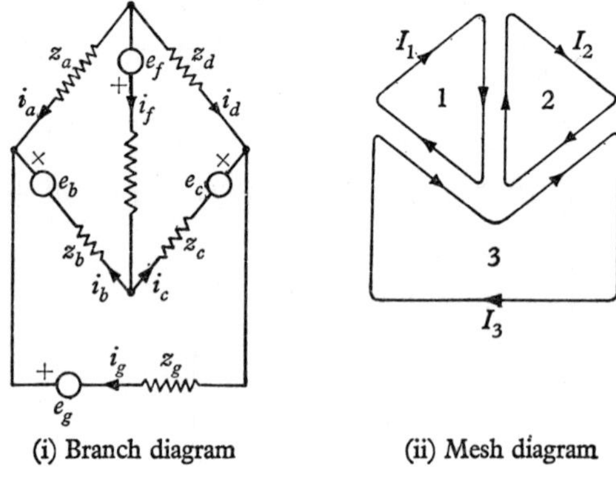

(i) Branch diagram

(ii) Mesh diagram

Fig. 99

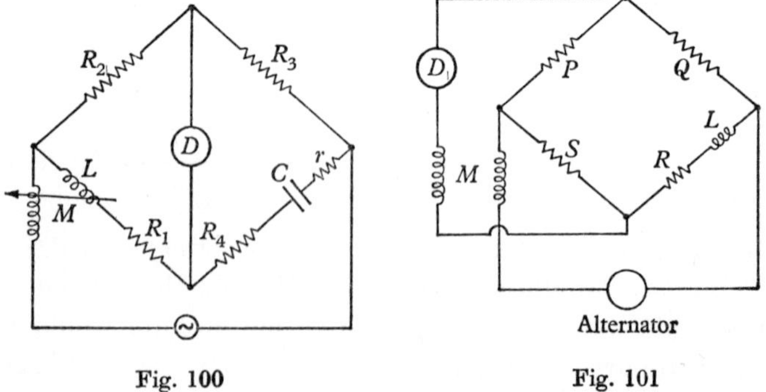

Fig. 100

Fig. 101

It is thought, however, that applying matrices to these simple circuits may have made clear the beginnings of the matrix analysis of electrical networks. It is hoped that the student will now be in a position to read text-books specializing in this topic.

Exercise 55

1. Write down the relations between the branch and mesh currents in the network shown in fig. 98. Deduce the transformation matrix. Write down the branch-impedance matrix and deduce the mesh-impedance matrix. Finally write down the branch-e.m.f. matrix and deduce the mesh-e.m.f. matrix.

2. Determine similar matrices to those in Question 1 for the bridge network shown in fig. 99.

Solve the following questions by matrix analysis:

3. A star-connected 3-phase load consists of impedances $0 - j4\ \Omega$, $6 + j0\ \Omega$ and $0 + j5\ \Omega$ connected to the red, yellow and blue phases respectively of a 3-phase, 3-wire, 220 volt supply. Calculate the current in any one of the three lines. [L.U.]

4. Derive the conditions of balance of the a.c. bridge shown in fig. 100 and explain the advantage of making R_2 zero.

This bridge, when used to measure the capacitance C and equivalent series resistance r of a capacitor, balanced under the following conditions:

$$M = 6693\ \mu\text{H.},\quad L = 19{,}220\ \mu\text{H.},\quad R_1 = 200\ \Omega,$$

$$R_2 = 0\ \Omega,\quad R_3 = 100\ \Omega \quad \text{and} \quad R_4 = 186{\cdot}9\ \Omega.$$

Calculate the capacitance and series resistance of the capacitor. [L.U.]

5. The circuit shown in fig. 101 is used to measure the frequency of the source of supply, D being a null detector. Obtain the conditions for balance, and hence derive an expression for ω (radians per second) which is independent of R. Determine the supply frequency when balance results for the component values:

$$P = Q = S = 100\ \Omega,\qquad L = 0{\cdot}1\ \text{henry},\qquad M = 0{\cdot}05\ \text{henry}.$$

6. The impedances of the three phases of a star-connected load (no neutral wire) are $5 + j20$, $12 + j0$ and $1 - j10$ in order. The line voltage is 400 volts. Find the line currents. [L.U.]

Note. In Questions 3 and 6 the voltage given is the line voltage. If e_R, e_Y, e_B are the phase voltages, the line voltages are $e_R - e_Y$, $e_Y - e_B$ and $e_B - e_R$. For example in Question 3, $e_R - e_Y = 220 + 0j$, $e_Y - e_B = -220(0{\cdot}5 + j0{\cdot}866)$, $e_B - e_R = 220\,(-0{\cdot}5 + j0{\cdot}866)$.

The student is advised to work in terms of e_R, e_Y, e_B and only replace by numerical values after finding the mesh-e.m.f. matrix.

Answers

1.

$$[C] = \begin{bmatrix} 1 & 0 & 0 \\ 0 & 0 & -1 \\ 0 & 1 & 0 \\ -1 & 1 & 0 \\ 0 & -1 & 1 \\ 0 & -1 & 0 \end{bmatrix},$$

$$[Z] = \begin{bmatrix} z_a + z_d & -z_d & 0 \\ -z_d & z_c + z_d + z_f + z_g & -z_f \\ 0 & -z_f & z_b + z_f \end{bmatrix},$$

$$[E] = \begin{bmatrix} e_a - e_d \\ e_d - e_f \\ -e_b + e_f \end{bmatrix}.$$

2.

$$[C] = \begin{bmatrix} -1 & 0 & 0 \\ 1 & 0 & -1 \\ 0 & -1 & 1 \\ 0 & 1 & 0 \\ 1 & -1 & 0 \\ 0 & 0 & 1 \end{bmatrix},$$

$$[Z] = \begin{bmatrix} z_a + z_f + z_b & -z_f & -z_b \\ -z_f & z_c + z_f + z_d & -z_c \\ -z_b & -z_c & z_b + z_c + z_g \end{bmatrix},$$

$$[E] = \begin{bmatrix} e_b + e_f \\ -e_c - e_f \\ -e_b + e_c + e_g \end{bmatrix}.$$

3. $i_R = 29{\cdot}68 - j11{\cdot}06, \qquad i_Y = -44{\cdot}04 - j19{\cdot}79,$
$i_B = 14{\cdot}36 + j30{\cdot}85.$

[Phase sequence taken as R-Y-B and line voltage $e_R - e_Y$ taken as reference vector $220 + j0$.]

4.

$$R_1R_3 - R_2R_4 - R_2r - \frac{M}{C} = 0,$$

$$\omega LR_3 + \frac{R_2}{\omega C} - \omega M(R_3 + R_4 + r) = 0.$$

If $R_2 = 0$, then the equations for balance are independent of the frequency of the applied voltage and they also give the values of C and r separately and directly.

$$C = 0{\cdot}33465\ \mu\text{F.}; \quad r = 0{\cdot}266\ \Omega.$$

5. $\omega^2 = \dfrac{QS - PR}{LM}, \quad LP - MP - MR - MS - MQ = 0.$

These give $$\omega^2 = \frac{MQS - P(LP - MP - MS - MQ)}{LM^2}.$$

$$\omega = 2000.$$

6. $0{\cdot}5 - j29{\cdot}5, \qquad 16{\cdot}1 - j11{\cdot}5, \qquad -16{\cdot}6 + j41{\cdot}0.$

Appendix

INVERSION AND CURRENT LOCI

In Chapter 5 examples were given of plotting impedance loci on Argand diagrams. This appendix gives a purely geometrical method for plotting vector-impedance loci, and also goes on to show how actual vector-current loci can be evolved.

A. 1. Definitions

In fig. 102 O is a fixed point and P any other point. If a point P' is taken on OP (produced if necessary) so that

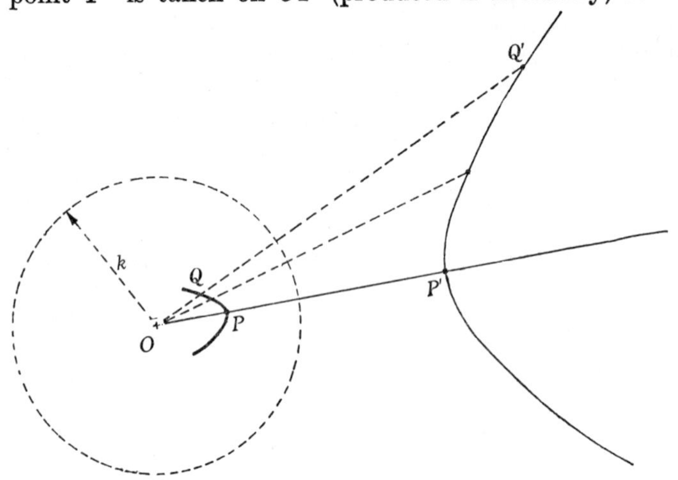

Fig. 102

$OP.OP' = k^2$, where k is a given constant, then P' is said to be the *inverse* of the point P with respect to the circle whose centre is O and radius k. O is called the *centre of inversion* and k the *radius of inversion*. k is also often called the constant of inversion.

If P' is the inverse of P, it is obvious that P is the inverse of P' with respect to the same centre and radius of inversion. The name 'inversion' is due to the fact that as $OP.OP' = k^2$, then $OP' = k^2/OP$; thus OP' varies inversely as OP.

If P describes a given curve and P' moves so as always to be the inverse of P, the locus of P' is a curve which is called the inverse, with respect to the circle of inversion, of the curve described by P. Fig. 102 shows an example of a curve and its inverse.

A. 2. Some useful theorems on inversion

I. *The inverse of a straight line with respect to a centre of inversion not on the line is a circle through the centre of inversion.*

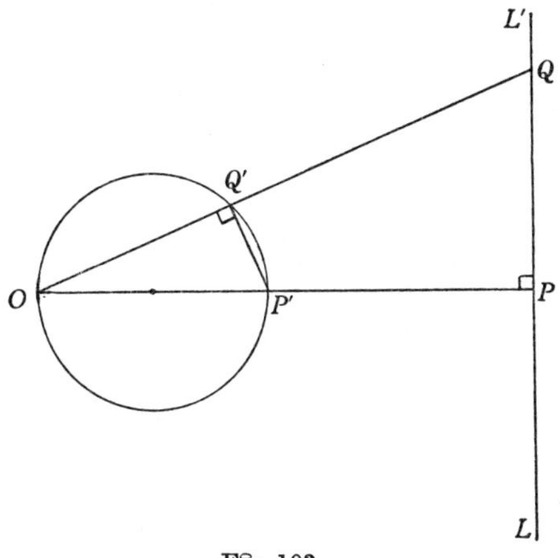

Fig. 103

Referring to fig. 103, let O be the centre of inversion, k the constant of inversion and LL' the given line. Draw OP perpendicular to LL'.

Take any other point Q on the line LL'.

On OP and OQ take points P', Q' such that $OP'.OP = k^2$ and $OQ'.OQ = k^2$, then P', Q' are the inverse points of P, Q.

Because $OP'.OP = OQ'.OQ$ $(= k^2)$, then $P'PQQ'$ is a cyclic quadrilateral. (Converse of theorem that products of segments of chords of a circle from an external point are equal.)

Thus $O\hat{Q}'P' = \hat{P}$ (exterior angle of cyclic quadrilateral = interior opposite angle).

But $\hat{P} = 90°$, by construction.

$$\therefore\ OQ'P' = 90°,$$

Now O and P' are fixed points, therefore the locus of Q', as Q moves along the line LL', is a circle passing through O and having OP' as diameter.

II. *The inverse of a circle with respect to a centre of inversion on its circumference is a straight line at right angles to the diameter through the centre of inversion.*

This is the converse of Theorem I above and follows easily from the fact that if Q' is the inverse of Q then Q must be

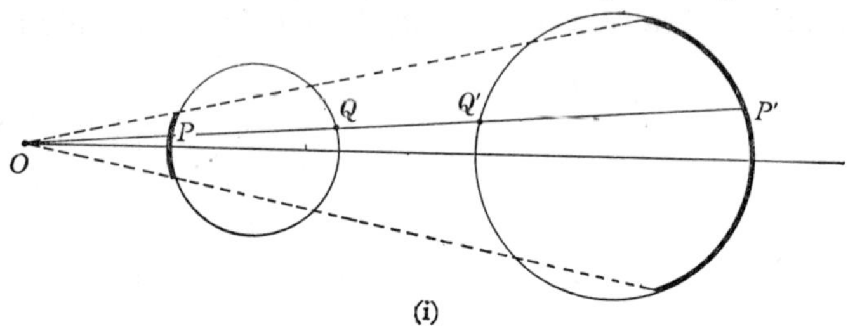

(i)

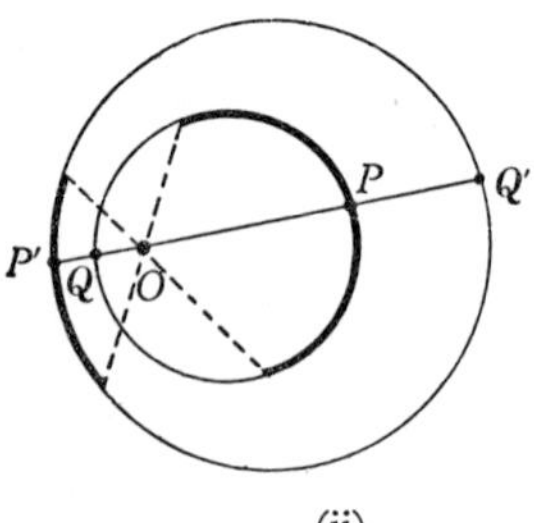

(ii)

Fig. 104

the inverse of Q'. Thus if the circle $OQ'P'$ is the inverse of the line LL', the line LL' is the inverse of the circle $OQ'P'$.

III. *The inverse of a circle with respect to a centre of inversion not on it is another circle.*

Let k be the radius of inversion and O the centre of inversion.

In fig. 104 let the smaller circle be the given circle in each case. In fig. 104 (i) O is outside this circle; in fig. 104 (ii) it is inside.

Draw a line OPQ to cut the given (small) circle in P and Q.

Let P' be the inverse of P and Q' the inverse of Q; then

$$OP.OP' = k^2. \tag{1}$$

Now $OP.OQ$ is constant (products of segments of chords of a given circle through a given point O).

Let $$OP.OQ = c. \tag{2}$$

From (1) and (2) $$\frac{OP.OP'}{OP.OQ} = \frac{k^2}{c},$$

i.e. $$\frac{OP'}{OQ} = \frac{k^2}{c} \quad \text{(constant).} \tag{3}$$

But as P (and Q) vary, the locus of Q is the given circle, therefore from (3) the locus of P' is a circle (similar figures).

Fig. 104 (i) and (ii) shows parts of the circles which are the inverses of each other in thick line.

A. 3. Application to current loci

In complex number notation a general impedance is denoted by $Z = R + jX$ and a general admittance by

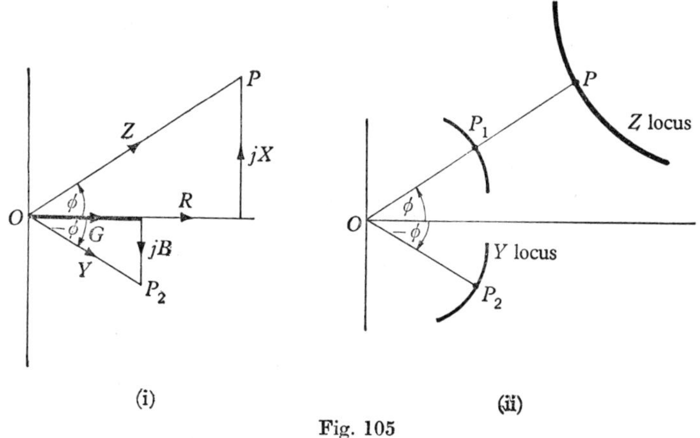

Fig. 105

$Y = G - jB$. The admittance, Y, of an impedance, Z, is given by $Y = 1/Z$.

If Z in polar form is $re^{j\phi}$, then Y in polar form is given by

$$Y = \frac{1}{Z} = \frac{1}{r} e^{-j\phi}.$$

In fig. 105 (i) the vector $\overrightarrow{OP}$, making angle ϕ with the initial line, represents Z, and $\overrightarrow{OP_2}$, making angle $-\phi$ with the initial line, represents Y.

Many cases arise in a.c. theory in which the (vector) impedance Z varies in a definite manner; that is, the point P has a definite locus. Corresponding to this there will be an admittance (vector) locus, the locus of P_2 (fig. 105 (ii)). The loci are related by the fact that $Z.Y = 1$; this means that $OP.OP_2 = 1$ and OP_2 makes an equal but opposite angle with the initial line to OP.

It is more convenient to plot the admittance locus (Y locus) as that of $G + jB$ instead of $G - jB$. This means that P_1 is the reflexion of P_2 in the initial line. If this is done the point P_1 is the inverse of P with respect to centre O and radius of inversion unity. (P_1 lies on the line OP and $OP_1.OP = 1$.) When the locus of P_1 has been plotted, if actual values of admittance are required the distances above the initial line are deliberately taken as negative imaginary instead of positive imaginary. (They are of course still read as positive imaginary for the Z locus.)

If the admittance locus is known the current locus can easily be obtained, usually merely by reading off lengths on a different scale, as the following example shows.

A. 4. Worked Examples

(1) Fig. 106 shows a circuit with two branches in parallel. The first branch consists of a variable resistance R in series

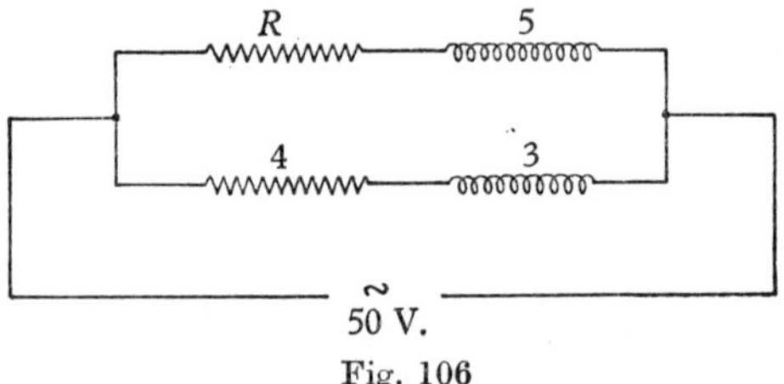

Fig. 106

with an inductive reactance of 5 ohms. The second consists of a resistance of 4 ohms in series with an inductive reactance of 3 ohms. The current locus will be drawn for an applied alternating voltage of 50 volts. From it the value of the current will be found when $R = 3$ ohms.

The complex impedances are given by

$$Z_1 = R + j5 \quad \text{and} \quad Z_2 = 4 + j3. \tag{1}$$

Scales are chosen for impedance and admittance as shown below. The scale for current can be deduced from that for admittance, using the fact that the impressed voltage is 50.

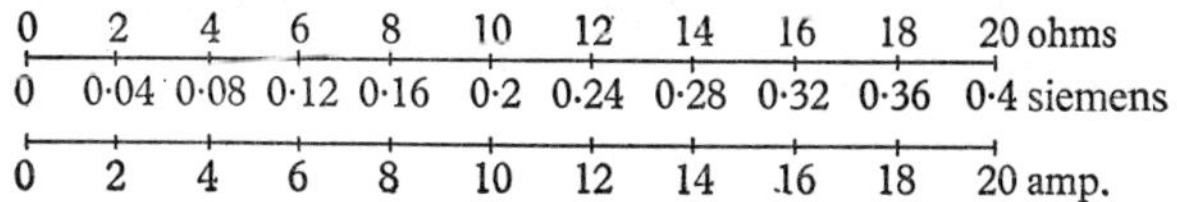

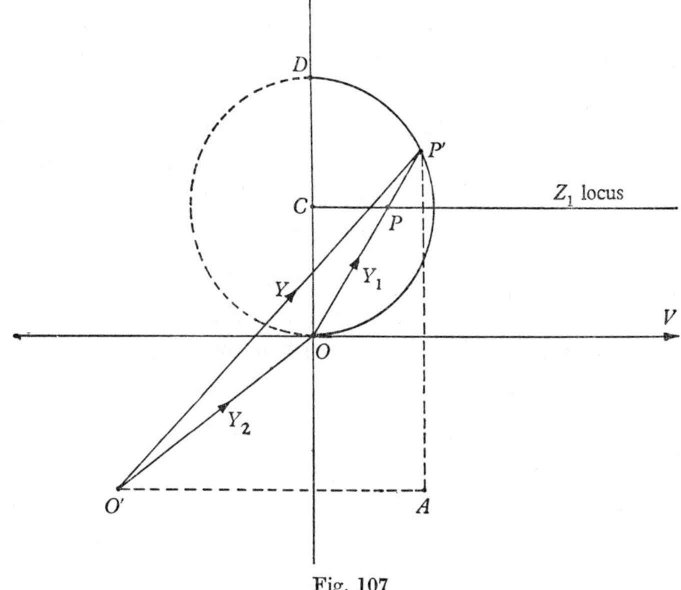

Fig. 107

Referring to fig. 107, a convenient point O is chosen as pole, and also a reference line OV to represent the direction of the voltage vector.

The locus of Z_1 as R varies is now drawn, making OC of length 5 on the ohm scale.

Now Y_1 ($= 1/Z_1$) is the inverse of this line with respect to centre O and radius of inversion 1.

Let $OD = 1/OC$, where OD is measured as $\frac{1}{5} = 0{\cdot}2$ on the siemen scale; then D is the inverse of C.

From §A. 2, Theorem I, the inverse of the Z_1 locus is the circle on OD as diameter, although as the Z_1 locus extends only to the right of C the inverse is only the semicircle shown in full line. This semicircle is therefore the locus for the admittance

Y_1 (remembering that for reading off susceptances negative values are plotted upwards).

Note that for convenience the siemen scale was chosen so that the centre of the circle for admittance lies on the extremity of the Z_1 locus.

$$\text{When} \qquad Z_1 = \overrightarrow{OP}, \quad Y_1 = \overrightarrow{OP'}. \tag{2}$$

$$\text{Now} \quad Y_2 = \frac{1}{Z_2} = \frac{1}{4+j3} = \frac{4-j3}{25} = 0{\cdot}16 - j0{\cdot}12. \tag{3}$$

The total admittance of the circuit, being a parallel circuit, is given by $Y = Y_1 + Y_2$, and the current is $50Y$ and can be read off on the ampere scale. As Y_2 is fixed in value it can best be added to Y_1 by selecting a new origin O' such that $\overrightarrow{O'O} = 0{\cdot}16 - j0{\cdot}12$; remembering that negative susceptances are plotted upwards. This means that O' is chosen at a distance 0·16 to the left of and 0·12 below O, both measured on the mho scale.

$\overrightarrow{O'P'}$ now represents the total admittance Y.

In fig. 107 P has been drawn to represent the required case when $R = 3$ ohms and $\overrightarrow{OP} = 3 + j5$. From measurement of $O'P'$ it is found that the total admittance in this case is 0·365 mho.

On the current scale, at voltage 50, this corresponds to a current of <u>18·25 A.</u>

If the complex admittance were needed it can be given by measuring $O'A$ and AP'.

In this case <u>$Y \simeq 0{\cdot}25 - j0{\cdot}26$.</u>

The angle which $O'P'$ makes with the reference line represents the angle of *lag*. In this case it is approximately 47°.

(2) A resistance of 8 ohms, an inductance of 4 millihenries, and a capacitance of 1000 picofarads are connected in series. An e.m.f. of constant amplitude 2 volts is introduced into the circuit, and its frequency is varied over a range including the resonant frequency. Draw the current locus and from it draw the resonance curve. [L.U.]

$R = 8\Omega$, $L = 4 \times 10^{-3}$ henries, $C = 1 \times 10^{-9}$ farad.

Let $\omega_0/2\pi$ be the resonant frequency.

Then $$\omega_0{}^2 = \frac{1}{LC} = \frac{1}{4} \times 10^{12} = 25 \times 10^{10}.$$

$$\therefore\ \underline{\omega_0 = 5 \times 10^5.} \tag{1}$$

It will be assumed that as the resonant frequency is high, the range over which the frequency is varied is small in comparison with it.

Let $\omega/2\pi$ be any frequency within this range; then

$$X = \omega L - \frac{1}{\omega C} \quad \text{and} \quad Z = R + jX. \tag{2}$$

Let $\omega = (\omega_0 + h)$, where h is small compared to ω_0. (3)

Then $$X = (\omega_0 + h)L - \frac{1}{(\omega_0 + h)C}$$

$$= \omega_0 L + hL - \frac{1}{\omega_0 C}\left(1 + \frac{h}{\omega_0}\right)^{-1}$$

$$\simeq \omega_0 L + hL - \frac{1}{\omega_0 C} + \frac{h}{\omega_0{}^2 C}.$$

But $$\omega_0 L = \frac{1}{\omega_0 C} \quad \text{and} \quad \frac{1}{\omega_0{}^2 C} = L,$$

therefore $$X \simeq hL + hL = 2hL,$$

i.e. $$\underline{X \simeq 8 \times 10^{-3}\, h \text{ ohms.}} \tag{4}$$

If $|Y|$ is the magnitude of the admittance in siemens, then the magnitude of the current vector is $2|Y|$ as the voltage amplitude is 2.

Fig. 108 shows suitable scales for impedance, admittance and current.

A suitable pole O is chosen and a reference line taken as the direction of the voltage vector.

OC is marked off as 8 ohms on the impedance scale. It is the impedance (purely resistive) at resonance when

0	8	16	24	32 ohms
0	0·0625	0·125	0·1875	0·25 siemens
0	0·125	0·25	0·375	0·5 amp.

Fig. 108

$\omega = \omega_0 = 5 \times 10^5$. As ω varies the Z locus will be the line ABC as shown in fig. 109. Corresponding to the point C, the admittance is $\frac{1}{8}$ siemen. As can be seen, the scale for

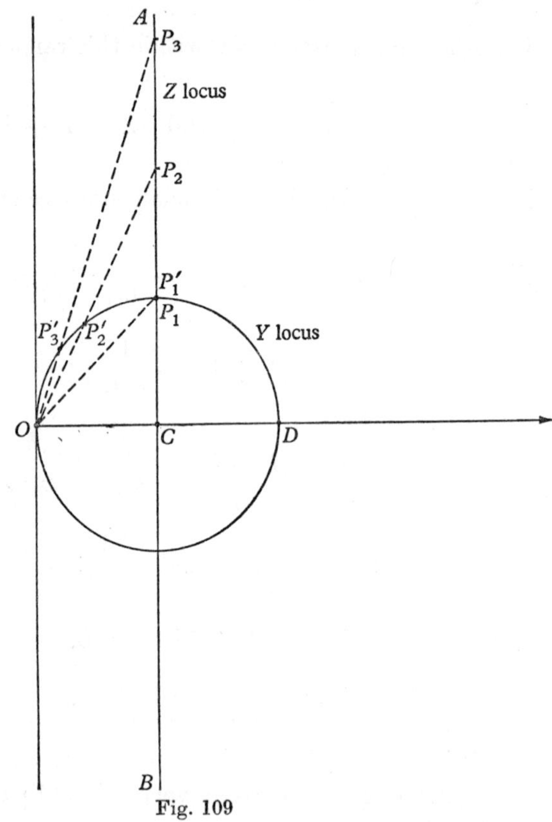

Fig. 109

admittance has been chosen so that D, the inverse of C, is such that $OD = 2OC$. Thus the admittance locus is the circle on OD as diameter. This is also the current locus if vector lengths are read off on the ampere scale.

The resonance curve can now be constructed using vector lengths OP' as current magnitudes and plotting them against values of $\frac{\omega}{\omega_0}$ for convenience. That is,

$$\frac{\omega}{\omega_0} = \frac{\omega_0 + h}{\omega_0} = 1 + \frac{h}{\omega_0}$$

will be plotted on the horizontal axis in fig.110. At resonance of course, $\omega/\omega_0 = 1$. Equation (4) is used to calculate values of h for varying values of X read off on CA on the ohm scale. Fig. 110 shows the form of the resulting resonance curve.

Only a few points have been taken in this case, but in practice as many points as necessary can be taken.

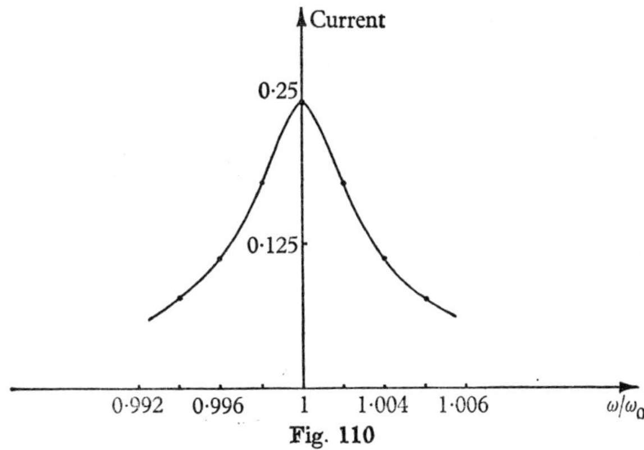

Fig. 110

Values for h/ω_0. (i) When $X = CP_1 = 8$ ohms (on impedance scale), from equation (4)

$$h = \frac{8 \times 10^3}{8}.$$

$$\therefore \frac{h}{\omega_0} = \frac{10^3}{5.10^5} = 0{\cdot}002.$$

The corresponding current is OP_1', measured on the ampere scale.

(ii) At P_2, $X = 16$ ohms and $h = 2 \times 10^3$, $h/\omega_0 = 0{\cdot}004$.

The current is OP_2' on the ampere scale

(iii) At P_3, $X = 24$ ohms and $h = 3 \times 10^3$, $h/\omega_0 = 0{\cdot}006$.

The current is OP_3' on the ampere scale.

Similar values are obtained by choosing points on CB.

Note. Analytical methods may of course be used in solving the problems which have here been solved graphically using vector methods. Although only giving approximate solutions, graphical methods do have the advantage of somewhat reducing the arithmetic involved in analytical methods.

Exercise 56

1. Draw a straight line and mark a point O whose perpendicular distance from the line is 2 in. Taking O as centre and 2 in. as the radius of inversion, mark a number of points on the inverse of the line. Sketch the full inverse curve.

2. Fig. 111 shows a linkage in which $OA = OB$, and $AP = PB = BQ = QA$. The point O is fixed. Prove that as the position of P varies,

(i) OPQ is a straight line,

(ii) P and Q are inverse points with respect to O as centre of inversion.

Note. This provides a mechanical device for constructing inverse curves; for if P describes a given curve, Q will describe the inverse of this curve. The linkage is called Peaucellier's Linkage, after its inventor.

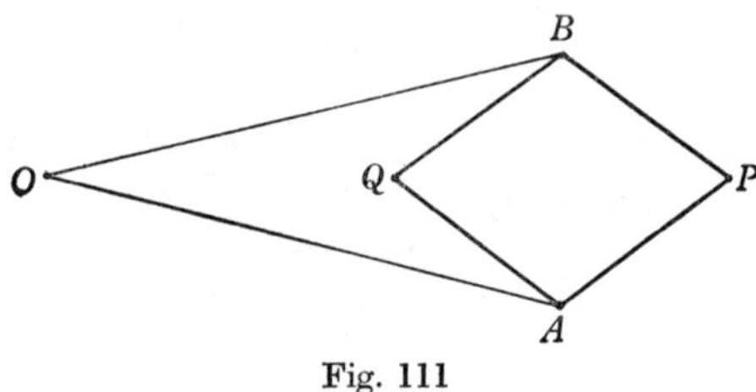

Fig. 111

3. A circuit consists of a resistance R in series with an inductive reactance of 4 Ω. Draw the vector current locus when 100 volts alternating is applied. What is the magnitude of the current when $R = 3$ Ω, and what is then the angle of lag?

4. A circuit has two branches in parallel; one contains a resistance R in series with a capacitative reactance of 5 Ω, and the other contains a resistance of 3 Ω in series with an inductive reactance of 4 Ω. Draw the vector current locus and find the magnitude of the current when $R = 4\ \Omega$. What is the phase angle for this value of R? For what value of R will the current and voltage be in phase? The applied voltage is 100 volts alternating.

5. A series circuit having an inductance of 0·2 henry and a variable resistance is connected to a supply of 230 volts at 50 Hz. Draw the vector locus of the current and from the locus find the magnitude of the current when the resistance has values 20, 50 and 100 ohms. [L.U.]

Answers

2. (i) $APBQ$ is a rhombus, therefore PQ bisects AB at right angles. $\triangle OAB$ is isosceles, therefore the perpendicular bisector of AB must also pass through O. Thus PQ passes through O.

(ii) Take $OB = l$ (constant) and $BQ = BP = a$ (constant).

Let $B\hat{O}Q = \alpha$ and $B\hat{Q}P = \theta$, both variable.

Then
$$\begin{aligned} OP.OQ &= (l\cos\alpha + a\cos\theta)(l\cos\alpha - a\cos\theta) \\ &= l^2\cos^2\alpha - a^2\cos^2\theta \\ &= (l^2 - a^2) - (l^2\sin^2\alpha - a^2\sin^2\theta). \end{aligned}$$

But in
$$\triangle OBQ: \frac{a}{\sin\alpha} = \frac{l}{\sin\theta}.$$

$$\therefore\ l^2\sin^2\alpha - a^2\sin^2\theta = 0.$$

Thus
$$OP.OQ = l^2 - a^2 = \text{constant}.$$

3. 20 A; 53° lag.

4. 22 A; 10° lag; $R = 2{\cdot}5\ \Omega$.

5. 3·5 A; 2·85 A; 1·95 A.

GENERAL MISCELLANEOUS EXERCISES

No hints about method are given in the answers to these questions as it is thought that by this stage the student should have confidence in his own ability.

1. Prove the results:

(i) $$\int_0^{\pi} \frac{d\theta}{4 - 3\sin\theta} = \frac{\pi}{\sqrt{7}} + \frac{2}{\sqrt{7}}\tan^{-1}\left(\frac{3}{\sqrt{7}}\right).$$

(ii) $$\int_0^{\infty} \frac{x\,dx}{(x+1)^2(x^2+4)} = \frac{2\pi}{25} + \tfrac{3}{25}\log_e 2 - \tfrac{1}{5}.$$

[L.U.]

2. (*a*) Evaluate $\displaystyle\int \log\frac{1}{t}\,dt$.

(*b*) If $I_n = \displaystyle\int e^{-x^3} x^{3n+2}\,dx$, show that

$$3I_n - 3nI_{n-1} + e^{-x^3}.\,x^{3n} = 0,$$

and use this result to evaluate $\displaystyle\int_0^1 e^{-x^3} x^8\,dx$. [L.U.]

3. Evaluate the following limits:

(i) $\displaystyle\lim_{x\to\frac{1}{2}\pi} \frac{\log_e \sin x}{(\pi - 2x)^2}$. (ii) $\displaystyle\lim_{\theta\to 0} \frac{\theta - \sin^{-1}\theta}{\sin^3\theta}$.

(iii) $\displaystyle\lim_{x\to 0} \frac{\tan x - \sin x}{\sin^3 x}$. (iv) $\displaystyle\lim_{x\to 0}\left[\frac{1}{\log_e(1+x)} - \frac{1}{x}\right]$.

4. Evaluate:

(i) $\displaystyle\int_0^{\frac{1}{2}\pi} \sin^7 x\,.\cos^5 x\,dx$. (ii) $\displaystyle\int_0^{\pi} \sin^4 x\,.\cos^6 x\,dx$.

(iii) $\displaystyle\int_{-\frac{1}{2}\pi}^{\frac{1}{2}\pi} \sin^3 x\,.\cos^6 x\,dx$. (iv) $\displaystyle\int_{-\pi}^{\pi} \sin^6 x\,.\cos^4 x\,dx$.

5. If $f(m, n) = \displaystyle\int_0^1 t^{m-1}(1-t)^{n-1}\,dt$ where m, n are positive, show that

$$f(m+1, n) + f(m, n+1) = f(m, n),$$
$$nf(m+1, n) = mf(m, n+1),$$

and express $f(m+1, n)$ and $f(m, n+1)$ in terms of $f(m, n)$.

Show, also, that if n is a positive integer,

$$f(m, n) = \frac{(n-1)!}{m(m+1)\dots(m+n-1)}. \qquad \text{[L.U.]}$$

6. Find the indefinite integrals:

(i) $\int e^{3x} \sin 5x\, dx.$ (ii) $\int \sqrt{\left(\frac{x-3}{x+2}\right)}\, dx.$

(iii) $\int \frac{d\theta}{3 - 2\sin\theta}.$

Evaluate $$\int_0^a x^3 \sqrt{(a^2 - x^2)}\, dx.$$

7. By the use of Taylor's theorem find correct to 5 decimal places:

(i) $\sin 29°$. (ii) $\tan 47°$. (iii) $\cos 59°$.

8. Using Maclaurin's theorem, show that

$$\log_e (1-x) = -x - \tfrac{1}{2}x^2 - \tfrac{1}{3}x^3 \dots .$$

Given that the equation $e^x = 2xe$ has a root near to e, by writing this root as $e(1-a)$, where a is small, show that a is given approximately by $a = \frac{e - 2 - \log_e 2}{e - 1}$ and hence evaluate this root to three decimal places. [I.E.E.]

9. Verify the following series by Taylor's theorem:

(i) $\log_e (a+x) = \log_e a + \frac{x}{a} - \frac{x^2}{2a^2} + \frac{x^3}{3a^3} - \dots .$

(ii) $\tan(x+h) = \tan x + h \sec^2 x + h^2 \sec^2 x \,.\, \tan x + \dots .$

10. Prove that if $y = \log_e(1+x)$ then $\frac{d^n y}{dx^n} = \frac{(-1)^{n-1}(n-1)!}{(1+x)^n}$

and hence obtain the expansion $\log_e(1+x) = \sum_{n=1}^{\infty} \frac{(-1)^{n-1} x^n}{n}$.

For what values of x is this result valid? Use it to show that if $|x|$ is sufficiently small, $\log_e (1 + e^x) = \log_e 2 + \frac{x}{2} + \frac{x^2}{8}$ if the fourth and higher powers of x are negligible. [I.E.E.]

11. Evaluate the integrals:

(i) $\int_0^5 x\sqrt{(25 - x^2)}\, dx$. (ii) $\int_4^8 \sqrt{(64 - x^2)}\, dx$.

(iii) $\int_0^1 \frac{x^2}{x^2 + 1} dx$. (iv) $\int_0^1 x^3 e^{2x}\, dx$.

12. Evaluate the integrals:

(i) $\int_0^\infty e^{-2x}\, dx$. (ii) $\int_1^\infty \frac{dx}{x\sqrt{(2x^2 + 1)}}$.

(iii) $\int_{-\infty}^\infty \frac{dx}{x^2 + 2x + 2}$. (iv) $\int_a^2 \frac{x^2\, dx}{\sqrt{(x^2 - 1)}}$ $(a > 1)$.

13. The function $f(x)$ is defined by $f(x) = \int_0^x \frac{\sin u\, du}{u}$. By interpreting the integral as an area, or by determining $f'(x)$ and $f''(x)$, show that $f(x)$ has a maximum value at $x = \pi$.

By expanding the integrand obtain the expansion of $f(x)$ in powers of x, and hence obtain an approximation to the maximum value by calculating $f(\sqrt{10})$ correct to two decimal places. [I.E.E.]

14. (*a*) Assuming that $\cos x + j \sin x = e^{jx}$, show that

$$\cos jx = \cosh x \quad \text{and} \quad \sin jx = j \sinh x.$$

If $x + jy = \sin (a + jb)$, find x and y in terms of a and b. Hence evaluate $\sin (\frac{1}{3}\pi + \frac{1}{4}j)$.

(*b*) Find $\log_e \left(\frac{3 - j}{2 + j}\right)$ in the form $a + jb$.

15. Find the value of $\sin^{-1} 3$ in the form $a + jb$.

16. If the complex variables $w = u + jv$, $z = x + jy$ are related by $w = \sin z$, find the loci in the w plane corresponding to (i) the line $x = \pi/4$, (ii) that part of the line $y = \log_e 2$ for which $0 \leqslant x \leqslant 2\pi$ in the z plane. Illustrate by sketches. [L.U.]

17. (i) If the equation

$$(1+\mu)L_1L_2x^3 + r(L_1+L_2)x^2 + \frac{L_2x}{C} + \frac{r}{C} = 0$$

has a purely imaginary root for x, the other symbols denoting constants, show that $L_2 = \mu L_1$ and solve the equation completely if this condition is satisfied.

(ii) Y and Z are complex numbers such that $1/Y = Z = R + j\omega L$ in which R and L are positive real constants. Determine the loci of the points representing Y and Z on an Argand diagram as ω varies from zero to infinity. [I.E.E.]

18. Evaluate $\cos(1+j2)$ in the form $a+jb$, obtaining the values of a and b correct to three decimal places. [I.E.E.]

19. If $z = x+jy$ and $w = u+jv$ are connected by $\left(\dfrac{z+c}{z-c}\right)^2 = \dfrac{w+2c}{w-2c}$, where c is real, prove that as the point representing z describes the circle $x^2+y^2=c^2$, the point representing w describes the segment of the real axis between $-2c$ and $+2c$, once in each direction. [I.E.E.]

20. What locus on the complex plane is represented by an equation $|z-a| = k$, where k is a positive constant, a is a complex constant, and z a variable complex number?

If $w = (2z+1)/(z+2)$, express $w - 5/4$ in terms of z and hence show that if the point representing z describes the imaginary axis, the point representing w will describe a circle with centre at $\frac{5}{4}$. What is the radius of this circle? [I.E.E.]

21. What is the locus on the z-plane of the point $z = \frac{1}{2}(1+e^{j\theta})$ as θ varies from 0 to 2π? Show that as z describes this locus the point $w = z/(1-z)$ describes the imaginary axis. Determine the values of z corresponding to $w = 0, \pm j$ and ∞. [I.E.E.]

22. Show that the quadratic equation $r^2 - 2r\cos\theta + 1 = 0$ is satisfied by $r = e^{j\theta}$ and by $r = e^{-j\theta}$. Express $\dfrac{r\sin\theta}{r^2-2r\cos\theta+1}$, regarded as a function of r, as a sum of partial fractions and hence derive the expansion

$$\frac{r\sin\theta}{r^2-2r\cos\theta+1} = r\sin\theta + r^2\sin 2\theta + r^3\sin 3\theta + \dots .$$

for r numerically less than 1. [I.E.E.]

23. A capacitor of 5 microfarads capacitance is shunted by a variable resistance R. Draw the vector locus of the impedance of the combination at 50 Hz as R varies from zero to infinity. Read off the impedance when the value of R is 500 ohms. [L.U.]

24. A coil having a resistance of 4 Ω and a reactance of 10 Ω, is connected in parallel with an identical coil which also has a variable resistance R in series with it. Draw the combined admittance locus as R varies, and find the combined admittance when $R = 16$ Ω.

25. A coil has a reactance of 20 Ω and a resistance of 4 Ω. It is connected in series with a resistance which can be varied between 0 and 50 Ω. Assuming a supply of 100 volts, draw the current vector locus.

26. Solve the following differential equations:

(i) $\dfrac{dy}{dx} + \dfrac{y}{x} = y^2 \log_e x.$

(ii) $y^2 + x^2 \dfrac{dy}{dx} = xy \dfrac{dy}{dx}.$

(iii) $\dfrac{dy}{dx} = \dfrac{y - 3x + 1}{2x + y - 4}.$

27. Solve the differential equations:

(i) $\dfrac{dy}{dx} = y \tan x - 2 \sin x.$

(ii) $\dfrac{dy}{dx} = \dfrac{x(1 + y^2)}{(1 + x^2)}.$ [I.E.E.]

28. Solve the differential equation, with the condition given:

$$(x^2 - y^2)\frac{dy}{dx} = xy; \quad y = 1 \text{ when } x = 0. \qquad \text{[L.U.]}$$

29. The amplitude a of the output of an oscillator varies with the time t according to the equation

$$2\frac{da}{dt} = \eta\left(a - \frac{a^3}{4}\right)$$

in which η is a positive constant. By considering the isoclines of this equation, or by solving the equation, determine and show on a sketch the general nature of the solution curves on the (t, a) plane. Show that $d^2a/dt^2 = 0$ when $a = 2/\sqrt{3}$. What conclusion may be drawn from this regarding the shape of the solution curves? (Only positive values of a need be considered). [I.E.E.]

30. What is meant by an isocline of a first-order differential equation? Show that the isoclines of the equation

$$(x + y)\frac{dy}{dx} = x - y$$

are straight lines through the origin, and that two of these lines have gradients equal to the value of dy/dx to which they correspond. What deduction may be made about these two lines? Indicate by a sketch the general form of the solution curves of this equation. [I.E.E.]

31. A solution curve of the differential equation $dy/dx = x + y^3$ is started at the point (x_0, y_0) in the first quadrant, and is followed in the direction of increasing x. Show that the curve remains in the first quadrant, and that, if it passes through (x_1, y_1) with $x_1 > x_0$, then

$$x_1 - x_0 = \int_{y_0}^{y_1} \frac{dy}{x + y^3}.$$

Use the fact that x is positive to show that

$$x_1 - x_0 < 1/2y_0^2 - 1/2y_1^2$$

and deduce that on this solution curve y will tend to infinity when $x < x_0 + 1/2y_0^2$. [I.E.E.]

32. (*a*) Solve the equation $\dfrac{d^3y}{dx^3} - 4\dfrac{dy}{dx} = \sin 2x$.

(*b*) Solve the equation $x^2\dfrac{d^2y}{dx^2} - 3x\dfrac{dy}{dx} + 4y = x^3$.

33. (i) Find the general solution of the differential equation

$$\frac{d^2y}{dx^2} + 4\frac{dy}{dx} + 5y = 3\sin x + x^2.$$

(ii) By applying the substitution $r = e^t$, or otherwise, solve the differential equation $r^2 \dfrac{d^2u}{dr^2} - 2u = r$, given that when $r = 1$, $u = 0$ and $du/dr = 0$. [L.U.]

34. Change the independent variable in the differential equation

$$(1 - x^2)\frac{dy}{dx} + xy = x^2$$

from x to θ by the substitution $x = \sin\theta$. Hence or otherwise obtain y as a function of x. [I.E.E.]

35. Change the independent variable in the equation

$$x\frac{d^2y}{dx^2} - \frac{dy}{dx} + 4x^3y = 0$$

by the substitution $x^n = u$.

Show that, with suitable choice of n, the new equation is linear with constant coefficients, and hence determine y as a function of x, given that $y = 1$ and $dy/dx = 0$ when $x = \sqrt{\pi}$. [I.E.E.]

36. Show that the equation

$$\frac{d^2y}{d\theta^2} + \tan\theta\frac{dy}{d\theta} + n^2\cos^2\theta\, y = 0$$

is reduced to the linear form with constant coefficients by the substitution $x = \sin\theta$. Hence obtain the solution for which $y = 0$ and $dy/d\theta = 1$ when $\theta = 0$. [I.E.E.]

37. Show that the solution of the equation

$$\frac{d^2y}{dt^2} + 2n\frac{dy}{dt} + n^2y = A\cos pt$$

for which y and dy/dt both vanish when $t = 0$ can be written

$$y = A\{\cos(pt - \phi) - e^{-nt}(nt + \cos\phi)\}/(n^2 + p^2),$$

where $$\tan\phi = \frac{2np}{(n^2 - p^2)}.$$ [L.U.]

38. Solve the following differential equations, with the conditions given:

(i) $\dfrac{d^2x}{dt^2} + 9x = t + \tfrac{1}{3}$; $x = \tfrac{1}{6}$, $\dfrac{dx}{dt} = \tfrac{1}{3}$, when $t = 0$.

(ii) $\dfrac{d^2y}{dx^2} + 9y = 4e^{3x}$; $y = 2$, $\dfrac{dy}{dx} = 3$, when $x = 0$.

(iii) $\dfrac{d^2y}{dx^2} - 2\dfrac{dy}{dx} - 3y = 2x - 1$; $y = \tfrac{1}{2}$, $\dfrac{dy}{dx} = -\tfrac{2}{3}$,

when $x = 0$.

(iv) $\dfrac{d^2i}{dt^2} - 6\dfrac{di}{dt} + 13i = 39$; $i = 5$, $\dfrac{di}{dt} = 2$, when $t = 0$.

(v) $\dfrac{d^2y}{dx^2} - 2\dfrac{dy}{dx} = 4x$; $y = 2$, $\dfrac{dy}{dx} = 1$, when $x = 0$.

(vi) $\dfrac{d^2i}{dt^2} - 2\dfrac{di}{dt} + 2i = 3\sin t$; $i = 0 = \dfrac{di}{dt}$, when $t = 0$.

(vii) $\dfrac{d^2i}{dt^2} + 9i = 2\cos 3t$; $i = 0 = \dfrac{di}{dt}$, when $t = 0$.

(viii) $\dfrac{d^2y}{dx^2} + 10\dfrac{dy}{dx} + 25y = e^{-5x} + 2\sin 5x$; $y = 0 = \dfrac{dy}{dx}$,

when $x = 0$.

(ix) $x^2\dfrac{d^2y}{dx^2} - x\dfrac{dy}{dx} + y + 2x = 0$; $y = 0$ when $x = 1$

and when $x = e$. [L.U.]

39. Determine the solution of the differential equation

$$\frac{d^2x}{dt^2} + 4\frac{dx}{dt} + 4x = 2\cos 2t,$$

which satisfies the initial conditions $x = dx/dt = 0$ at $t = 0$. Distinguish between the steady and transient parts of the solution, and show that for $t > 0$ the greatest numerical value of the transient is e^{-1} times the amplitude of the steady solution. Indicate by a sketch the graphs of the complete solution and its two parts, over the interval $t = 0$ to $t = \pi$. [I.E.E.]

40. Solve the equation

$$\frac{d^2x}{dt^2} + 2\frac{dx}{dt} + 2x = 10 \cos \omega t.$$

Determine the amplitude of the steady oscillation for large values of t, and show that this never exceeds 5, whatever the frequency. [I.E.E.]

41. Solve the following differential equations by Laplace transform methods, subject to the stated initial conditions:

(i) $\dfrac{d^2y}{dt^2} - \dfrac{dy}{dt} + 4y = 3 \sin 2t; \quad y = 3, \dfrac{dy}{dt} = 0$ at $t = 0$.

(ii) $\dfrac{d^2y}{dt^2} + 2\dfrac{dy}{dt} + y = 2te^{-t}; \quad y = 0, \dfrac{dy}{dt} = 1$ at $t = 0$.

(iii) $\dfrac{d^2y}{dt^2} + 4\dfrac{dy}{dt} + 4y = 4e^{jt}; \quad y = 0 = \dfrac{dy}{dt}$ at $t = 0$.

(iv) $\dfrac{d^2x}{dt^2} - y = 0, \dfrac{d^2y}{dt^2} + 4x = 0; \ y = x = 2; \dfrac{dy}{dt} = \dfrac{dx}{dt} = 0,$ when $t = 0$.

(v) $\dfrac{d^2x}{dt^2} + 2\dfrac{dx}{dt} - y = 1, \dfrac{d^2y}{dt^2} - 3\dfrac{dx}{dt} + 2y = 0;$

$$y = x = \frac{dy}{dt} = \frac{dx}{dt} = 0 \quad \text{at} \quad t = 0.$$

42. Solve the simultaneous equations

$$3\frac{dx}{dt} = -4x + 2y, \quad 3\frac{dy}{dt} = x - 5y$$

with the initial conditions $x = 3$, $y = 0$ at $t = 0$. Sketch the graphs of x and y against t for $t > 0$ and determine the maximum value of y and the corresponding values of x and t. [I.E.E]

43. Solve the simultaneous differential equations

$$\frac{d^2x}{dt^2} + x + y = 0, \quad 4\frac{d^2y}{dt^2} - x = 0,$$

subject to the conditions that, when $t = 0$, $x = 2a$, $y = -a$, $\dfrac{dx}{dt} = 2b$, $\dfrac{dy}{dt} = -b$, and show that the solution is then purely periodic. [L.U.]

44. Solve the equations

$$\frac{dx}{dt} + y = \sin t, \quad \frac{dy}{dt} + x = \cos t,$$

subject to the conditions $x = 2$, $y = 0$, when $t = 0$. [L.U.]

45. x and y are functions of t, satisfying the simultaneous equations

$$\frac{dx}{dt} - 2y = \cos t, \quad \frac{dy}{dt} + 2x = 2 \sin t.$$

Determine the solution satisfying the initial conditions $x = 1$, $y = 0$ at $t = 0$, and obtain the r.m.s. values of x and y for this solution. For what initial value of x will the solution for y be identically zero, and what is then the solution for x?

[I.E.E.]

46. Variables x and y satisfy the simultaneous differential equations

$$\frac{dx}{dt} + \frac{dy}{dt} - x + y = e^{2t}, \quad \frac{dx}{dt} + y = 2e^{2t}.$$

If at $t = 0$, $x = 0$ and $y = 2$, show that immediately after $t = 0$, x becomes positive. [I.E.E.]

47. x and y are functions of t satisfying the simultaneous differential equations

$$\frac{dx}{dt} + 3x + 5y = \cos 3t, \quad 2x - \frac{dy}{dt} - y = 0.$$

Initially, that is at $t = 0$, x and y are both zero. Show that the initial value of dx/dt is 1 and hence obtain x in terms of t.

[I.E.E.]

48. The displacement x of an oscillating system at time t satisfies the differential equation:

$$\frac{d^2x}{dt^2} + 2a\frac{dx}{dt} + (a^2 + \omega^2)\, x = h\, e^{-at} \cos \omega t.$$

Determine the solution for which x and dx/dt both vanish at $t = 0$, and show that in the subsequent oscillation the maximum amplitude is $h/2a\omega e$. [I.E.E.]

49. The charge q on the capacitor in a given electric circuit satisfies the equation

$$L\frac{d^2q}{dt^2} + R\frac{dq}{dt} + \frac{q}{C} = V,$$

where $R^2 > \frac{4L}{C}$ and V is constant. At the instant $t = 0$, when the switch is closed, q and $I \left(= \frac{dq}{dt}\right)$ are both zero. Show that for $t > 0$,

$$q = VC\left[1 + \frac{\beta}{\alpha - \beta}e^{-\alpha t} - \frac{\alpha}{\alpha - \beta}e^{-\beta t}\right],$$

where $-\alpha$ and $-\beta$ are the roots of the quadratic equation

$$Lx^2 + Rx + \frac{1}{C} = 0.$$

Show further that the current reaches a maximum value at time

$$t = \frac{1}{\beta - \alpha}\log_e\frac{\beta}{\alpha}.$$ [I.E.E.]

50. Using Laplace transform methods, solve the following problem:

A constant voltage V_0 is applied at $t = 0$ to a series L-C-R circuit. Assuming that $q = 0 = i$ at $t = 0$, show that the current at time t is given by

$$i = \frac{V_0}{nL}e^{-kt}\sin nt \quad \text{for} \quad n^2 > 0,$$

$$= \frac{V_0}{L}te^{-kt} \quad \text{for} \quad n^2 = 0,$$

where $$k = \frac{R}{2L} \quad \text{and} \quad n^2 = \frac{1}{LC} - \frac{R^2}{4L^2}.$$

51. An electrical circuit consists of an inductance L, resistance R and capacitance C in series. A constant e.m.f. E is applied in series with the circuit at time $t = 0$ when the current i and potential v across the capacitor are zero. Obtain the differential equation for v.

Find the values of v and i at time t, given that $L = 1/5C\omega^2$ and $R = 2/5C\omega$, and prove that the greatest value of v is $E(1 + e^{-\frac{1}{3}\pi})$. [L.U.]

52. A circuit consists of inductance L and capacitance C in series. An alternating electromotive force $E \sin nt$ is applied to the circuit commencing at time $t = 0$, the initial current and charge on the capacitor being zero. Prove that the current at time t is given by

$$I = \frac{nE}{L(n^2 - \omega^2)}\{\cos \omega t - \cos nt\}, \quad CL\omega^2 = 1.$$

Prove also that if $n = \omega$, the current at time t is

$$\frac{Et \sin \omega t}{2L}.$$ [L.U.]

53. An e.m.f. $E \cos pt$ is switched across an electric circuit consisting of a coil of inductance L henries, resistance R ohms, in series with a capacitor of capacitance C farads. State the differential equation for the charge q in the capacitor at any time t and the form of the general solution when the resistance is just sufficient to prevent natural oscillations.

If $L = 0{\cdot}0001$, $R = 2$, find the value of C for this condition to hold, and calculate the amplitude of the steady current for an imposed peak voltage of 100 at 50 Hz. [L.U.]

54. Solve the differential equations

$$7\frac{di_1}{dt} + 4\frac{di_2}{dt} + 50i_1 = E,$$

$$2\frac{di_1}{dt} + 5\frac{di_2}{dt} + 20i_2 = 0,$$

given that E is constant and $i_1 = i_2 = 0$ when $t = 0$.

Sketch a network which would give these equations.

55. Sketch a circuit which satisfies the equations

$$L\frac{di_1}{dt} - Ri_2 = E, \quad Ri_2 + \frac{q}{C} = 0, \quad i_1 + i_2 = \frac{dq}{dt}.$$

If the voltage E is constant, find the differential equation satisfied by q. Show that q oscillates only if $L < 4CR^2$.

If this condition is satisfied and if $q = 0 = i_1$ at $t = 0$, show that

$$q = EC - \frac{E}{2Rp} e^{-t/2RC} (\sin pt + 2RCp \cos pt),$$

where $p^2 = \frac{1}{LC} - \frac{1}{4R^2C^2}$.

56. Two points A, B are joined by a wire of resistance R without self-inductance; B is joined to a third point C by two wires each of resistance R, of which one is without self-inductance and the other has a coefficient of self-inductance L. If the ends A, C are kept at a potential difference $E \cos \omega t$ and if there are no mutual inductances, prove that the current in AB is

$$\frac{E}{R}\sqrt{\left(\frac{4R^2 + L^2\omega^2}{9R^2 + 4L^2\omega^2}\right)} \cos (\omega t - \alpha),$$

where
$$\tan \alpha = \frac{RL\omega}{(6R^2 + 2L^2\omega^2)}.$$

Find also the difference of potential of B and C. [L.U.]
(*Note.* The required current is the steady-state current.)

57. Find the general solutions of the differential equation $\frac{d^2x}{dt^2} + 2h\frac{dx}{dt} + x = 0$ for the following cases: (i) $h > 1$, (ii) $h = 1$, (iii) $h < 1$.

The above equation, with $0 < h < 1$, represents the motion of a galvanometer. To determine h, the degree of damping, the ratio ϵ of the amplitudes of consecutive half-swings is found. Show that

$$h^2 = \frac{(\log \epsilon)^2}{\{\pi^2 + (\log \epsilon)^2\}}.$$ [C.U.]

58. A ballistic galvanometer has an undamped period of 10 sec. Its deflexion for a steady current of 1 microampere is 250 mm. when the scale is at a distance of 1 metre. What would be its throw for a charge of 1 microcoulomb?

59. The rectangular coil of a moving-coil instrument is wound on an aluminium former of resistivity $2{\cdot}67 \times 10^{-8}$ Ωm, and of cross-sectional area 2 mm^2. The sides of the former, each 30 mm in length, move in a uniform flux density of 0·1 tesla. The width of the former is 20 mm, the end members with the axis of rotation through their centres being effectively outside the permanent magnet field.

Deduce the damping torque exerted on the former when it is rotating with an angular velocity of 2 rad/s. [L.U.]

60. The coil of a moving-coil galvanometer has 400 turns and is suspended in a uniform field of strength 0·1 tesla. It moves against a control torque of 2×10^{-7} newton metre per radian. The coil is rectangular and is 20 mm wide and 30 mm high; its moment of inertia is $2{\cdot}5 \times 10^{-7}$ kg m^2.

If the galvanometer resistance is 250 ohms, calculate a value for the resistance which, when connected across the galvanometer terminals, will give critical damping. Assume the damping to be entirely electromagnetic.

61. If ϕ is a function of x and y, where $x = e^u \cos v$, $y = e^u \sin v$, show that

$$\frac{\partial \phi}{\partial u} = x\frac{\partial \phi}{\partial x} + y\frac{\partial \phi}{\partial y}, \quad \frac{\partial \phi}{\partial v} = x\frac{\partial \phi}{\partial y} - y\frac{\partial \phi}{\partial x},$$

and hence prove the relation

$$\frac{\partial^2 \phi}{\partial u^2} + \frac{\partial^2 \phi}{\partial v^2} = (x^2 + y^2)\left(\frac{\partial^2 \phi}{\partial x^2} + \frac{\partial^2 \phi}{\partial y^2}\right).$$ [L.U.]

62. (i) If $z = \dfrac{\sin(x+t)}{\sin(x-t)}$, prove that

$$\frac{\partial}{\partial t}\left(\frac{1}{z}\frac{\partial z}{\partial t}\right) = \frac{\partial}{\partial x}\left(\frac{1}{z}\frac{\partial z}{\partial x}\right).$$

(ii) If u and v are functions of r and θ which satisfy the equations

$$\frac{\partial u}{\partial r} = \frac{1}{r}\frac{\partial v}{\partial \theta} \quad \text{and} \quad \frac{\partial v}{\partial r} = -\frac{1}{r}\frac{\partial u}{\partial \theta},$$

prove that

$$r^2\frac{\partial^2 u}{\partial r^2} + r\frac{\partial u}{\partial r} + \frac{\partial^2 u}{\partial \theta^2} = 0.$$ [I.E.E.]

63. Evaluate du for the function $u = e^{\frac{1}{2}x} \cos(x - \phi)$ when

$$x = 0,\ \phi = \tfrac{1}{2}\pi,\ dx = 0{\cdot}2 \text{ and } d\phi = -0{\cdot}3.$$

64. If $u = 1/r$, where $r = \sqrt{(x^2 + y^2 + z^2)}$, show that

$$\left(\frac{\partial u}{\partial x}\right)^2 + \left(\frac{\partial u}{\partial y}\right)^2 + \left(\frac{\partial u}{\partial z}\right)^2 = \frac{1}{r^4}.$$

65. If $V = x^n f(Y, Z)$, where $Y = y/x$ and $Z = z/x$, prove that

$$x\frac{\partial V}{\partial x} + y\frac{\partial V}{\partial y} + z\frac{\partial V}{\partial z} = nV. \qquad \text{[L.U.]}$$

66. If $z = f(x, y)$ and $x = \frac{1}{2}(u^2 - v^2)$, $y = uv$, show that

(i) $$u\frac{\partial z}{\partial v} - v\frac{\partial z}{\partial u} = 2\left(x\frac{\partial z}{\partial y} - y\frac{\partial z}{\partial x}\right).$$

(ii) $$\frac{\partial^2 z}{\partial u^2} + \frac{\partial^2 z}{\partial v^2} = (u^2 + v^2)\left(\frac{\partial^2 z}{\partial x^2} + \frac{\partial^2 z}{\partial y^2}\right). \qquad \text{[L.U.]}$$

67. If

$$S = \frac{2V}{y} - \frac{2\pi}{3}xy + \pi y\sqrt{(x^2 + y^2)},$$

where V is constant, show that S is a minimum when

$$y = \frac{\sqrt{5}}{2}x \quad \text{and} \quad x = \sqrt[3]{\frac{24V}{25\pi}}.$$

68. Show that the expression $x^2 + y^2 + 2\lambda xy + x^2y^2$, where λ is a positive constant, has a minimum value at $x = 0$, $y = 0$ if $\lambda \leqslant 1$, and that if $\lambda > 1$ it has a minimum value at each of the points $x = -y = \pm\sqrt{(\lambda - 1)}$. [L.U.]

69. Six equal uniform conducting wires form a tetrahedron $ABCD$. A given current i is introduced at A and removed at a point P of BC such that $BP = \lambda . BC$. Show that the current in AD is independent of the position of P, and show that the ratio of the current in BP to that in CP is

$$\left(\frac{3 - 2\lambda}{1 + 2\lambda}\right). \qquad \text{[C.U.]}$$

70. Solve, by determinants, the equations:

(i)
$$\begin{aligned} 4x + 3y + 5z &= 11, \\ 9x + 4y + 15z &= 13, \\ 12x + 10y - 3z &= 4. \end{aligned}$$

(ii)
$$\begin{aligned} 2x + y - 2z &= 3, \\ x - 5y + 9z &= 4, \\ 5x - 2y + 3z &= 9. \end{aligned}$$

71. Using Relaxation methods, solve the following equations correct to 4 decimal places:

$$\begin{aligned} 9{\cdot}32 i_1 - 1{\cdot}5\, i_2 + 0{\cdot}63 i_3 &= 5, \\ 2{\cdot}1\, i_1 + 8{\cdot}7\, i_2 - 1{\cdot}3\, i_3 &= 4{\cdot}5, \\ 1{\cdot}7\, i_1 - 2{\cdot}41 i_2 + 10{\cdot}7\, i_3 &= 7{\cdot}4. \end{aligned}$$

72. Use the mesh-current method for solving for the branch currents in the two networks shown in figs. 112, 113. Solve the equations by use of determinants.

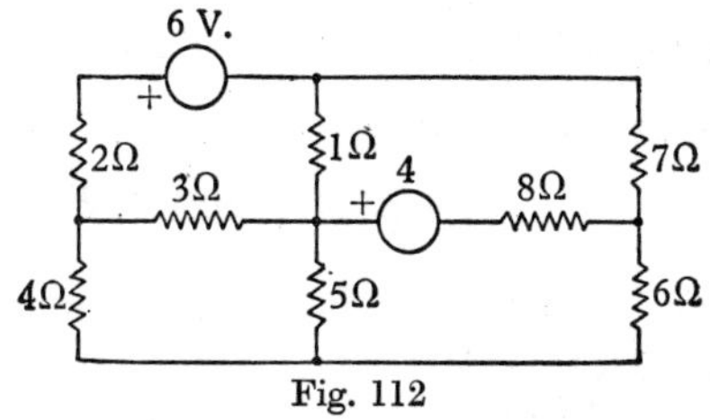

Fig. 112

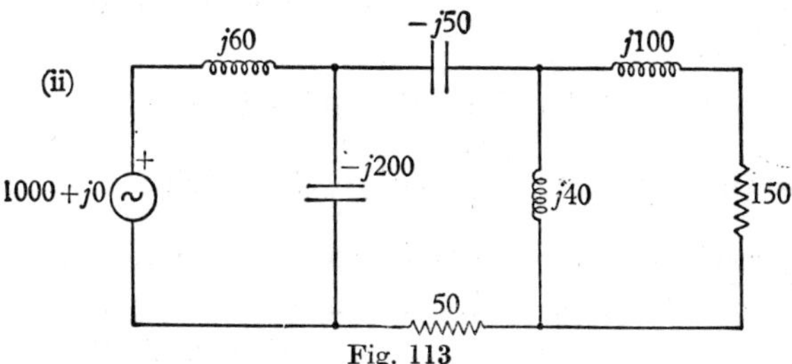

Fig. 113

73. Show that a particular solution as a power series of the differential equation

$$2x^2\frac{d^2y}{dx^2}+3(x^2-x)\frac{dy}{dx}+3y=0$$

is of the form

$$y=Ax(1-3x+3x^2-\frac{9}{5}x^3+\dots),$$

and find the first four terms of another solution of this equation. [L.U.]

74. Find the general solution of the differential equation

$$(1-x^2)\frac{d^2y}{dx^2}-7x\frac{dy}{dx}-9y=0$$

in series of ascending powers of x.

By applying the ratio test, or otherwise, prove that the series converges for $-1<x<1$. [L.U.]

75. If $y=e^{nx+j(\omega t+\alpha)}$ satisfies $\frac{\partial^4 y}{\partial x^4}=k\frac{\partial^2 y}{\partial t^2}$, where n, ω, α, k, are real constants, prove that $n=\pm(1+j)m$ or $\pm(1-j)m$, where $4m^4=k\omega^2$. Deduce that

$$e^{mx}\sin(mx+\omega t+\alpha)\quad\text{and}\quad e^{-mx}\sin(mx+\omega t+\alpha)$$

also satisfy the given equation. [L.U.]

76. Show that $u=Ae^{mx}\cos(\omega t+mx)+Be^{-mx}\cos(\omega t-mx)$ is a solution of $\frac{\partial^2 u}{\partial x^2}=2\frac{\partial u}{\partial t}$, where A, B and ω are constants, provided that $m^2=\omega$.

Find the values of the constants, given the conditions (i) $m>0$, (ii) u remains finite as $x\to\infty$, (iii) $u=\cos t$ when $x=0$. [L.U.]

77. What must be the value of β if $v=A\cos(\omega t-\beta x)$ satisfies the equation: $\frac{\partial^2 v}{\partial t^2}=c^2\frac{\partial^2 v}{\partial x^2}$? Sketch the graphs of v against x when β is positive for each of the values $t=n\pi/2\omega$ with $n=0$, 1, 2, 3, and 4 in turn, drawing the graphs in succession with the x-scales corresponding, so that the variation of v with t for a fixed value of x can be followed from one graph to the next. What is the period and wavelength and what is the significance of the constant c? [I.E.E.]

78. $V = f(z)$ with $z^2 = x^2/4t$, so that V is a function of x and t.

Determine $\partial V/\partial t$ and $\partial^2 V/\partial x^2$, and show that, if $\partial V/\partial t = \partial^2 V/\partial x^2$, then $f(z)$ must satisfy

$$\frac{d^2 f(z)}{dz^2} + 2z \, . \, \frac{df(z)}{dz} = 0.$$

Hence show that $f(z)$ must be of the form $A \int e^{-z^2}\, dz + B$.

[I.E.E.]

79. Assuming that $u = \frac{1}{r} F(r) \cos(\omega t + \alpha)$ is a solution of the partial differential equation

$$\frac{\partial^2 u}{\partial r^2} + \frac{2}{r}\frac{\partial u}{\partial r} = \frac{1}{c^2}\frac{\partial^2 u}{\partial t^2},$$

where ω, α and c are constants, and $F(r)$ is a function of r only, obtain the ordinary differential equation satisfied by $F(r)$ and give the general solution for $F(r)$.

Given that, for all values of t, (i) u is finite at $r = 0$, (ii) $\partial u/\partial r = 0$ at $r = a$, and that u is not identically zero, prove that $\omega a/c = \beta$ must satisfy the equation $\beta = \tan \beta$. [L.U.]

80. The function $y = t^n e^{-x^2/4t}$ satisfies the partial differential equation $\partial y/\partial t = \partial^2 y/\partial x^2$ if n is suitably chosen. Find the value of n, and show that with this value of n, y has a maximum value at $x = 0$ for any fixed positive value of t, and determine the values of x for which y has a value e^{-1} times its maximum value. Sketch the graph of y against x for small and large positive values of t. [I.E.E.]

81. Show that $V = r^n \cos 2\theta$ will satisfy the equation

$$\frac{\partial^2 V}{\partial r^2} + \frac{1}{r}\frac{\partial V}{\partial r} + \frac{1}{r^2}\frac{\partial^2 V}{\partial \theta^2} = 0$$

if n has either one of two values. Find a particular solution of this equation which equals $\cos 2\theta$ when $r = a$, and which vanishes when $r = 2a$ for all values of θ. [I.E.E.]

82. The function $V = r^n . f(\theta)$ is to satisfy the partial differential equation

$$r\frac{\partial}{\partial r}\left(r\frac{\partial V}{\partial r}\right) + \frac{\partial^2 V}{\partial \theta^2} = 0,$$

n being a constant and $f(\theta)$ a function of θ. Obtain and solve the equation to be satisfied by $f(\theta)$. If r and θ are polar co-ordinates on a plane, find particular solutions of this partial differential equation valid (*a*) inside, and (*b*) outside the circle $r = 1$, which are finite throughout their respective regions, including $r = \infty$ in (*b*), and which in each case reduce to $\cos 3\theta - \cos \theta$ on the boundary $r = 1$. [I.E.E.]

83. A telephone line, 20 km long, has the following constants per loop kilometre:

$$R = 30\,\Omega, \quad L = 3\,\text{mH}, \quad C = 0{\cdot}056\mu\text{F}, \quad \text{and} \quad G = 10\,\mu\text{S}.$$

Calculate (*a*) the value of the impedance which would correctly terminate the line, (*b*) the attenuation of the line in d.b., (*c*) the phase of the voltage at the receiving end relative to the sending-end, when the line is correctly terminated, and (*d*) the velocity of propagation along the line. The frequency is $5000/2\pi$ Hz. [L.U.]

84. An open-wire transmission line has the following primary constants:

$$L = 3\,\text{mH/km}, \quad C = 0{\cdot}01\,\mu\text{F/km},$$
$$R = 60\,\Omega/\text{km} \quad \text{and} \quad G = 1\mu\text{S/km}.$$

Ignoring negligibly small quantities, calculate the approximate values of the characteristic impedance, the propagation constant and the velocity of wave propagation of this line at a frequency of 100 kHz. Derive any formulae used from the differential equations for a uniform line. [L.U.]

85. Expand $f(x) = e^x$ in a sine series in the interval $0 < x < \pi$.

86. Expand $f(x) = x \sin x$ in a cosine series in the interval $0 < x < \pi$. Show that if $x = \frac{1}{2}\pi$,

$$\tfrac{1}{2}\pi = 1 + \frac{2}{1.3} - \frac{2}{3.5} + \frac{2}{5.7} \cdots.$$

87. A function of x, $f(x)$, has period 2π and $f(x) = 2x$ for $0 \leqslant x \leqslant \frac{1}{2}\pi$ and $f(x) = 2(\pi - x)$ for $\frac{1}{2}\pi \leqslant x \leqslant \pi$. If its Fourier series contains sine terms only, sketch the graph of the function in the range $0 \leqslant x \leqslant 2\pi$ and find the general term of the series. [L.U.]

88. The function $f(t)$ has period 1 and is defined by $f(t) = t^2 - t + \frac{1}{6}$ for $0 \leqslant t < 1$. Sketch the graph of $f(t)$ for t between -1 and $+2$, and show that its Fourier expansion is

$$f(t) = \sum_{n=1}^{\infty} \frac{\cos 2n\pi t}{n^2\pi^2}.$$

Deduce the sum of the series $\sum_{n=1}^{\infty} \frac{1}{n^2}$. [I.E.E.]

89. Assuming that $f(x)$ can be expanded in the form

$$f(x) = a_1 \sin\frac{\pi x}{l} + a_2 \sin\frac{2\pi x}{l} + \dots + a_n \sin\frac{n\pi x}{l} + \dots,$$

the expansion being valid for x between 0 and l, show how the coefficients $a_1, a_2, \dots, a_n, \dots$ may be determined. Obtain the expansion when $f(x) = x(l - x)$. [I.E.E.]

90. $f(t)$ is an even periodic function of t, of period $2\pi/\omega$. For $0 < t < \pi/\omega$, $f(t) = 2\omega t/\pi$. Sketch the graph of $f(t)$ against t between $t = -2\pi/\omega$ and $t = +2\pi/\omega$, and obtain its Fourier expansion. [I.E.E.]

91. A sawtooth wave of period T is defined by $f(t) = \dfrac{ht}{T}$ for $0 < t < T$, h being a constant.

(i) Sketch the graph of $f(t)$ for $-2T < t < 2T$.

(ii) Show that $f(t) - \frac{1}{2}h$ is an odd function.

(iii) Determine the Fourier expansion of $f(t)$. [I.E.E.]

92. The Fourier series of a function $f(x)$ of period 2π does *not* contain sines of multiples of x. Also $f(x) = \cos x$ when $0 \leqslant x \leqslant \dfrac{\pi}{2}$ and $f(x) = 0$ when $\dfrac{\pi}{2} \leqslant x \leqslant \pi$. Sketch the graph of $f(x)$ for one complete period and show that the Fourier series is

$$\frac{1}{\pi} + \tfrac{1}{2}\cos x + \frac{2}{\pi}\sum_{p=1}^{\infty} \frac{(-1)^{p+1}}{4p^2 - 1}\cos 2px. \qquad \text{[L.U.]}$$

93. If $V = X \cos t$, where X is a function of x only, determine X so that V may satisfy the equation $\frac{\partial^2 V}{\partial x^2} - c^2 \frac{\partial^2 V}{\partial t^2} = 0$, and so that $V = 0$ when $x = 0$ for all values of t. [I.E.E.]

94. Derive the relationships used for evaluating the amplitudes of the several frequency components of a given recurrent non-sinusoidal waveform.

Estimate the approximate amplitude of the fundamental 50 Hz component of a current waveform, one complete cycle of which is supplied by:

t(milliseconds)	0	1·67	3·33	5	6·67	8·33	10	11·67	13·33	15	16·67	18·33	20
i(milliamperes)	37	49	46	30	12	1	3	9	12	10	11	21	37

[L.U.]

95. Explain how three unbalanced currents in a 3-phase system may be resolved analytically and graphically into balanced positive, negative and zero phase-sequence components. Determine these components for the currents

$$I_R = 0 + j10 \text{ amperes}, \qquad I_Y = \text{zero},$$
$$I_B = 0 - j10 \text{ amperes},$$

and draw the vector diagrams. [L.U.]

96. In a 3-phase system the three line currents are: $I_R = 20 + j30$; $I_Y = 10 - j40$; $I_B = -15 + j60$. Determine the values of the positive, negative and zero phase-sequence components. Draw vector diagrams showing them to scale.

97. An electric current flows in a cable of negligible leakance and inductance. At a distance x from one end, show that the voltage v and current i satisfy the differential equations $-\frac{\partial v}{\partial x} = Ri$, $-\frac{\partial i}{\partial x} = C\frac{\partial v}{\partial t}$, where R and C have their usual meanings. Deduce that $\frac{\partial^2 v}{\partial x^2} = RC\frac{\partial v}{\partial t}$. A cable of this type is of length l and is initially in a steady state with $v = \frac{V_0 x}{l}$. At $t = 0$ both ends are suddenly grounded and kept so (i.e. $v = 0$ when $x = 0$ and $x = l$ for all $t > 0$).

Find series for the values of v and i at distance x at time t.

98. Sketch the parabola $x^2 = 4ay$ and the straight line $2y = 4a - x$ and find the coordinates of their points of intersection.

Express as repeated integrals (i) the area enclosed by the parabola and the straight line and (ii) the moment of this area about the y-axis.

Evaluate these integrals and hence show that the centroid of the area is at a distance a from the y-axis. [L.U.]

99. Show in a diagram the region over which the integral

$$\int_0^1 dx \int_0^{\sqrt{(x-x^2)}} \frac{4xy\, e^{-(x^2+y^2)}}{x^2+y^2}\, dy$$

extends. Transform to polar co-ordinates, and hence or otherwise show that the integral has the value e^{-1}. [L.U.]

100. Perform the matrix multiplication:

$$\begin{bmatrix} 1 & 3 & -2 \\ 4 & 2 & 7 \end{bmatrix} \cdot \begin{bmatrix} 1+j & 4 \\ 3 & 1+j2 \\ 2-j & 2-j3 \end{bmatrix} \cdot \begin{bmatrix} 4 & -2 \\ 3 & 1 \end{bmatrix}.$$

101. Solve the equations in Question 70, using matrices.

102. Solve the network of fig. 113 using the Kron mesh method (i.e. using matrix analysis).

103. Solve for the current distribution in the circuit shown in fig. 114, using matrix analysis.

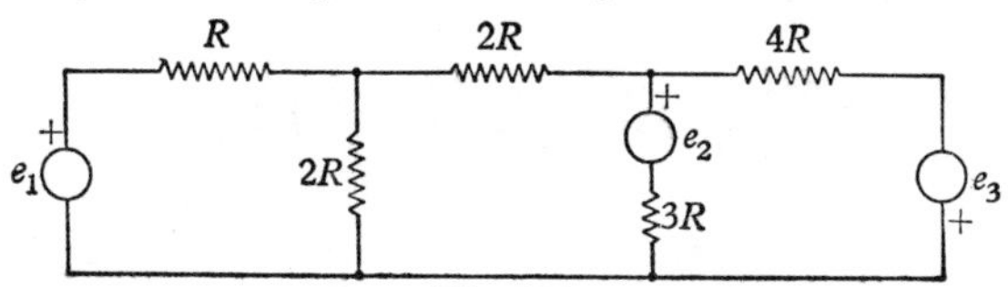

Fig. 114

104. Find the solutions of the wave equation

$$\frac{\partial^2 y}{\partial x^2} = \frac{1}{c^2}\frac{\partial^2 y}{\partial t^2}$$

for which $y = 0$ at $x = 0$ and at $x = l$ for all values of t. Find the particular solution for which, in addition, at $t = 0$

$$y = 0 \quad \text{and} \quad \frac{\partial y}{\partial t} = A \sin\left(\frac{3\pi x}{l}\right).$$ [C.U.]

105. Transform the equation $x^2 \dfrac{d^2y}{dx^2} - (1 - 3x)\dfrac{dy}{dx} + y = 0$ by putting $x = 1/t$, $y = ute^{-t}$. Hence, or otherwise, find a solution which remains finite when x tends to infinity. [C.U.]

106. If $y = F(x) \cos \omega t$, where ω is a constant greater than 1, is a solution of the equation

$$\frac{\partial^2 y}{\partial x^2} + 2\frac{\partial y}{\partial x} = \frac{\partial^2 y}{\partial t^2},$$

find the ordinary differential equation satisfied by $F(x)$. Obtain the general solution of this equation and hence a solution of the equation for y in the above form. Given that this solution satisfies the conditions $y = 0$ at $x = 0$ and at $x = l$, for all values of t, and does not vanish identically, prove that ω must have one of the values given by

$$\omega^2 = 1 + n^2\pi^2/l^2, \quad \text{where } n = 1, 2, \ldots .$$ [L.U.]

107. Two capacitors of capacitances C_1 and C_2 each have one plate earthed. The two insulated plates are connected at time $t = 0$ by a wire of resistance R and self-inductance L. At this instant the first capacitor has charge Q and the second capacitor is uncharged. Show that the current I flowing in the wire satisfies the equation

$$L\frac{d^2I}{dt^2} + R\frac{dI}{dt} + \left(\frac{1}{C_1} + \frac{1}{C_2}\right) I = 0.$$

Hence show that, if $\dfrac{1}{C_1} + \dfrac{1}{C_2} = \dfrac{R^2}{4L}$, then $I = (Qt/LC_1)e^{-Rt/2L}$.

[L.U.]

108. If $F(D)$ is linear with constant coefficients, a particular integral of $F(D)y = e^{ax}$ is given by $y = \dfrac{e^{ax}}{F(a)}$, when $F(a) \neq 0$ (see § 7·341). If $F(a) = 0$, a particular integral is given by $\dfrac{x^n e^{ax}}{F^n(a)}$, where $F^n(a)$ is the first derivative of $F(D)$ which is *not* zero when a is substituted for D. Use this to find particular integrals for the equations:

(i) $(D^2 - 2D - 3)y = e^{3x}$,

(ii) $(D^2 + 16)y = \cos 4x$, taking $\cos 4x$ as the real part of e^{j4x},

(iii) $(D - 2)(D^2 + 4)y = \sin 2x$, taking $\sin 2x$ as the imaginary part of e^{j2x},

(iv) $(D + 2)^2 y = e^{-2x}$.

Verify the results by substitution in the actual equations.

ANSWERS TO GENERAL MISCELLANEOUS EXERCISES (p. 458)

2. (*a*) $t - t\log_e t + C$; (*b*) $\frac{2}{3} - 5/3e$.

3. (i) $-\frac{1}{8}$; (ii) $-\frac{1}{6}$; (iii) $\frac{1}{2}$; (iv) $\frac{1}{2}$.

4. (i) $\frac{1}{120}$; (ii) $\frac{3\pi}{256}$; (iii) 0; (iv) $\frac{3\pi}{128}$.

5. $f(m+1, n) = \dfrac{m}{m+n} f(m,n)$;

$$f(m, n+1) = \frac{n}{m+n} f(m,n).$$

6. (i) $\dfrac{e^{3x}}{34}(3\sin 5x - 5\cos 5x) + C$;

(ii) $\sqrt{(x^2 - x - 6)} - \frac{5}{2}\cosh^{-1}\left(\dfrac{2x-1}{5}\right) + C$;

(iii) $\dfrac{2}{\sqrt{5}}\tan^{-1}\left(\dfrac{3\tan\theta/2 - 2}{\sqrt{5}}\right)$; $\frac{2}{15}a^5$.

7. (i) 0·48481; (ii) 1·07237; (iii) 0·51504.

8. 2·679.

10. $-1 < x \leqslant 1$.

11. (i) $\frac{125}{3}$; (ii) $32\left(\dfrac{\pi}{3} - \dfrac{\sqrt{3}}{4}\right)$; (iii) $(1 - \pi/4)$;

(iv) $\frac{1}{8}(e^2 + 3)$.

12. (i) $\frac{1}{2}$; (ii) $\log_e\left(\dfrac{1+\sqrt{3}}{\sqrt{2}}\right)$; (iii) π;

(iv) $\frac{1}{2}\{\cosh^{-1} 2 - \cosh^{-1} a + 2\sqrt{3} - a\sqrt{(a^2 - 1)}\}$.

13. $x - \dfrac{x^3}{3.3!} + \dfrac{x^5}{5.5!} - \dfrac{x^7}{7.7!} + \ldots$; 1·85.

14. (*a*) $0·893 + j0·126$;

(*b*) $\frac{1}{2}\log_e 2 - j\dfrac{\pi}{4}$ (principal value).

15. $(4n+1)\dfrac{\pi}{2} + j\log_e(3+\sqrt{8})$ (general value, where $n = 0, \pm 1, \pm 2 \ldots$).

16. (i) That branch of the hyperbola $u^2 - v^2 = \frac{1}{2}$ for which u is positive.

(ii) The ellipse $\dfrac{u^2}{25} + \dfrac{v^2}{9} = \dfrac{1}{16}$.

17. (i) $-\dfrac{r}{\mu L_1}, \pm \dfrac{j}{\sqrt{(CL_1 + \mu CL_1)}}$.

(ii) Z locus is the line $x = R$, $0 \leqslant y \leqslant \infty$;

Y locus is the semicircle, centre $(1/2R, 0)$, radius $1/2R$, in the fourth quadrant.

18. $2{\cdot}033 - j3{\cdot}052$.

20. Circle, centre at point representing a, radius k. Radius $\frac{3}{4}$.

21. Circle, centre $(\frac{1}{2}, 0)$, radius $\frac{1}{2}$. 0, $\frac{1}{2}(1+j)$, $\frac{1}{2}(1-j)$, 1.

22. $\dfrac{j}{2}\left\{\dfrac{1}{(1 - re^{-j\theta})} - \dfrac{1}{(1 - re^{j\theta})}\right\}$.

23. $309 - j243$.

24. $0{\cdot}075 - j0{\cdot}106$.

26. (i) $\dfrac{1}{xy} + \frac{1}{2}(\log_e x)^2 = C$; (ii) $y = Ae^{y/x}$;

(iii) $\frac{1}{2}\log(3x^2 + xy + y^2 - 8x - 5y + 9) + \dfrac{3}{\sqrt{11}}\tan^{-1}\left\{\dfrac{3y - 4x - 2}{\sqrt{11}(2x + y - 4)}\right\}$.

27. (i) $y = \dfrac{1}{2}\dfrac{\cos 2x}{\cos x} + \dfrac{C}{\cos x}$ or $y = \cos x + A\sec x$;

(ii) $y = \tan\{\log_e A\sqrt{(1+x^2)}\}$.

28. $y = e^{-x^2/2y^2}$.

29. General solution is $a^2 = \dfrac{4\,Ce^{nt}}{1 + Ce^{nt}}$. At $a = 2/\sqrt{3}$, the curves change concave to convex.

30. The lines are $y = (-1 \pm \sqrt{2})x$, and are asymptotes to the solution curves, which are rectangular hyperbolas.

32. (a) $y = A + Be^{2x} + Ce^{-2x} + \frac{1}{16}\cos 2x$;

(b) $y = (A + B \log_e x)x^2 + x^3$.

33. (i) $y = e^{-2x}(A \cos x + B \sin x) + \frac{3}{8}(\sin x - \cos x)$

$$+ \frac{x^2}{5} - \frac{8}{25}x + \frac{22}{125};$$

(ii) $u = \dfrac{r^2}{3} + \dfrac{1}{6r} - \dfrac{r}{2}$.

34. $y = x - \sqrt{(1 - x^2)} \sin^{-1} x + A\sqrt{(1 - x^2)}$.

35. $y = -\cos x^2$.

36. $y = (1/n) \sin (n \sin \theta)$.

38. (i) $x = \dfrac{1}{54}(7 \cos 3t + 4 \sin 3t) + \dfrac{t}{9} + \dfrac{1}{27}$;

(ii) $y = \frac{1}{9}(16 \cos 3x + 7 \sin 3x) + \frac{2}{9}e^{3x}$;

(iii) $y = -\frac{5}{72}e^{3x} - \frac{5}{24}e^{-x} - \frac{2}{3}x + \frac{7}{9}$;

(iv) $i = 2e^{3t}(\cos 2t - \sin 2t) + 3$;

(v) $y = 1 + e^{2x} - x^2 - x$;

(vi) $i = \frac{3}{5}e^t(\sin t - 2 \cos t) + \frac{3}{5}(\sin t + 2 \cos t)$;

(vii) $i = \dfrac{t}{3}\sin 3t$; (viii) $y = \frac{1}{25}(1 + 5x)e^{-5x}$

$$+ \frac{x^2}{2}e^{-5x} - \frac{1}{25}\cos 5x;$$

(ix) $y = x \log_e x(1 - \log_e x)$.

39. $x = -\frac{1}{2}t\,e^{-2t} + \frac{1}{4}\sin 2t$.

40. $x = A\,e^{-t}\sin (t + \alpha) + \dfrac{10}{\sqrt{\{(2 - \omega^2)^2 + 4\omega^2\}}}\cos (\omega t - \beta)$,

where A, α are arbitrary and $\beta = \tan^{-1}\dfrac{2\omega}{2 - \omega^2}$.

Amplitude: $$\frac{10}{\sqrt{\{(2 - \omega^2)^2 + 4\omega^2\}}} = \frac{10}{\sqrt{(\omega^4 + 4)}}.$$

41. (i) $y = \frac{3}{2} e^{t/2} \cos \frac{\sqrt{15}t}{2} - \frac{3}{2\sqrt{15}} e^{t/2} \sin \frac{\sqrt{15}t}{2}$

$+ \frac{3}{2} \cos 2t$;

(ii) $y = te^{-t} + \frac{1}{3} t^3 e^{-t}$;

(iii) $y = -\frac{4}{5}(2-j)te^{-2t} + \frac{4}{25}(3-j4)e^{jt}$

$- \frac{4}{25}(3-j4)e^{-2t}$;

(iv) $x = 2 \cos t \cosh t + \sin t \sinh t$,

$y = 2 \cos t \cosh t - 4 \sin t \sinh t$;

(v) $x = 3e^{-t} + 2t - 4 + e^{-t/2} \cos \frac{\sqrt{3}}{2} t +$

$\sqrt{3}\, e^{-t/2} \sin \frac{\sqrt{3}}{2} t, y = -3e^{-t} + 3 - 2\sqrt{3}\, e^{-t/2} \sin \frac{\sqrt{3}}{2} t.$

42. $x = 2e^{-t} + e^{-2t}, \quad y = e^{-t} - e^{-2t}$.
Max. y is $\frac{1}{4}$, when x is $\frac{5}{4}$ and $t = \log_e 2$.

43. $x = 2a \cos \frac{1}{\sqrt{2}} t + 2\sqrt{2}\, b \sin \frac{1}{\sqrt{2}} t,$

$y = -a \cos \frac{1}{\sqrt{2}} t - \sqrt{2}\, b \sin \frac{1}{\sqrt{2}} t.$

44. $x = 2 \cosh t, \quad y = \sin t - 2 \sinh t$.

45. $x = \cos 2t + \sin t, y = -\sin 2t$; r.m.s. for $x = 1$, r.m.s. for $y = 1/\sqrt{2}$. For $y \equiv 0, x = 0$ at $t = 0$, then $x = \sin t$.

46. $x = -(\cos t + 2 \sin t) + e^{2t} \simeq \frac{5t^2}{2}$ for small t.

47. $x = e^{-2t}(\frac{1}{6} \sin 3t - \frac{1}{4} \cos 3t) + \frac{1}{4} \cos 3t$.

48. $x = \frac{ht}{2\omega} e^{-at} \sin \omega t$.

51. $LC \frac{d^2v}{dt^2} + RC \frac{dv}{dt} + v = E$;

$v = E\{1 - e^{-\omega t}(\cos 2\omega t + \frac{1}{2} \sin 2\omega t)\}$;

$i = \frac{5}{2} \omega CE\, e^{-\omega t} \sin 2\omega t$.

53. $C \simeq 100\,\mu\text{F}$; approx. 3·14 amperes.

54. $i_1 = E\left\{\frac{1}{50} - \frac{1}{210} e^{-\frac{10t}{3}} - \frac{8}{525} e^{-\frac{100t}{9}}\right\}$;

$$i_2 = \frac{E}{105}\left(e^{-\frac{100t}{9}} - e^{-\frac{10t}{3}}\right).$$

56. $E \cos \omega t - \dfrac{E\sqrt{(4R^2 + L^2\omega^2)}}{\sqrt{(9R^2 + 4L^2\omega^2)}} \cos(\omega t - \alpha)$.

57. (i) $x = e^{-ht}(Ae^{kt} + Be^{-kt})$, where $k = \sqrt{(h^2 - 1)}$;

(ii) $x = (A + Bt)e^{-t}$;

(iii) $x = e^{-ht}(A \cos pt + B \sin pt)$,

where $p = \sqrt{(1 - h^2)}$.

58. 155 mm. (approx.)

59. $53{\cdot}9 \times 10^{-7}$ newton metres.

60. 1035 Ω (approx.).

63. 0·5.

70. (i) $x = -5, y = 7, z = 2$; (ii) $x = 2, y = 5, z = 3$.

71. $i_1 = 0{\cdot}5667$, $i_2 = 0{\cdot}4867$, $i_3 = 0{\cdot}7112$.

72. (i) $-1{\cdot}231$, $-1{\cdot}216$, $-0{\cdot}779$, $-0{\cdot}437$, $-0{\cdot}118$, $-0{\cdot}320$, $0{\cdot}014$, $-0{\cdot}334$;

(ii) $14{\cdot}11 - j10{\cdot}69$, $4{\cdot}24 + j1{\cdot}79$, $9{\cdot}88 - j12{\cdot}48$, $3{\cdot}09 - j0{\cdot}25$, $6{\cdot}79 - j12{\cdot}23$, $-9{\cdot}88 + j12{\cdot}48$.

73. $Bx^{3/2}(1 - \frac{3}{2}x + \frac{9}{8}x^2 - \frac{9}{16}x^3 + \ldots)$.

74.
$$y = A\left(1 + \frac{3^2}{2!}x^2 + \frac{3^2 \cdot 5^2}{4!}x^4 + \frac{3^2 \cdot 5^2 \cdot 7^2}{6!}x^6 + \ldots\right)$$
$$+ B\left(x + \frac{4^2}{3!}x^3 + \frac{4^2 \cdot 6^2}{5!}x^5 + \ldots\right).$$

76. $B = 1 = \omega = m$, $A = 0$.

77. Period $2\pi/\omega$; wavelength $2\pi c/\omega$; c is the velocity of the wave.

78. $\dfrac{\partial V}{\partial t} = -\dfrac{z}{2t} \cdot f'(z)$; $\dfrac{\partial^2 V}{\partial x^2} = \dfrac{1}{4t} \cdot f''(z)$.

79. $F''(r) + \dfrac{\omega^2}{c^2} F(r) = 0;$

$$F(r) = A \cos \frac{\omega}{c} r + B \sin \frac{\omega}{c} r.$$

80. $n = -\tfrac{1}{2}; \quad x = \pm 2\sqrt{t}.$

81. $n = \pm 2; \quad V = \dfrac{1}{15}\left(\dfrac{16a^2}{r^2} - \dfrac{r^2}{a^2}\right) \cos 2\theta.$

82. $f(\theta) = A \cos n\theta + B \sin n\theta.$

(a) $r^3 \cos 3\theta - r \cos \theta$; (b) $(1/r^3) \cos 3\theta - (1/r) \cos \theta$.

83. (a) $297{\cdot}5 - j176{\cdot}5$, (b) 9·14 db.,

(c) 1·63 rad (lagging), (d) 61,330 km/s.

84. 547·6 Ω, $0{\cdot}054 + j3{\cdot}44$, 182,600 km/s.

85. $\displaystyle\sum_{n=1}^{\infty} \frac{2n}{\pi(1+n^2)} \{1 - e^{\pi} \cos n\pi\} \sin nx.$

86. $\displaystyle 1 - \tfrac{1}{2} \cos x + \sum_{n=2}^{\infty} \frac{2}{(n^2-1)} \cos (n+1)\pi \cos nx.$

87. $\dfrac{8}{\pi n^2} \sin n \dfrac{\pi}{2} \sin nx.$

88. $\pi^2/6.$

89. $\displaystyle\sum_{n=1}^{\infty} \frac{8l^2}{(2n-1)^3 \pi^3} \sin \frac{(2n-1)\pi x}{l}.$

90. $\displaystyle 1 - \frac{8}{\pi^2}\left\{\cos \omega t + \frac{1}{3^2} \cos 3\,\omega t + \frac{1}{5^2} \cos 5\,\omega t + \ldots\right\}.$

91. $\displaystyle f(t) = \frac{h}{2} - \sum_{n=1}^{\infty} \frac{1}{n\pi} \sin \frac{2n\pi t}{T}.$

93. $X = A \sin cx.$

94. 19·8.

95. $I_{R1} = -2{\cdot}89 + j5, \quad I_{Y1} = 5{\cdot}77,$

$I_{B1} = -2{\cdot}89 - j5; \quad I_{R2} = 2{\cdot}89 + j5,$

$I_{Y2} = -5{\cdot}77, \quad I_{B2} = 2{\cdot}89 - j5;$

$I_{R0} = I_{Y0} = I_{B0} = 0.$

96. $I_{R1} = 36{\cdot}37 + j13{\cdot}88$, $I_{Y1} = -6{\cdot}15 - j38{\cdot}43$,
$I_{B1} = -30{\cdot}19 + j24{\cdot}55$, $I_{R2} = -21{\cdot}37 - j0{\cdot}55$,
$I_{Y2} = 11{\cdot}17 - j18{\cdot}22$, $I_{B2} = 10{\cdot}21 + j18{\cdot}78$,
$I_{R0} = I_{Y0} = I_{B0} = 5 + j16{\cdot}67$.

97. $$v = \sum_{n=1}^{\infty} (-1)^{n+1} \frac{2V_0}{n\pi} e^{-n^2\lambda t} \sin \frac{n\pi x}{l},$$

$$i = \sum_{n=1}^{\infty} (-1)^{n} \frac{2V_0}{lR} e^{-n^2\lambda t} \cos \frac{n\pi x}{l}, \quad \text{where } \lambda = \frac{\pi^2}{l^2CR}.$$

100. $$\begin{bmatrix} (33 + j48) & (-9 + j6) \\ (192 - j63) & (-16 - j11) \end{bmatrix}.$$

103. $$\frac{20e_1 - 4e_2 + 3e_3}{46R}, \quad \frac{-2e_1 + 5e_2 + 2e_3}{23R},$$

$$\frac{6e_1 + 8e_2 + 17e_3}{92R}, \quad \frac{14e_1 - 12e_2 + 9e_3}{92R}, \quad \frac{26e_1 + 4e_2 - 3e_3}{92R}.$$

104. $$y = \sum_{n=1}^{\infty} \left(B_n \cos \frac{nc\pi t}{l} + C_n \sin \frac{nc\pi t}{l}\right) \sin \frac{n\pi x}{l}.$$

Particular solution required is $y = \dfrac{Al}{3\pi c} \sin \dfrac{3\pi ct}{l} \sin \dfrac{3\pi x}{l}$.

105. $t^2 \dfrac{d^2u}{dt^2} - (t - 1) \dfrac{du}{dt} = 0.$ $\quad y = \dfrac{C}{x} e^{-1/x}$.

106. $F''(x) + 2F'(x) + \omega^2 F(x) = 0.$
$y = e^{-x}\{A \cos \sqrt{(\omega^2 - 1)}x + B \sin \sqrt{(\omega^2 - 1)}x\} \cos \omega t.$

108. (i) $\dfrac{x}{4} e^{3x}$; (ii) $\dfrac{x}{8} \sin 4x$; (iii) $\dfrac{x}{16} (\cos 2x - \sin 2x)$;

(iv) $\dfrac{x^2}{2} e^{-2x}$.

INDEX

(Numbers refer to the pages)